ASPECTS OF INDUSTRY IN ROMAN YORKSHIRE AND THE NORTH

Edited by Pete Wilson and Jennifer Price

I0892407

Oxbow
2002

First published in the United Kingdom in 2002. Reprinted in 2016 by
OXBOW BOOKS
10 Hythe Bridge Street, Oxford OX1 2EW

and in the United States by
OXBOW BOOKS
1950 Lawrence Road, Havertown, PA 19083

© Oxbow Books and the individual authors, 2016

Paperback Edition: ISBN 978-1-84217-078-6
Digital Edition: ISBN 978-1-78570-417-8 (ePub)

A CIP record for this book is available from the British Library

All rights reserved. No part of this book may be reproduced or transmitted in any form or by any means, electronic or mechanical including photocopying, recording or by any information storage and retrieval system, without permission from the publisher in writing.

For a complete list of Oxbow titles, please contact:

UNITED KINGDOM
Oxbow Books
Telephone (01865) 241249, Fax (01865) 794449
Email: oxbow@oxbowbooks.com
www.oxbowbooks.com

UNITED STATES OF AMERICA
Oxbow Books
Telephone (800) 791-9354, Fax (610) 853-9146
Email: queries@casemateacademic.com
www.casemateacademic.com/oxbow

Oxbow Books is part of the Casemate Group

CONTENTS

PREFACE

Many profound changes in material culture took place when Britain became a province of the Roman empire. The range of materials and artefacts in circulation became broader, a multitude of crafts for local and regional markets were developed, and the organisation and scale of production and the patterns of distribution and consumption increased in complexity.

In the area of study, Yorkshire and the Roman North, evidence for the role of the army in craft production is often identifiable from finds in legionary and auxiliary forts, their satellite settlements, the canabae and vici, and other industrial complexes. It is also apparent that craft production to serve civil markets took place in towns and in a variety of rural settlements. Much less is known about the organisation of these activities although they presumably existed side by side with the military production networks. The quality and quantity of information available for the study of the production processes varies considerably. Some activities, such as pottery making or jet-working, produce a wide range of waste materials which survive well in the archaeological record, while others, such as metal- or glass- or leather- or stone-working often leave very few traces apart from the finished artefacts, and are consequently rather difficult to interpret.

The day-school which provided the inspiration for this volume was concerned with the crafts and materials of Roman Yorkshire, and some of the papers subsequently prepared for publication have focused in this region, while others have also included information from elsewhere in the Roman North. The ten papers explore a range of current research issues. They have been grouped in three sections which have considerable overlap between them. The first examines the range of craft activity and industrial production in the military and urban settlements at York (Cool), and in two contrasting rural areas, the North York Moors (Wilson) and Holme on Spalding Moor in the Wolds (Halkon). The second and third are studies of individual materials and production processes, which have been divided into high-temperature activities - pottery (Swan), glass (Price), copper alloys (Dungworth) and non-ferrous metals (Bayley), and low-temperature activities - leather (van Driel Murray), jet (Allason-Jones) and stone (Gaunt and Buckland). Some of these topics have long-standing traditions of research, while others have received rather less scholarly attention until recent years. We hope that the papers with their bibliographies will focus attention on the contribution that Yorkshire and the North has made to the current state of knowledge of Romano-British craft production, and that they will stimulate future research in the region and elsewhere in the province.

Pete Wilson and Jennifer Price

ACKNOWLEDGEMENTS

The production of this volume would not have been possible without the contributions of many people, who have provided knowledge, support and facilities in a variety of ways. The work had its origins in a one-day conference on *Aspects of Industry in Roman Yorkshire* held in York in 1994, an event organised jointly by the Roman Antiquities Section of the Yorkshire Archaeological Society and the Yorkshire Philosophical Society, for which the Roman Antiquities Section provided the programme and invited the speakers and the Philosophical Society arranged the venue, in the Tempest Anderson Hall at the Yorkshire Museum. Seven papers were given, and the conference was chaired by Dr John Peter Wild, who encouraged the publication of the proceedings. In the event, it has not been possible to include all the papers given at the conference but additional contributions have been offered and the geographical coverage has been expanded to include the Roman North as well as Yorkshire.

During the production of the volume, Yvonne Beadnell and Trevor Woods of the Department of Archaeology, University of Durham and John Vallender of the English Heritage Centre for Archaeology Graphics Studio have produced illustrations, the York Archaeological Trust and the Yorkshire Museum have made available figures for the papers by Cool and Allason-Jones respectively, Jill Wilson has converted the manuscripts into a book, and Oxbow Books has agreed to publish it. In addition, the comments of the anonymous reader of the manuscripts have enhanced the overall consistency and quality of the work.

We are very grateful to everyone who has contributed to the volume, but our greatest debt is to the authors of the papers for their initial work and for their support and patience during the unforeseen events which have caused delays in completion.

Pete Wilson and Jennifer Price

List of contributors

Lindsay Allason-Jones, MLitt
Museum of Antiquities,
Department of Archaeology,
University of Newcastle,
Newcastle upon Tyne NE1 7RU

Justine Bayley, PhD
English Heritage Centre for Archaeology,
Fort Cumberland,
Fort Cumberland Road,
Eastney,
Portsmouth P04 9LD

Paul Buckland, PhD
Department of Archaeology,
Northgate House,
West Street,
Sheffield S1 4ET

Hilary Cool, PhD
16 Lady Bay Road,
West Bridgford,
Nottingham NG2 5BJ

David Dungworth, PhD
English Heritage Centre for Archaeology,
Fort Cumberland,
Fort Cumberland Road,
Eastney,
Portsmouth P04 9LD

Geoff Gaunt, PhD
Department of Archaeology,
University of Bradford,
Bradford, BD7 1DP

Peter Halkon, MA
Department of History,
University of Hull,
Hull, HU6 7RX

Carol van Driel-Murray, PhD
Instituut voor Pre- en Protohistoriche Archeologie,
Nieuwe Prinsengracht,
1018 VZ Amsterdam

Jennifer Price, PhD
Department of Archaeology,
University of Durham,
South Road,
Durham, DH1 3LE

Vivien Swan, DLitt
Department of Archaeology,
University of Durham,
South Road,
Durham, DH1 3LE

Pete Wilson, PhD
English Heritage Centre for Archaeology,
Fort Cumberland,
Fort Cumberland Road,
Eastney,
Portsmouth P04 9LD

CRAFT AND INDUSTRY IN ROMAN YORK

H E M COOL

Introduction

This paper seeks to summarise the evidence for manufacturing activities in Roman York. At the outset it should be noted that for this aspect of life in the fortress and *colonia,* as for many others, our knowledge is limited and heavily biased towards areas where large-scale excavation has taken place. Though this bias is true of many towns, it is especially marked at York. Over much of the core area of Roman deposits, successive generations of occupation have led to a considerable overburden so that in some places the natural ground surface is up to ten metres below the modern one (Ottaway 1993, 13) thus militating against casual finds. The principal areas of archaeological activity in the modern period have been to the east of the fortress in the area termed by the Royal Commission the *canabae* (RCHME 1962, xxxiv) and to the south-west of the Ouse in the area traditionally termed the *colonia.* The interior of the fortress itself has seen relatively little excavation.

Much of the detailed evidence on which this paper is based is unpublished, but I have been greatly aided by being able to consult two manuscripts in advance of publication by former colleagues at the York Archaeological Trust. The pottery of Roman York has been the subject of a synthetic overview by Dr J Monaghan (Monaghan 1997) and the non-ceramic finds from the excavations at the General Accident site, Tanner Row and Rougier Street have been studied by Mr A D Hooley (Hooley *forthcoming*). Both of these works are so fundamental to what follows that it is appropriate to signal my indebtedness at the outset as well as at the end of the paper.

The main source of information about the manufacturing activities that took place in Roman York has come from excavations conducted by the York Archaeological Trust since 1973, and this paper considers the evidence that had been recovered up to 1994. The publication policy followed for those excavations was to publish by fascicules as study of the various categories of evidence for a site was completed. In order to simplify the referencing in the discussions that follow, publications that relate to the stratigraphy of a site are given here. The publications cited in the body of the text for these excavations refer to the publication of the artefactual material itself. If no reference is cited then the material is unpublished. The definitive site reports currently available are those for the excavations at Church Street (Whitwell 1976), Skeldergate and Bishophill (Carver *et al* 1978), the Ebor Brewery site, 21-33, Aldwark (Magilton 1986) and 9 Blake Street (Hall 1997). Final reports are still awaited for the other sites but summaries of the structural remains of some are available in publications dealing with other aspects of the archaeology. Those for the excavations at the General Accident site, Tanner Row and Rougier Street are included in Perrin 1990, for 16-22 Coppergate in Bayley 1992 and for Purey Cust Nuffield Hospital in Cool *et al* (1995). Preliminary notes on some sites are available in the *Roman Britain in* series published annually in *Britannia.* Sites in this category include the Adam's Hydraulics site Peaseholme Green (Frere 1992, 273), the Foss Island Road/Lawrence Street site (Frere 1990, 327), the County Hospital site at Fossbank (Frere 1983, 294), Jewbury (Frere 1984, 282) and Wellington Row (Frere 1990, 325). Useful summaries of the stratigraphic evidence for otherwise unpublished sites are also available in the pottery synthesis (Monaghan 1997). The finds other than pottery from Wellington Row have not been studied in any detail but have been the subject of an assessment exercise (Ottaway and Cool 1995) which provides some hints of manufacturing activity. The locations of the principal York sites mentioned in the text are shown on Figure 1.

Manufacturing activity can be divided into two categories, those that were carried out at high temperatures and those that were not. In general, items made of fired clay, metal and glass fall into the former category, and those of material such as wood, bone, textiles, leather and stone into the latter. In this paper the evidence for manufacturing activities will first be discussed according to this division, and then an overview will be presented.

High temperature industries

When the Ninth Legion arrived at York in the early 70's AD there would have been a pressing need to establish industries for the production of iron objects, tiles and pottery to provide structural elements for the buildings and vessels for everyday use. Other than the products themselves, however, this activity has left little direct evidence.

A timber legionary fortress required vast quantities of nails many of which are likely to have been made by legionary smiths on site. The scale of production needed can be gauged by a discovery at Inchtuthil. When this fortress was de-commissioned, nearly 10 tons of nails estimated to be equivalent of in excess of one million nails were disposed of by burial. These were primarily the short nails used for attaching the cladding to the buildings; most of the large structural nails were missing and had probably been salvaged for re-cycling (Manning 1985a, 291). Huge though this deposit was, it was far from being the total number of nails required in the fortress as Shirley (1996, 122; see also Taylor 1999) has calculated that over 1.5 million nails would be required for fixing the roof tiles, shingles and external cladding there. At York the only evidence of this smithing activity in the fortress contemporary with its building is a smith's punch and a minute quantity of hammer scale from the *praetentura* site at 9 Blake Street (Cool *et al* 1995, 1591-2, fig 752 no 6279).

Blake Street produced another smith's punch and a fragment

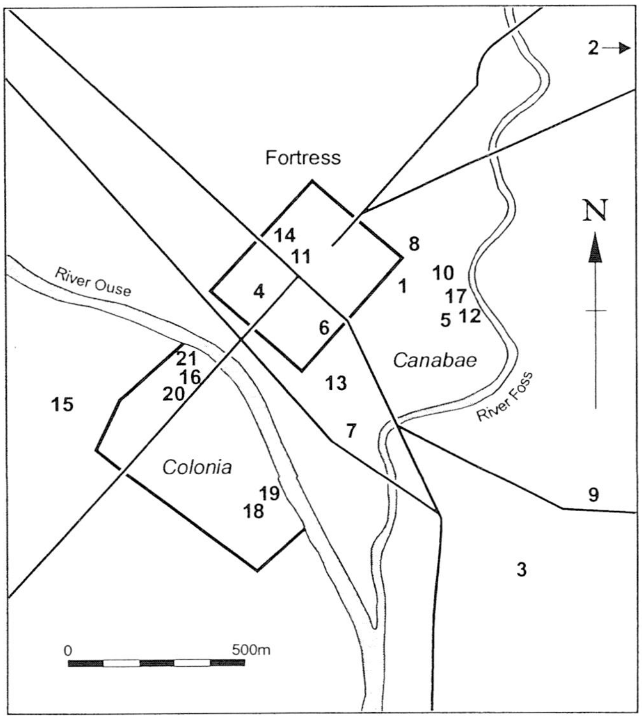

Key
1 21-33, Aldwark (Ebor Brewery site)
2 Appletree Farm, Heworth Without
3 Paragon Street
4 9, Blake Street
5 Borthwick Institute
6 Church Street
7 16-22 Coppergate
8 Fossbank (County Hospital)
9 Foss Island Road/Lawrence Street
10 Jewbury
11 The Minster
12 Peasholme Green (Adam's Hydraulic site)
13 Parliament Street
14 Purey Cust Nuffield Hospital
15 Railway Station
16 Rougier Street
17 St Cuthbert's Church
18 St Mary Bishophill Senior
19 Skeldergate
20 Tanner Row (General Accident site)
21 Wellington Row

Figure 1. Map showing the location of the principal sites in York mentioned in the text

of iron scrap in mid-second to third-century contexts (*ibid* 1995, fig 572 nos 6280 and 6284), but most of the evidence for smithing in the Roman period that has been recognised hitherto has come from the *colonia*. Excavations on the site of the church of St. Mary, Bishophill Senior produced evidence in the form of `objects in various states of completion', forging and fuel slags and hammer scale (Biek 1976, 68) associated with the debris of copper alloy workshop. The workshop was the earliest structure on the site and was active during the later second century (Ramm 1976, 39). The largest group of smith's tools and waste from the *colonia* were recovered from the General Accident site, Tanner Row and Rougier Street (Hooley *forthcoming*). At those sites smithing slag, hearth linings, iron scrap and smithing tools were found in late second-century contexts. The smithing slags and hearth lining were not found in primary deposits, but re-deposited in ones such as drainage channel fills, road make-up and road surface spreads. There were no concentrations of this material and this, together with the large size of some of the pieces of slag, suggests that the material may have derived from the cleaning or destruction of a single smithy in close vicinity to the excavated area. The distribution pattern of the hearth linings and the slags has led Hooley (*forthcoming*) to postulate that the material represents the periodic clearance and renewal of the hearth linings of a smithy, rather than a single depositional episode.

The General Accident site also produced seven smithing tools from contemporary mid- to late second-century contexts, including punches, a chisel (Figure 2.1), a drift (Figure 2.2) and a handle most probably from a hearth shovel (Figure 2.3). The presence of the drift is noteworthy. Modern smiths who wish to make a hole in an iron bar will pierce it with a punch and then enlarge it with a drift, which is a tool with its widest point below the head. Roman tools like this are so rare that it seems Roman smiths habitually used the punch both to initiate and widen the hole (Manning 1985b, 9). In discussing these tools, Hooley (*forthcoming*) noted that not only was this unusual tool present, but the tools were carefully made and well-finished, all factors suggesting a smithy with an above average tool stock.

Tiles would have been another structural item needed in large quantities when the timber fortress was being built in the early Flavian period, and from time to time thereafter as the fortress was rebuilt and the civilian settlement developed. Tile wasters have been found to the south-west of the fortress in excavations at the Borthwick Institute (King 1974, 213), 21-33 Aldwark (Monaghan 1997, 1071) and the Adams Hydraulic site at Peaseholme Green (*ibid*, 1075). In the same area, at St Cuthbert's Church, there are reports of stacks of tiles being found in the first half of the nineteenth century (RCHME 1962, 65) and these too may relate to manufacture (Monaghan 1997, 1065). The St Cuthbert's Church tiles include examples with stamps of the both the sixth legion and the ninth legion, while those from the Borthwick Institute have stamps of the ninth legion. No kiln structures relating to the tile wasters have been found, but the regular appearance of the wasters in this area suggest that military tile production was centred here in both the first and second centuries and possibly later as well.

The provision of adequate supplies of coarse pottery for the garrison would have been a major concern when the Ninth Legion arrived as there was no local pottery industry that could supply the soldiers. This problem was solved at York, as it was at several other first-century legionary fortresses in Britain, by the production of pottery by, or under the supervision of, the legion itself. At York this development is the origin of the local pottery called Ebor/Eboracum ware (Perrin 1977; Swan this volume). The production of this

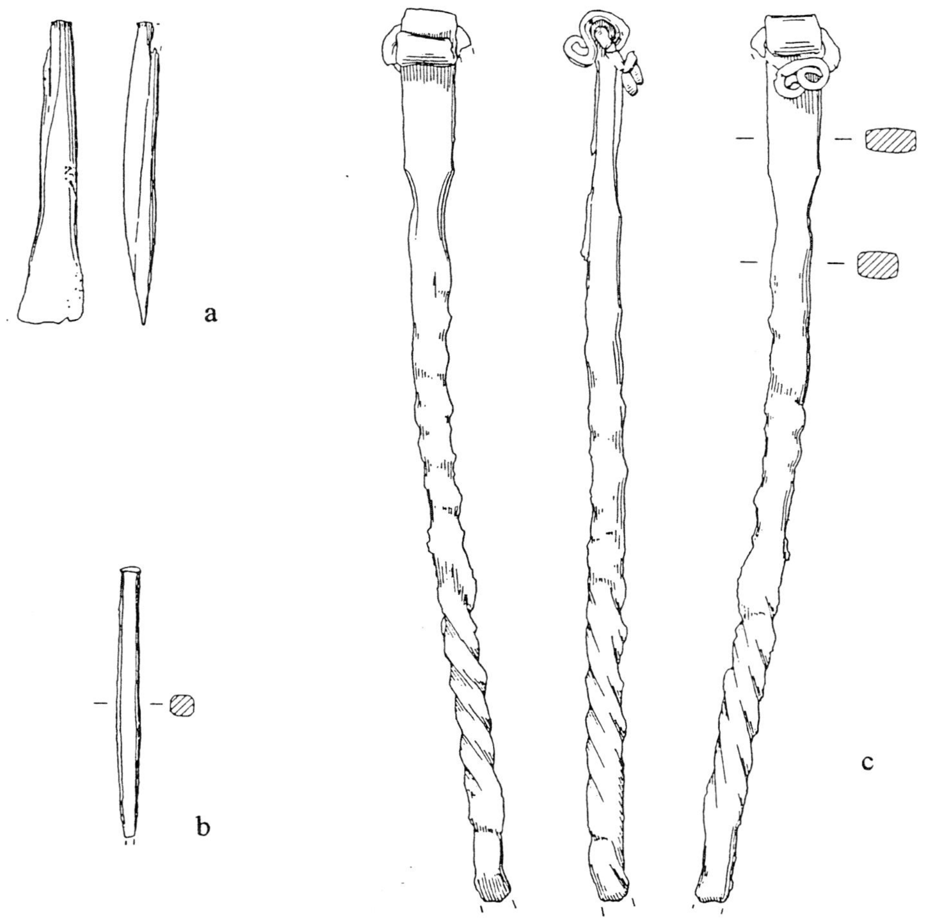

Figure 2. Smithing tools from the General Accident site: a) chisel; b) drift; c) handle, probably from a hearth shovel. (Scale 1:2)

pottery must have started almost as soon as the Ninth Legion arrived as fragments are found in the earliest levels at 9 Blake Street in Period 1 contexts dated to the early 70's (Monaghan 1993, 705). The location of the kilns that produced Ebor ware have to be deduced, in the main, from the position of redeposited pottery wasters and kiln material. One focus of activity is again the area to the south-west of the fortress where tile wasters have been found, and no doubt the tiles and the pottery vessels were products of a single industry. At 21-33 Aldwark wasters and kiln material had been dumped on the site by *c* AD 225, as the material was incorporated in a road built on the site at that time (Monaghan 1997, 1069-70). An earlier road had been constructed in the later second century, and there no evidence was found to suggest that this material had been present when it had been built suggesting that the dumping was taking place towards the end of the second century or early in the third. A third road, broadly contemporary with the second, also used kiln material as hard core, but Monaghan notes that the material in two roads was not the same and suggests this might be explained if there were two loosely contemporary kiln dumps in the vicinity and that the workmen quarried different ones for the two different roads.

Some 150m to the south-east of the site at 21-33 Aldwark, more Ebor ware with `wasting characteristics' was recovered at the Adams Hydraulic site on Peaseholme Green (*ibid*, 1075). This pottery can be dated to the first quarter of the third century and belongs to the range of Ebor ware that is influenced by North African potters (Perrin 1981, 59). Again this material was not found in a primary deposit but redeposited in a ditch. More evidence of the manufacture of Ebor ware in this area may be provided by the pottery found at the Borthwick Institute (King 1974) that lies between and some 50 to 100m to the south of the 21-33 Aldwark and Adams Hydraulics sites. Re-examination of the pottery (Monaghan 1997, 1065) suggests that it is of mid-second-century date and contains no true wasters, but that it may represent the clearance of a warehouse or store close to the kilns.

This area to the south-east of the fortress was clearly a major focus of pottery and tile production. The sites discussed must have been towards the northern edge of the kiln area as in his survey of the pottery Monaghan found very little wasters or kiln material from the sites immediately to the north such as the County Hospital, Fossbank and Jewbury (*ibid*, 1095). How far to the south the kiln area extended is unclear because the area has seen little archaeological activity. Certainly it does not appear to stretch to the vicinity of 16-22 Coppergate as no wasters are reported from the very large, if mainly residual, Roman pottery assemblage from that site (*ibid*, 1077-85).

One other centre of potting activity is known for certain and there are hints that there may have been a third. The former consists of the kilns at Appletree Farm, Heworth Without (Swan 1992, 1 with references to early work; Monaghan 1993, 708; 1997, 1138-42). These are the only kilns that have been excavated in the immediate vicinity of York and lie 3km north-east of the fortress. These kilns were active in the Hadrianic-Antonine period, and produced vessels that fall into the Ebor ware range. The third focus of activity is much more speculative and is represented by two sites lying nearly 1km to the south-east of the fortress across the River Foss. At Paragon Street a few Ebor ware wasters were found, while at the Foss Island Road/Lawrence Street site some wasters of probably third-century greywares were recovered (Monaghan 1997, 1092; see also Swan - this volume).

One final piece of evidence for pottery manufacture that should be briefly noted is the mould for a samian bowl of

Form 37 (Stanfield and Simpson 1958, 16, pl 16 no 206) recorded as coming from the site of the present Railway Station (RCHME 1962, 63 monument 48). It has long been pointed out that there is no other evidence for the manufacture of samian pottery at York in either the form of wasters or the very distinctive kilns (Stanfield and Simpson 1958, 16); and Dr Monaghan (1997, 948) is surely correct when he identifies this as an antiquarian find and nothing to do with industry in York during the Roman period.

The evidence for the working of non-ferrous metals is widely scattered through York but is not very extensive. Within the fortress a small amount of evidence was recovered from 9 Blake Street where fragments from a crucible used to melt impure silver were recovered from a context associated with the first-century timber fortress (Cool *et al* 1995, 1592), and the same site produced a few offcuts of copper alloy (*ibid*). The bronze working evidence reported from below the centurion's quarters of Barrack 2 at the Minster (Phillips and Heywood 1995, 182) may also relate to the early fortress as it predates the mid-second-century building of the barrack. In the area of the *canabae* there is evidence for the melting of both copper alloy and silver in the Coppergate/Piccadilly area. Bayley (1992, 761) notes the presence of a fragment of a typical Roman crucible from a watching brief in the area and other fragments of probably Roman crucibles in post Roman contexts from the excavations at 16-22, Coppergate (*ibid*, 830). Part of a limestone mould for the manufacture of a pewter plate mould was recovered from a sewer trench in Parliament Street (Tweddle 1986, 199, fig 94.689). The dates of the crucibles and mould cannot be suggested from the archaeological contexts, but a late Roman date seems most likely for the mould. There is a growing body of evidence for the manufacture of pewter vessels in Britain in the early Roman period, but the majority still appear to be a late Roman phenomenon (Beagrie 1989, 175).

A very specialised branch of the non-ferrous metal industry is represented by the seven coin moulds found at 21-33 Aldwark (Magilton 1986, 39-40). The moulds were made using coins of Commodus and Severus of AD 193-97, but only obverses were represented indicating that they may have been dumped from a coining operation taking place elsewhere. The only examples found stratified in Roman contexts came from ones broadly contemporary with the laying out of the roads which incorporated the kiln waste and wasters discussed above, and which may be dated to the third century. The date of AD 193-97 provided by the coins which were being copied is, of course, only a *terminus ante quem* for the copying activity. These moulds form part of a wider group which have been found widely scattered throughout Roman Britain, mainly copying coins of Septimius Severus to Severus Alexander (Boon 1974, 111). They are thought to relate to a copying episode of the 230's or 240's (*ibid*, 113).

The status of this activity is interesting. Can it be viewed as part of the legitimate industry of Roman York, or are the moulds evidence of criminal activity? Making coins of copper alloy was not illegal until the fourth century provided they were not used to defraud but rather were used to provide small change when it was in short supply (*ibid*, 114). A classic example of cast coinage possibly not being illegal is the coins made in *c* AD 122 which Walker (1988, 291-92) has proposed were the product of a lawful mint, and part of the official activity aimed at increasing the volume of small bronze coins in the province. The status of the coins produced in the 230's or 240's is obviously different as the volume is much smaller, but it does occur at a time when supplies of small change to Britain from the central mints was low (*ibid*, 300). Perhaps the Aldwark coin moulds represent a local, and possibly legitimate, response to the lack of small change which was needed to oil the wheels of commerce.

There is also evidence for the working of copper alloys south of the Ouse in the *colonia*. Part of a metal workshop apparently using both iron and copper alloys and active in the late second century was recovered at the site of the church of St Mary, Bishophill Senior (Ramm 1976, 39). This produced a crucible and the remains of two superimposed boxes containing copper alloy waste similar to those recovered at Verulamium (Frere 1972, 18-9). At the latter site these were interpreted as the remains of trays positioned below a workbench to catch filings produced by working at a lathe so that the metal could be recycled. Similar trays from Catterick have been shown to have been filled by an activity such as sweeping up filings (Bayley *et al* 2002), so that they may not automatically imply the use of a lathe. The combination of smithing waste in both iron and copper alloy suggests that the workshop may have been producing items similar to the iron key inlaid with a copper alloy strip found in an early second-century context at 9 Blake Street (Cool *et al* 1995, fig 750 no 6272). The waste boxes may suggest the production of lathe-turned items such as vessels. Taken together the evidence suggests a workshop producing high-class products.

At the General Accident site/Tanner Row a crucible for the melting of an alloy that was probably brass was recovered together with a fragment from a mould that had probably been used for the manufacture of copper alloy objects of unknown form (Budd 1987a, 2). Both of these were found in late second-century contexts, as was a flake from a copper alloy crucible found at the neighbouring site at Rougier Street (Hooley *forthcoming*). At the latter site crucibles used for a copper alloy that was probably leaded gunmetal were found in fourth-century contexts, though in one of the cases the context contained much residual material and the crucible may have derived from earlier activity. A small quantity of copper alloy slag and spillage was found at both sites in contexts ranging from the later second to fourth centuries. At neither of these sites was the evidence for working copper alloy in a primary context or directly related to a workshop. Given the amount of dumping that took place at the General Accident site, the presence of scattered fragments of crucibles, moulds and slag need not be indicative of metal working in the vicinity. Indeed it has been suggested that the composition of the pottery assemblage might suggest that one component of the dumped material might have been derived from the fortress (Perrin 1990, 252).

The same problem of origin besets the evidence for gold working at the General Accident site. This consists of three heating trays for melting small amounts of gold (Budd 1987b, 1), probably with the aid of a blow pipe, a tiny gold globule and two gold wire offcuts (Hooley *forthcoming*). Where the material comes from stratified contexts, it belongs to the late second century. The material was not found as part of a coherent group but it could all come from a workshop specialising in working up small quantities of gold to make jewellery from sheet and wire. The type of product might be the sort of jewellery found at Rhayader, Powys and Southfleet, Kent (Cool 1986) where the ornaments are built up by soldering wire and, most interestingly in the General Accident context, small gold globules onto sheet gold backings

One of the most interesting groups of industrial waste from York consists of glass-melting pots and the semi-reacted batch material from 16-22 Coppergate. This was initially thought to be of Viking date (Bayley 1987, 249) because the bulk of the material was found stratified in ninth to eleventh-century contexts. Subsequent analysis has shown that a small amount was found securely stratified in Roman contexts, and this distribution pattern is typical of Roman material at Coppergate where the earliest layers appear to have been much disturbed. Given the extreme rarity of evidence for the manufacture of glass in the north-western provinces during the Roman period, this material has been the subject of an extensive programme of scientific and typological investigation which is fully reported elsewhere and will only be briefly summarised here (Cool *et al* 1999; Jackson *et al* 1998).

Approximately 3kg of pottery fragments retaining melted glass were recovered from a minimum of 36 vessels (Figures 3 and 4). This is by far the largest group of glass-melting pots or crucibles to have been recovered from any site in Roman Britain. They are made of normal Ebor Ware fabric and while some seem to be a specialised form of bowl, possibly made especially for the glass workers, others are typical domestic forms which on typological grounds would appear to have been gathered together in the later second or earlier third century. The glass being melted in them was a soda glass similar to the typical 'Roman' soda glass composition but with slightly raised levels of iron, alumina and potassium. The first two elements might suggest that unrefined sand was being used, and the third that some plant ash was being incorporated.

The first stage in the manufacture of glass from the raw ingredients of sand and soda is a fritting process whereby the ingredients are mixed together and heated so that they can react in the solid state, and gasses and impurities can be driven off (see Turner 1956, 293T-295T for a discussion of the process). At Coppergate fragments of part-fused quartz with or without glass on top and blocks of glass with a white frothy quartz layer on top or interleaved with the glass were recovered in addition to the glass-melting pots. Examination under a scanning electron microscope has shown that this material consists of fragments of unreacted silica in a glassy matrix. This is precisely the material that might be expected if an episode of fritting had gone wrong and the temperature had been allowed to rise too much. Fragments of this semi-reacted material also adhered to some of the glass-melting pots showing that both the fritting and the melting were connected.

The number of glass-melting pots recovered suggests glass-making and melting was being carried out on a relatively large scale. What the glass was subsequently used for is not clear, but it might have been intended for window glazing as during the later second century when the fortress was being rebuilt there would have been a high demand for glass for glazing. A small number of cylindrical moiles and

Figure 3. Glassmaking crucibles from Coppergate.

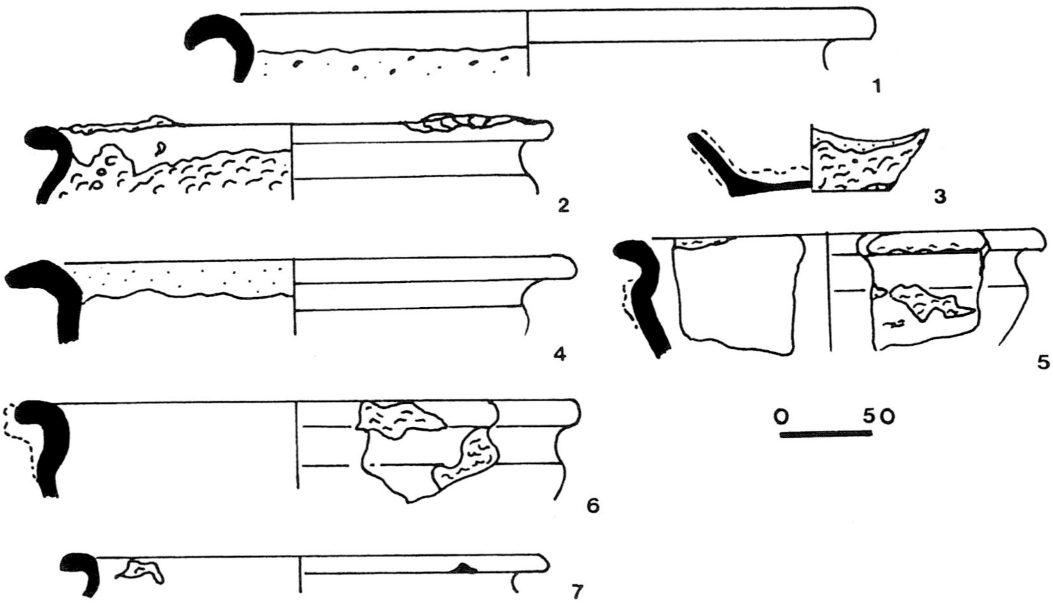

Figure 4. A selection of the glassmaking pots made in Ebor ware. Nos. 1-3 are examples of ordinary domestic vessels. Nos. 4-7 are examples of the specialist form probably deliberately made for the glass-workers (after Cool et al 1999, fig 2). (Scale 1:4)

clipped and pinched fragments of glass of the sort that are normally associated with glass blowing sites (see for example Price and Cool 1991, 25), were recovered from the Coppergate site. There are hints, however, that the glass melting and the glass blowing may not have been connected because of differences in colour and chemical composition. Certainly the glass blowing debris does not reflect the blowing of vessels on anything like the scale of the glass making and melting demonstrated by the glass-melting pots.

Recently (1999) more glass melting pots and blowing waste which appears very similar to that from Coppergate have been recovered during excavations at the Royal York Hotel (Site Code YORYM 1999.840). Analysis of the material has not yet been carried out, so it is unknown whether this represents glass making or only glass working.

The evidence for glass melting or vessel production elsewhere in York is very slight. There is a single fragment of the sort of pinched waste that can be associated with either bead manufacture or the blowing of vessels from 9 Blake Street (Cool *et al* 1995, fig 752 no 6253). As this came from the material that was dumped on the site from an unknown source after *c* AD 280, however, its original date and source is unknown. It is clear that much of the Romano-British glass vessel industry used recycled vessel glass as its raw material (Price and Cool 1991, 27). In previous publications (*ibid*, 24; Cool *et al* 1995, 1592) it has been suggested that melted fragments from the General Accident site, Tanner Row had their origin in the preliminary stages of melting such cullet and could, therefore, be viewed as evidence for industrial activity. Re-examination of this material, in association with information about the contexts from which it came, has raised considerable doubts about this interpretation. The melted fragments from the General Accident site are widely scattered and do not necessarily suggest deliberate melting. A much larger amount of melted material was recovered from the neighbouring Rougier Street site (more than half a kilogram) and this was concentrated in a late second-century make-up layer. The other contents of the make-up, however, also showed considerable signs of burning and charring, and might be interpreted as the remains of a building which collapsed *in situ*. In such circumstances the melting of the glass might be accidental and cannot be used as evidence of industrial activity.

Tesserae are another item made of glass that might be indicative of industrial activity. As noted in the discussion of the examples from 9 Blake Street (Cool *et al* 1995, 1592), though they may have been used for wall or floor mosaics, evidence for this is rare from Romano-British sites. In Britain they have rarely been found on sites where such luxurious mosaics are to be expected but have been found on sites where metalworking was taking place. There is, therefore, a distinct possibility that their presence is indicative of an activity such as enamelling or bead-making. In addition to the examples from Blake Street, others have been recovered from the Purey Cust Nuffield Hospital site (*ibid*, fig 775 no 6523), the Minster (Price 1995, 353 fig 146 nos 93-4), and Wellington Row (Ottaway and Cool 1995, 35). Non-ferrous metalworking might also be suggested at the last-mentioned site by crucibles and bright flecks visible in the X-radiographs of the ironwork, as the latter can be indicative of non-ferrous metal waste (Jane Cowgill *pers comm*). Full investigation of all this evidence will be needed before it can be interpreted with confidence, but it is worth noting that it may be an example of a late Roman industry at York as it was concentrated in the late fourth to fifth-century dumps at that site.

All of these industries would have required considerable supplies of fuel, and it is interesting to note that this may, in part, have been in the form of coal. The largest quantities of coal yet reported were from the Trentholme Drive cemetery where it was clearly being used to fire the crematorium (Wenham 1968, 21), but the coal at the General Accident site, Tanner Row and Rougier Street, Wellington Row and 1-9 Micklegate (Dearne and Branigan 1995, 98) was presumably present for industrial purposes. The reason for coal to be present at Blossom Street where it was found in pits of Severan date is unclear (Wenham 1965, 542). Given the presence of a cemetery in the area (Frere 1990, 326), a funerary rather than an industrial reason may be more likely.

Low temperature craft activities

The manufacturing activity that is most frequently associated with York is the working of jet, and a picture is sometimes presented of a thriving export trade to the Rhineland (see for example Salway 1981, 658; Ottaway 1993, 84, but see also Allason-Jones this volume). Finished artefacts made of jet or a jet-like substance are common finds at York, but direct evidence of manufacturing activity is not particularly common and the recent excavations have added only a little to that collected in the Royal Commission survey (RCHME 1962, 142). The evidence at that time consisted of the blocks of jet and unfinished pins recovered during the building of the present railway station in 1873, a fragment possibly from the manufacture of bangle from The Mount, and a similar unprovenanced fragment now in the Yorkshire Museum. A possible core produced while lathe-turning of a bracelet, also unprovenanced and in the Yorkshire Museum, had been reused as a counter as may have a similar disc found at 9 Blake Street (Cool *et al* 1995, fig 757 no 5818). This reuse has turned them into `finished' articles and thus their presence on sites cannot be used as direct evidence of manufacturing activity there. A recent survey of the jet in the Yorkshire Museum has reinterpreted the unfinished pins from the Railway Station as blanks for beads (Allason-Jones 1996, 50 nos. 327-31), and has added an unfinished finger ring from a grave at Heslington to the corpus of early finds (*ibid*, 50 no 323).

A jet slab retaining one cut edge (Cool *et al* 1995, 1649 no 5819; Allason-Jones 1996, 51 no 339) which may be evidence of manufacture was also found at Blake Street but, with the exception of this, the new evidence is concentrated in the *colonia*. Allason-Jones (1996, 52 no 342) has suggested that one block from Rougier Street was a possible bead blank, and that another block from the General Accident site, Tanner Row was a blank for a handle or massive pin (*ibid*, 52 no 340). Hooley (*forthcoming*) has identified two small fragments at Rougier Street and the General Accident site, Tanner Row as waste flakes from a manufacturing process. Given the amount of sieving that took place at the General Accident site (Hall and Kenward 1990, 295), it seems likely that more flakes of this nature would have been recovered had the manufacture of jet items been one of the activities regularly carried out on the site. The flakes were found in a later second and an early to mid-third-century contexts respectively, and therefore though they may not indicate an on-site industry, they do confirm the suggestion that jet-working was being carried out in York before the late Roman period (RCHME 1962, 142). Pebbles of jet or a jet-like substance and some roughly squared blocks have also been recovered from Wellington Row and these may have been present on site as raw material. Some of these too were found in early third-century contexts. A fragment of unworked jet was also found in the fourth-century backfill of a well in Skeldergate (MacGregor 1978, 60 no 276)

As well as producing evidence of the manufacture of jet ornaments, the building works for the railway station in 1873, also produced unfinished bone pins and a piece of `ivory' interpreted as being a polishing tool (RCHME 1962, 63 monument 48). It has been suggested that this area was used for industrial activity prior to the development of the late Roman cemetery (*ibid*) but to what extent the jet and bone-working debris can be interpreted as being in primary deposits as opposed to having been dumped from elsewhere is unknown, as is also their chronological relationships to both each other and the cemetery. Other evidence of the working of bone and similar materials in York is very scarce though this must have been a very common activity. There is a single bone point rough-out at the General Accident site, Tanner Row in a late second-century context (Hooley *forthcoming*), and fragments of worked antler that appear to be the by-products of manufacture from the Church Street sewer (MacGregor 1976, 23 no 144) and from Blake Street in a late Roman context (Cool *et al* 1995, 1598).

The sites to the south-east of the Ouse are frequently water-logged and this has resulted in the recovery of much leather including debris from manufacture. At the General Accident site, Tanner Row considerable quantities of offcuts from the manufacture of shoe soles were found in the late second to early third-century dumps (Hooley *forthcoming*),

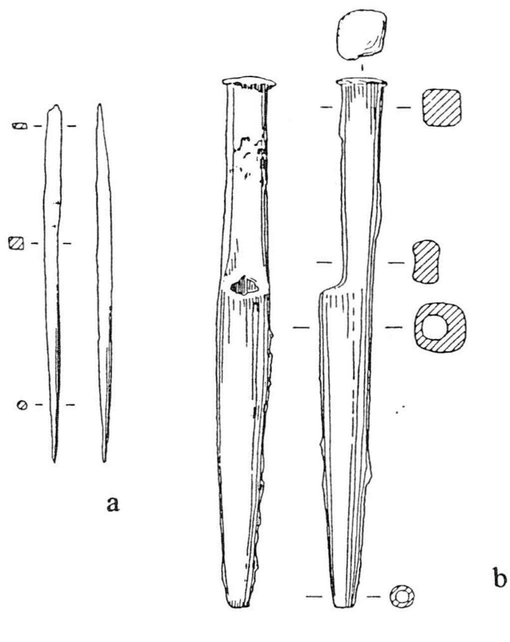

Figure 5. Leather-working tools from the General Accident site: a) awl; b) hole punch. (Scale 1:2)

and similar material was found in contemporary deposits at the nearby Wellington Row site. An interesting feature of the large quantities of dumped leatherwork at the General Accident site is that much appears to be the result of recycling damaged military material such as tent panels. Damaged areas had been cut out and discarded and presumably the tent panels themselves had been patched and repaired. Several discarded, worn items had regular shapes and curves cut out from them as if areas had been recycled before final discard (Hooley *pers comm*). The General Accident site also produced two iron tools associated with leather-working from these deposits, an awl (Figure 5.a) and a hole punch (Figure 5.b) of unusually high quality (Hooley *forthcoming*). Leather-working clearly went on elsewhere in York as well, for example at Blake Street (Cool *et al* 1995, fig 753 nos 6282-3) but away from the river we have to rely on the presence of tools to indicate this rather than the leather itself.

Textile production is very poorly attested in York and relatively few of the associated tools such as spindle whorls have been recovered. Amongst those that have, however, there are some of special note because they represent rarely encountered types. These include the two bone spindles and the wooden whorl from the General Accident site, Tanner Row (Hooley *forthcoming*) and the bone spindle from a very late Roman context at St. Mary, Bishophill Senior (Ramm 1976, 59). Another item of some interest both in itself and from its context, is the ivory weaving tablet found in a centurion's barrack at the Minster (Walton 1995). Ivory items are not common in Roman Britain and so this weaving tablet would have been likely to have come from an expensively furnished work basket. As the production of braid using such tablets has not often been considered a military activity, perhaps this item reflects the presence of the centurion's wife or companion in his quarters.

Much of the stone- and woodworking carried out in Roman York would have been connected with the building trades, and thus lies beyond the scope of this paper. It is to be expected that carpenters and stonemasons manufacturing objects would also have been active in the city throughout the Roman period. They have, however, left no direct traces of their activities in the form of workshops, waste or tools. Indirect evidence for the activities of the masons can be found in the geology of the stones found in York. Buckland (1988, 239) notes that all the references to "gritstone" in the Royal Commission survey (RCHME 1962) that he could check were to rocks of Upper Carboniferous Millstone Grit whose nearest outcrops are about 25km from York (see also Gaunt and Buckland – this volume). This was obviously the preferred building stone in York and very large quantities were transported to the city. It was also used for altars (for example) RCHME 1962, 115-19 nos 30-31, 34-35, 40, 42, 45-47), dedication tablets (for example RCHME 1962, 119 nos 52-55), statues (for example RCHME 1962, 120 nos 58-59), tombstones (for example RCHME 1962, 121-28 nos 71-73, 77-82, 92-99, 101, 102) and sarcophagi (for example RCHME 1962, 128 nos 103-09). Though it is always possible that these were carved elsewhere and brought to York as finished items, more economical hypothesis is that they were commissioned and carved locally. Certainly items such as altars, tombs and sarcophagi carved with the names of particular individuals living and/or buried at York would, at the very least, have to be finished locally. The statue of Arimanes dedicated by Volusious Irenaeus (for example RCHME 1962, 20 no 58, pl 46) is such an unusual piece that it implies close liaison between the donor and the craftsman, again most easily carried out if both resided in the same place. The evidence that survives thus points to thriving stone masons and sculptors yards in York.

An overview of manufacturing activity in Roman York

As will be obvious from the forgoing sections, there is quite a substantial body of evidence for manufacturing activity in Roman York but equally it is apparent that it suffers from three basic problems when we try to build up a coherent picture. The first is the spatial bias mentioned at the outset of this paper, i.e. there is evidence where there have been excavations and these have been concentrated in certain areas. The second is that there is a distinct chronological bias as well. Where it can be dated, most of the evidence belongs to the second and earlier third century. The manufacturing activity that must have gone on during the earliest years of the fortress is rarely represented. There is an almost equal lack of knowledge about later third and fourth-century activity, though there are indications that this would be corrected, in part at least, were the large quantities of artefactual material recovered from the Wellington Row site to be studied in detail. The third problem is one of re-deposition. The majority of the evidence has not been found in primary deposits but forming parts of dumps or make-up layers. The only examples of sites retaining both parts of the structures and the waste products are the Heworth kiln (see p3) and the metalworking workshop on the site of the church of St Mary, Bishophill Senior (see p 4). An assumption can be made that the redeposited debris has not moved far from its place of origin to its place of discovery, but it should always be born in mind that it is only an assumption. Material might have been transported some distance if the aim was to level previously unoccupied areas before building. We may note, for example, Perrin's suggestion (1990, 252) that some of the later second-century accumulations at the General Accident site Tanner Row comprised material brought over from the fortress. This seems to be supported by the significant amount of military leatherwork in these accumulations. These include a graffito on a corner reinforcing panel of a tent attributing it to the century of Sollius Iulianus (Hassall and Tomlin 1987, 373 no 31). This man is known to have been active on Hadrian's Wall towards the end of Hadrian's reign, and this highlights the fact that not only may dumped material be moved but it may also considerably predate the dumping activity. The evidence of the pottery and the leatherwork does pose the question of whether or not some of the evidence for industrial activity at the Tanner Row site relates to the early years of the fortress and not to the early years of occupation in the *colonia*. It is not possible to answer this question, but it does emphasise the problems of interpretation that can be present with redeposited material.

The spatial bias against evidence from the interior of the

fortress is a particular problem. Legions were not just fighting units, they contained a large number of craftsmen and the fortresses would have been centres of manufacturing activity. Vegetius provides a useful list of these people and their activities.

'The legion, furthermore, has wood-workers, masons, carpenters, blacksmiths, painters and all other workers for constructing buildings...... the legion has men prepared to make new weapons, vehicles, and other kinds of military devices or to repair the damaged ones. They also had workshops for making shields, breastplates, and bows, in which also arrows, missiles, helmets, and all types of armour were made. For this was a special concern, that whatever seemed necessary to the army was never lacking in the camp.'

Vegetius 2.xi (p 83)

It is likely that a considerable amount of such activity took place within the fortress. At the fortress of the Second *Augusta* at Caerleon, for example, the remarkable group of material found in a third-century rampart building at Prysg Field suggests that it functioned as both a weapons and armour store and perhaps as a workshop (Bishop and Coulston 1993, 200). Earlier there is evidence that the working of precious metals was carried out in the quarters of the *praefectus castrorum* (Zienkiewicz 1993, 76). Within the fortress at York, the only evidence of such activity is the tiny quantities of metalworking debris at 9 Blake Street (see p 1) and the reported presence of bronze working at the Minster (Phillips and Heywood 1995, 98, see also 182). The presence of a *fabricae* in the vicinity of the Purey Cust Nuffield hospital can be postulated from the damaged military equipment found there (Cool *et al* 1995, 1632), but hard evidence of its presence is lacking.

Manufacturing activity by legionaries, or under their supervision, need not have been confined within the walls of the fortress, and it seems very likely that one of the main industrial centres for the ninth legion and later for the sixth, was the area to the south-west of the fortress in the Aldwark/Coppergate area. The northern part of this area was clearly the location of fired clay industries directly associated with both legions (p 2). In the south there is evidence of non-ferrous metal working, the making of glass and the blowing of glass vessels. Much legionary military equipment was made of copper alloys and so the presence of such an industry under military control is not surprising. The military associations of the collection and working of glass are less often commented on, but just as pottery industries may have been established so that in Vegetius's words `whatever seemed necessary to the army was never lacking in the camp', the possibility that from time to time the legion undertook the production of glass windows and vessels should not be ruled out. Certainly the location of the Coppergate glass industry so close to the fortress and in an area that might be suggested to be under the control of the military is not unique. A very large collection of broken vessel glass (cullet), presumably gathered for remelting and reworking, was found buried in the *canabae* of the legionary fortress at Nijmegen, Netherlands (Isings 1980); and the cullet deposit in Guildhall Yard, London also has a similar relationship to the fort in London (John Shepherd *pers comm*).

It would thus be possible to postulate that this area to the south-west of the fortress was given over to military manufacturing activities in the first to early third centuries and, furthermore, that the area was zoned with fired clay industries which required more space to the north and other activity to the south. In the light of this suggestion it is worth re-examining the ironworking evidence recovered from 16-22 Coppergate (Ottaway 1992, mf 2.A4 nos 1465-1615 and tables 29 and 30). The possibility that some of this, or much else of the ironwork assemblage other than nails and tacks, could predate the ninth century has been discounted on the grounds that very little ironwork was found in Roman contexts on the site (*ibid*, 465 and 478). This could not take into consideration the fact that the majority of the Roman finds from the site are found residually, and relatively few in securely stratified Roman contexts. Over 40% of the Roman vessel glass, for example, was found in the ninth-century contexts, and, as already noted, the majority of the late second to early third-century glass-making evidence was also found in these deposits. Clear evidence for residual Roman ironwork in these deposits is provided by such items as the calthrop (*ibid*, 718, fig 313.3944). In addition it is hard to see from what source other than a Roman one the scrap found in fifth to ninth-century contexts (*ibid*, mf 2:A4 nos 1465-71) could have been derived, as theoretically the site was deserted at this time (Bayley 1992, 743). In these circumstances it seems quite probable that some of the iron smithing evidence recovered dates to the Roman rather than the Anglo-Scandinavian period, and that this is another industry that should be added to the legionary industrial zone.

If we are correct in seeing this area of York as a military industrial complex, it casts an interesting sidelight on the coin moulds from 21-33 Aldwark (see p 4). These suffer from the problem of being redeposited; but it is an interesting coincidence that they should be recovered in this area and it may hint that the coining activity was not illegal. It is hard to imagine that forgers would happily have set up an illicit activity anywhere near such an official area (see also Swan this volume).

At present there is no similar focus for civilian manufacturing activity in York. The pottery production at Heworth Without is thought to have had its origins as a civilian establishment, (Swan 1992, 1) though possibly the civilians in question were veterans (Monaghan 1997, 871). Manufacturing was also clearly being carried out within the *colonia* and some appears to have been of high quality goods. At present, for example, such evidence as there is for the manufacture of jet or shale items in York itself is concentrated in the *colonia* area or immediately outside of it. The probably high quality of the products of the metal workshop on the site of St Mary's church, Bishophill Senior has already been noted (see p 4), and a similar high standard might be hinted at by quality of the smithing tools found at the General Accident site, Tanner Row.

In conclusion, though we can begin to produce in outline a picture of manufacturing activity at York, we are far from having the full story; and it is to be hoped that future work, including the study of material already excavated, will provide some additional colour to this sketch.

Endnote

This paper was written in 1995. It was revised in 1998 to take account of material published subsequently and the results of the Coppergate Glass-making Project.

Acknowledgements

I would like to thank Dr P Addyman, the director of the York Archaeological Trust, for permission to use currently unpublished material from excavations by that organisation in this article. I am also grateful to Dr P Ottaway of the same organisation for providing the contextual information for the glass from Tanner Row and Rougier Street. I was invited to look at the glass working debris by Dr A Vince and Mr N Pearson. Ms L Allason-Jones discussed her survey of the York jet with me in advance of its publication, and I have also benefited from useful discussions with my fellow-workers on the Coppergate glass working project, Drs C Jackson and J Monaghan. The paper has also benefited from the comments of an anonymous referee. To all of these I extend my grateful thanks. I have already signalled my indebtedness to Mr A D Hooley and to Dr Monaghan for permission to use their work in advance of publication. It is, however, a great pleasure to thank them again here as they not only allowed me access to their manuscripts but also commented on this paper. It has been improved by those comments but any misinterpretations remain, of course, mine alone. Figures 2, 3 and 5 are courtesy of the York Archaeological Trust.

Bibliography

Vegetius Stelten, L F (ed) 1990 *Flavius Vegetius Renatus Epitoma Rei Militaris*. New York: Peter Lang

Addyman, P V, 1974 Excavations in York, 1972-1973 first interim report. *Antiq J* **54**, 200-31

Allason-Jones, L, 1989 *Roman Jet in the Yorkshire Museum*. York: Yorkshire Museum

Bayley, J, 1987 Viking glassworking: the evidence from York. *Annales du 10e Congrés de l'Association Internationale pour l'Histoire du Verre*. Amsterdam: l'Association Internationale pour l'Histoire du Verre, 255-70

---- , 1992 *Anglo-Scandinavian Non-ferrous Metalworking from Coppergate*. Archaeol York **17/7**. London: CBA

----, Canti, M, Campbell, G, and Dungworth, D, 2002 Metalworking at Catterick in P R Wilson *Cataractonium. Roman Catterick and its Hinterland. Excavations at Research 1958-1997*. Part II CBA Research Report **129**, 164-66.

Beagrie, N, 1989 The Romano-British Pewter Industry. *Britannia* **20**, 169-91

Biek, L, 1976 The metalworking residues. In Ramm (1976), 68

Bishop, M C, and Coulston, J C N, 1993 *Roman Military Equipment from the Punic Wars to the Fall of Rome*. London: Batsford

Boon, G C, 1974 Counterfeit coins in Roman Britain. In J Casey and R Reece (1974) *Coins and the Archaeologist*. Oxford: Brit Archaeol Rep, Brit Ser **4**, 95-170

Brinklow, D, Hall, R A, Magilton, J R, and Donaghey, S, 1986 *Coney Street, Aldwark and Clementhorpe Minor Sites and Roman Remains*. Archaeol York **6/1**. London: CBA

Buckland, P C, 1988 The stones of York. Building materials in Yorkshire. In J Price and P R Wilson (eds) *Recent Research in Roman Yorkshire*. Oxford: Brit Archaeol Rep, Brit Ser**193**, 237-87

Budd, P, 1987a *Crucible fragments from Tanner Row, York*. Ancient Monuments Laboratory Rep **18/87**. London: English Heritage

---- , 1987b *Examination of Some Crucible Fragments from Tanner Row and Rougier Street, York*. Ancient Monuments Laboratory Report **82/87**. London: English Heritage

Carver, M O H, Donaghey, S, and Sumpter, A B, 1978 *Riverside Structures and a Well in Skeldergate and Buildings in Bishophill*. Archaeol York **4/1**. London: CBA

Cool, H E M, 1986 A Romano-British gold workshop of the second century. *Britannia* **17**, 231-7

---- , Jackson, C M, and Monaghan, J, 1999 Glass-making and the Sixth legion at York. *Britannia* **30**, 147-61

---- , Lloyd-Morgan, G, and Hooley, A D, 1995 *Finds from the Fortress*. Archaeol York **17/10**. York: CBA

Dearne, M J, and Branigan, K, 1995 The use of coal in Roman Britain. *Antiq J* **75**, 71-105

Frere, S S, 1972 *Verulamium Excavations Vol. I*. Oxford: Rep Res Comm Soc Antiq London **28**

---- , (ed) 1983 Roman Britain in 1982. I. Sites explored. *Britannia* **14**, 280-335

---- , (ed) 1984 Roman Britain in 1983. I. Sites explored. *Britannia* **15**, 266-332

---- , (ed) 1990 Roman Britain in 1989. I. Sites explored. *Britannia* **21**, 304-64

---- , (ed) 1992 Roman Britain in 1991. I. Sites explored. *Britannia* **23**, 256-308

Hall, A R, and Kenward, H K, 1990 *Environmental Evidence from the Colonia*. Archaeol York **14/6**. London: CBA

Hall, R A, 1997 *Excavations in the Praetentura: 9 Blake Street*. Archaeol York **3/4**. York: CBA

Hassall, M W C, and Tomlin, R S O, 1987 Roman Britain in 1986. II. Inscriptions. *Britannia* **18**, 360-77

Hooley, A D, *forthcoming Roman finds from the General Accident site, Tanner Row and Rougier Street*. Archaeol York **17/-**

Isings, C, 1980 Glass from the *canabae legionis* at Nijmegen. *Ber Rijksdienst Oudheidkund Bodenmerz* **30**, 281-346

Jackson, C M, Cool, H E M, and Wager, E C W, 1998 The manufacture of glass in Roman York. *J Glass Stud* **40**, 55-61

King, E, 1974 Roman kiln material from the Borthwick Institute, Peaseholme Green: a report for the York Excavation Group. In Addyman (1974), 213-17

MacGregor, A, 1976 *Finds from a Roman Sewer System and an Adjacent Building in Church Street*. Archaeol York **17/1**. London: CBA

---- , 1978 *Roman Finds from Skeldergate and Bishophill*. Archaeol York **17/2**. London: CBA

Manning, W H, 1985a The iron objects. In L F Pitts and J K St Joseph (1978) *Inchtuthil: the Roman Legionary Fortress Excavations 1952-65*. London: Britannia Mono **6**, 289-99

---- , 1985b *Catalogue of the Romano-British Iron Tools, Fittings and Weapons in the British Museum*. London: British Museum

Magilton, J R, 1986 A Roman building and Roman roads in Aldwark. In Brinklow *et al* (1986), 33-48

Monaghan, J, 1993 *Roman Pottery from the Fortress: 9 Blake Street*. Archaeol York **16/7**. London: CBA

---- , 1997 *Roman Pottery in York*. Archaeol York **16/8**. York: CBA

Ottaway, P, 1992 *Anglo-Scandinavian Ironwork from 16-22 Coppergate*. Archaeol York **17/6**, London: CBA

----, 1993 *Roman York*. London: Batsford/English Heritage

----, and Cool, H E M, 1995 *Early York: Development and Change. Sites South-west of the Ouse in the First Millennium*. Unpublished Assessment and Project Design presented to English Heritage 13th March 1995

Perrin, J R, 1977 Legionary Ware in York. In J Dore and K Greene (eds) (1977) *Roman Pottery Studies in Britain and Beyond*. Oxford: Brit Archaeol Rep, Supp Ser **30**, 101-10

----, 1981 *Roman Pottery from the Colonia: Skeldergate and Bishophill*. Archaeol York **16/2**. London: CBA

----, 1990 *Roman Pottery from the Colonia 2: General Accident and Rougier Street*. Archaeol York **16/4**. London: CBA

Phillips, D, and Heywood, B, 1995 *Excavations at York Minster Volume 1. From Roman Fortress to Norman cathedral*. London: HMSO

Price, J, 1995 Roman glass. In Phillips and Heywood (1995), 346-71

----, and Cool, H E M, 1991 The evidence for the production of glass in Roman Britain. In D Foy and G Sennequier (eds) (1991) *Ateliers de Verriers de l'antiquité à la période pré-industrielle*. Rouen: Assoc Française for l'Archéologie du Verre, 23-9

RCHME 1962 = Royal Commission on Historical Monuments England *An Inventory of the Historical Monuments in the City of York 1. Eburacum: Roman York*. London: HMSO

Ramm, H G, 1976 The church of St Mary, Bishophill Senior, York, excavations 1964. *Yorks Archaeol J* **48**, 35-68

Salway, P, 1981 *Roman Britain*. Oxford: Clarendon Press

Shirley, E A M, 1996 The building of the legionary fortress at Inchtuthil, *Britannia* **27**, 111-28

Stanfield, J A, and Simpson, G, 1958 *Central Gaulish Potters*. London: Oxford Univ Press, Univ Durham Publ

Swan, V G, 1992 *Legio* VI and its men: African legionaries in Britain. *J Roman Pottery Stud* **5**, 1-33

Taylor, D J A, 1999 A note on, 'The building of the legionary fortress at Inchtuthil', *Britannia*, **30**, 297-99

Turner, W E S, 1956 Studies in ancient glasses and glassmaking processes. Part V. Raw materials and melting processes. *J Soc Glass Technology* **40**, 277T-300T

Tweddle, D, 1986 *Finds from Parliament Street and Other Sites in the City Centre*. Archaeol York **17/4**. London: CBA

Walker, D R, 1988 The Roman coins. In B Cunliffe (ed) (1988) *The Temple of Sulis Minerva at Bath. Volume 2. The Finds from the Sacred Spring*. Oxford: Oxford Univ Comm Archaeol Mono **16**, 281-358

Walton, P, 1995 An ivory weaving tablet. In Phillips and Heywood (1995), 428-29

Wenham, L P, 1965 Blossom Street excavations York, 1953-1955. *Yorks Archaeol J* **41**, 525-53

----, 1968 *The Romano-British Cemetery at Trentholme Drive, York*. London: HMSO, Ministry Public Buildings Works Archaeol Rep **5**

Whitwell, J B, 1976 *The Church Street Sewer and an Adjacent Building*. Archaeol York **3/1**. London: CBA

Zienkiewicz, J D, 1993 Excavations in the *Scamnum Tribunorum* at Caerleon: the Legionary Museum site 1983-5. *Britannia* **24**, 27-140

CRAFT AND INDUSTRY ON THE NORTH YORK MOORS IN THE ROMAN PERIOD

P R WILSON

Introduction

'Few sites in the area have shown evidence of manufacture ...' The late Don Spratt used these words to introduce his consideration of Roman period industrial activity in his excellent volume *Prehistoric and Roman Archaeology of North-East Yorkshire* (Spratt 1993, 160). The apparent dearth of evidence was reflected in the brevity of his discussion which ran to 250-300 words.

The difficulty of recognising the Roman period in the area is a fundamental obstacle to any study that seeks to examine the period in isolation. This problem has been discussed elsewhere (Wilson 1995) and will not be considered at length here. However a brief summary is necessary to provide a background to what follows.

The pottery tradition of the Moors is essentially highly conservative with hand-made calcite-gritted ware dominating the ceramic assemblages from the late Iron Age through the whole of the Roman period (but see Evans (1995, 48-50) with regard to Iron Age 'calcite-gritted' wares). Therefore recognition of Roman period sites is difficult. With a few exceptions it is only with the appearance of commercially produced East Yorkshire ceramics, notably Crambeck ware, in the later third and fourth centuries that Roman occupation is clearly evidenced.

For the purposes of this paper the North York Moors will be regarded as extending from the Tabular Hills in the south to include the Cleveland Hills and northern outliers such as Eston Moor. The eastern limit will be the coast and the study area will extend to take in the Hambleton Hills in the south-west (Figure 1).

Crafts, industry and the Roman army (Figure 2)

The known Roman period military sites in the study area are limited and despite the well-attested evidence for craft and industrial activities, even on short-lived military sites elsewhere, the evidence from the study area is meagre. At Lease Rigg fort Hayes and Rutter (1964, 70) report one piece of iron slag from the site. Although Hartley notes more in his first interim report he locates it all to topsoil and post-Roman features, or regards it as having been introduced into Roman levels in mole-runs. He concludes that 'there is nothing now to suggest metalworking in the Roman period' (Hartley 1978, 32). However his second interim report on the site includes

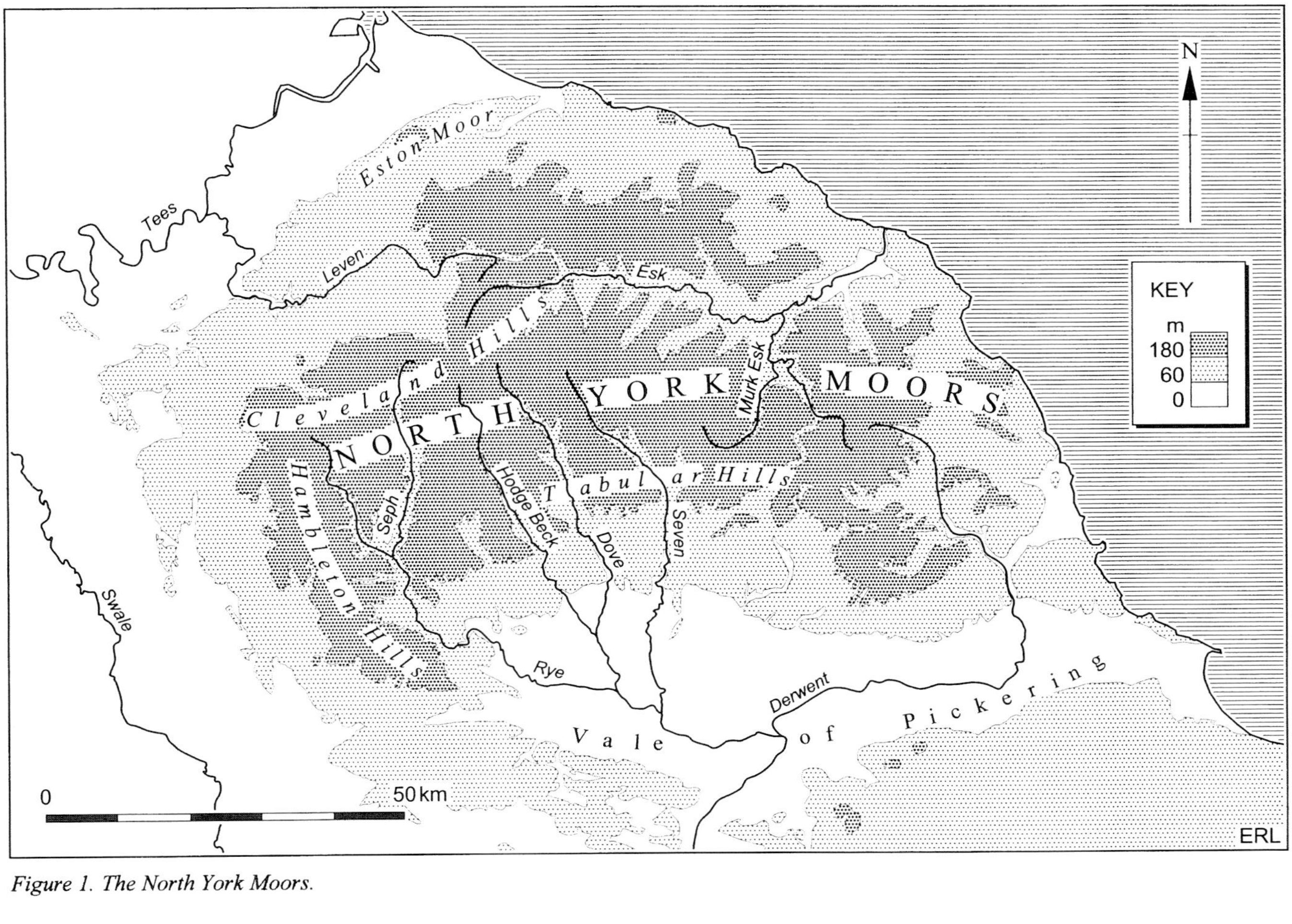

Figure 1. The North York Moors.

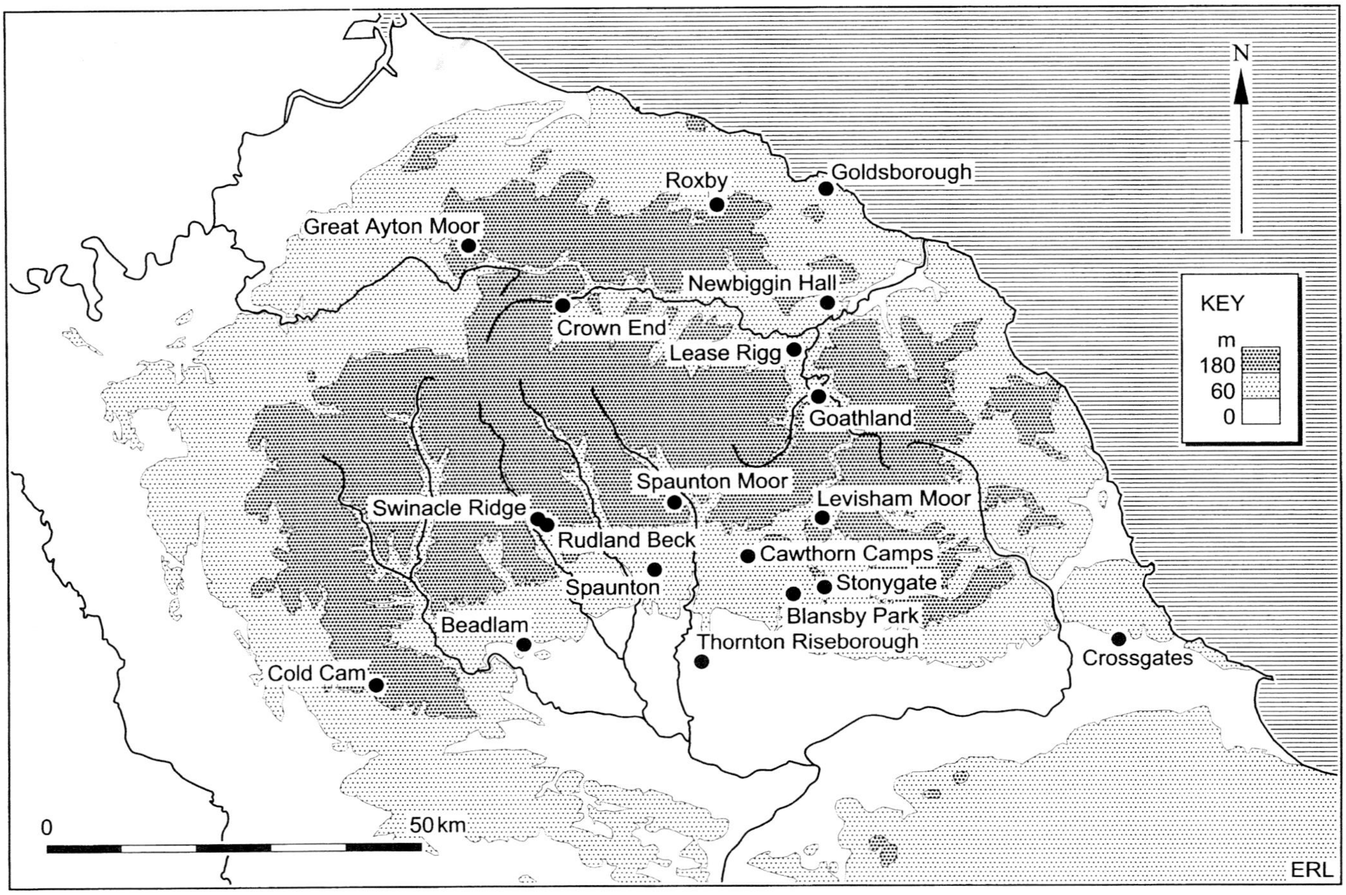

Figure 2. Sites mentioned in the text.

reference to `several pits, including what appears to be the stokehole and flue of a furnace .... No conclusive evidence of purpose was found, though intense heat suggests an industrial process' (Hartley 1980, 10). Given that the site suffered from extensive plough damage the suspicion may be entertained that the iron slag recorded from the site could have originated in Roman levels since redistributed by the plough, and could possibly have derived from industrial/craft processes associated with the recorded furnace, or other undiscovered features. Even so this slag is perhaps more likely to derive from smithing and not represent anything other than the routine repair and maintenance that might be expected on any fort site.

Still at Lease Rigg Hartley (1980, 10) refers to a phase of reconstruction in stone, albeit one only represented by a single heavily-robbed building (K). Hayes and Rutter refer to the discovery of building stone in their trenches B4 and B5 towards the south-eastern corner on the site, the former producing `a large squared block' resting on the lowest humic layer. This represents a fortuitous survival and attests to utilisation of something other than freestone in the stone buildings on the site and therefore suggests the possibility of Roman army masons working on the site.

The site of Cawthorn Camps produced little evidence of craft or industrial activity. Despite a reference to the use a of `squared stone' in a post-hole in the first interim report (Richmond 1925, 334), nothing further of relevance to the present inquiry was reported, not has any evidence derived from recent work on the site (Wilson and Lee 2000-2001). Even allowing for the possibility that further Roman military sites could be found in the region, notably the putative third fort towards the northern end of Wade's Causeway (Hartley 1993, 162), on the evidence available it would seem reasonable to assume that the appearance of the Roman army in the region provided little direct stimulus to developments in craft activity. Given that the initial military occupation of the area appears to have lasted for no more than about 50 years this is perhaps not surprising. The coastal signal stations may be regarded as the second phase of the Roman military presence in the study area and these sites utilised considerable quantities of worked stone (Wilson 1991).

If developments in craft and industry cannot be seen as a result of the appearance of the Roman army, can they be seen at all? In short the answer is `not really', at least if changes directly attributable to the incorporation of the area into the Roman province are what is sought. However the recognition of pre-Roman craft activity is also difficult, although a number of activities and associated artefact types have been recognised.

Crafts and the civilian population (Figure 2)

Querns

Quern production in the region has been discussed by various workers (Hayes 1974; 1976; Hayes *et al* 1980, 304-06; Briggs 1988, 298-300), and we are shortly to benefit from the publication of the research undertaken as part of the

Yorkshire Quern Survey (Heslop *in prep*).

The beehive quern is suggested as being current in the study area between *c* 200 BC and AD 200 (Hayes *et al* 1980, 306-07) and despite the recognition of beehive querns in early Iron Age contexts at Thorpe Thewles north of the Tees Valley, there is no evidence at present that necessitates any revision of their dating in the area of the North York Moors. However querns may also serve to demonstrate the conservatism of the region with the continued use of saddle querns into the Roman period, as for example from the fourth-century aisled building at Old Pasture Field, Spaunton (Whitaker 1967, 22; Hayes 1974, 25). However saddle querns, unlike beehive querns, can be used for 'crushing nuts, pounding roots, powdering cosmetics' (Heslop 1988, 61 - following Barker 1985, 12) and their continued use might simply demonstrate a range of activities beyond grinding grain.

Saddle querns and beehive querns may provide our first evidence relating to modes of production and, if so, it is likely that there was little change between the Iron Age and Roman periods. Where the production, lithology and distribution of saddle querns has been studied, for example at Lodsworth, West Sussex, it has been suggested that those who required them "would take away raw material which would be fashioned to the desired shape at home, and it is possible that every household met its own requirements."

This model would neatly explain the observed facts, namely:

1. *saddle querns are irregular and made by different techniques even on a single site.*
2. *they are found at no great distance from the quarry, a day or, at the most, two days return journey on foot, and*
3. *no saddle querns have been found roughed out at the quarry despite the most diligent search.' (Peacock 1987, 76).*

Such a model of self-sufficient home production of saddle querns would broadly accord with the evidence from the North York Moors. Although of the five quern quarries recognised in the study area, only the one on Spaunton Moor (Hayes *et al* 1980, 304-05) has produced saddle querns.

The concept of 'home production', along with that of 'local specialist production' will be pursued below. For the purposes of this study a 'home producer' is seen as someone working only to supply the needs of their family and immediate community and 'local specialist producer' someone who would supply their immediate community and those in a wider area.

Beehive querns derived from the channel sandstone of the Goathland quarry site are considered to travel up to 20km by Hayes *et al* (1980, 305-06). This would be well within the one or two days return journey discussed by Peacock in relation to saddle querns and could represent further self-sufficient home production.

Other rock sources local to the study area known to have been utilised for the production of beehive querns include Crinoid Grit, with a so-called 'factory' recognised on Spaunton Moor (Hayes *et al* 1980, 304-05). Outcrops of Corallian limestone and Moor Grit were also exploited. The availability of the different stone sources local to the study area varies considerably, with Moor Grit being restricted to very limited outcrops, Crinoid Grit being more common, but with outcrops restricted to the highest parts of the Moors, while channel sandstone, although inconsistent as a deposit within the earlier elements of the Middle Jurassic formation, outcrops frequently in the sides of the dales that radiate out of the Moorland block, with Corallian limestone and calcareous sandstone being common in the Tabular Hills and Hambleton Hills.

Briggs (1988) has argued that erratics, many originally derived from the Pennines, were a significant if not dominant source of raw material for querns, whereas Hayes *et al* (1980, 302) consider that glacial erratics were only 'occasionally selected for querns'. The exploitation of erratics may seem a likely model for the 'home' producer, seeking only to supply their own needs. However the production of beehive querns is more technically demanding than the production of saddle querns and this, coupled with the use of 'exotic' igneous rock, or Millstone Grit boulders, may be taken to point to specialist production.

In contrast the distribution of the local stone sources used for the recorded beehive querns suggests that, despite the greater technical skills required for the production of beehive querns, it is not certain that this would lead to the establishment of local specialist production in preference to home production over all of the study area. It seems clear that querns were transported to those areas beyond the Moorland block where stone was not readily available, for example the Vale of Pickering where querns of Crinoid Grit and channel sandstone are both found (Hayes *et al* 1980, table 8). This distribution suggests the existence of people who either produced finished querns for use outside the immediate area, or quarried stones for finishing elsewhere, during the currency of the beehive quern which extends into the Roman period by 100-150 years on conventional dating. The main evidence for this happening is the concentration of querns of Crinoid Grit in the Tabular Hills, where a local source of channel sandstone was readily available to the people of the Spaunton area (Figure 3).

Possible evidence for specialist production of beehive querns does not mean that local stone sources would not have been utilised for home production, although as suggested above the production of beehive querns would have required a degree of skill. The discovery of isolated roughouts of beehives, such as two from Blansby Park (Hayes 1988, 47), may point to home production. Both of these stones were Calcareous Grit, the local Corallian stone and may point to the utilisation of what was to hand by the immediate community. It is perhaps unreasonable in the areas were the stone types in common usage for beehive querns are found, to ascribe all, or even most of the production to the known 'factory' sites. The assumption of usage of stone resources by the local population begs questions of ownership, tenancy and associated rights. However it is possible that to date we have in general only recognised those quarries that were subject to long-term or large scale use, leaving evidence in the form of broken, or unfinished querns; or perhaps only those quarries that were not exploited for stone for other purposes such as

Figure 3. Beehive quern rough-out in a field wall on the Spaunton/Hutton-le-Hole parish boundary. (Photo J Price)

building, or those that were not reused after the Roman period. By the nature of the material quarries are not readily dateable, nor, in the main, is the debris from non-specialist exploitation.

Therefore it is perhaps best to see the production of beehive querns, both before and after the Roman conquest of the study area, as possibly operating on two levels. Firstly home production in those areas with suitable stone, serving at least part of the local need, although perhaps constrained by the skill requirement inherent in the artefact type. Secondly local specialist production possibly serving part of the immediate population that did not undertake home production and also a wider region, perhaps directed especially to those areas without suitable local stone sources.

There is a possibility that a somewhat similar situation might have pertained during the earlier Iron Age with saddle querns, or the stone to produce them, being exported to areas where stone is not easily available. However saddle querns are less well-represented in the archaeological record, although a possible example of such an export is a saddle quern from Thornton Riseborough (Hayes 1988, 45) in the Vale of Pickering. Even if such movement of saddle querns was taking place it need not indicate that home production was not the norm in the areas where stone was available. Furthermore not only were saddle querns less demanding to produce but little preference seems to have been shown with regard to stone type and virtually any rock type appears to have been utilised in conjunction with a rubbing stone (Challis and Harding 1975 i, 23).

The flat rotary quern appears during the Roman period and production in the study area is known in Bransdale where Crinoid Grit was exploited at Rudland Beck and Swinacle Ridge (Hayes *et al* 1980, 305). It is likely that the Spaunton Moor `factory' was also quarried for flat rotary querns. Although none was found there many were utilised in the structure of the nearby sixteenth-century glass furnace (Hayes 1974, 36).

Metal-working

Moving on from querns two groups of sites become of particular importance, those on Levisham Moor (Hayes 1983) and at Roxby (Inman *et al* 1985). The possible longevity of these sites, and the apparently conservative nature of the pottery assemblages from them has been discussed elsewhere (Wilson 1995, 72-73). However for the purposes of this consideration, and for that of pottery production that follows, it is sufficient to emphasise the difficulty of distinguishing late Iron Age from Roman period occupation with any certainty except where clearly Romanised ceramics are present. By the later third and fourth centuries hard-fired Roman ceramics were in use, but in relatively small quantities. For the earlier Roman period pottery in the native tradition appears to have been the norm with infrequent occurrences of mortaria, and even less commonly samian.

At Levisham Moor a roundhouse was recorded in Enclosure D that appears to have been entirely given over to ironworking. It contained three phases of furnace, as well as associated features and large quantities of iron slag. At Roxby two of the roundhouses, numbers 2 and 3, produced evidence of smithing and smelting respectively, suggesting a locally sophisticated industry. However the character of that `industry' is unclear, given the limited number of recorded furnaces it may be that the Roxby site was not engaged in anything more than home production, and that the two roundhouses formed part of what appears likely to have been a primarily agricultural community. Evidence for Iron Age iron smelting is also recorded from less fully explored sites in the area, for example the enclosure at Crown End, Westerdale (Harbord and Spratt 1975).

Local diversity is apparent between the Levisham Moor and Roxby sites, although the processes used at the two sites were similar. Both provided evidence of domed furnaces typical of later Iron Age smelting (Cleere and Crossley 1985, 41, 52-53), but apparently utilised different types of ore. In the case of the Roxby site the ore was derived from outcrops of ironstone within 1500m of the site (Inman *et al* 1985, microfiche 1.32-33); while at Levisham Moor bog iron ore, probably from Rosedale, was being exploited (Hayes 1983, 53). The impression being that by the later Iron Age, at least in the area of the North York Moors and Tabular Hills, the industry was well-enough established at a home production level to allow individual practitioners, or groups of practitioners to go their own way, with perhaps only limited dependence on outside expertise or supplies. However, a distinction should be drawn between smithing and smelting, the former being commonplace on sites of all type, while the latter appears as (probably) a specialist activity.

Although the pottery assemblages from both Levisham Moor Enclosure A and roundhouses 2 and 3 at Roxby were consistent with a late Iron Age date, they were also quite limited. The low level of clearly Romanised ceramics from the Levisham and Roxby sites is noticeable, although later Roman material is known and the sites could be occupied through the Roman period (Wilson 1995, 74-75, and below). It is possible that the iron-working components of these sites could extend into the Roman period. Looking beyond the study area most excavated Roman sites produce some evidence of metalworking of one type, or another, much presumably relates to home production in the form of blacksmithing for the immediate needs of the community. This would suggest that the limited evidence for smithing from sites in the study area may reflect the relative lack of recent excavations that would incorporate as a matter of routine the sort of sieving and sampling programmes that would allow the facilitate the recovery of hammer-scale and other waste. The widespread discovery of evidence for metalworking away from the study area suggests continuity of practice from the Late Iron Age into the Roman period, at least on minor rural sites. Cleere (1976, 127) has drawn attention to the inappropriateness of a hard distinction between the iron maker and the smith in the Roman period and it is probable that in the study area the Roman period `home producer' utilised the abundant local resources of the North York Moors, as had his Late Iron Age predecessor. This suggestion may be supported by the discovery of a *tuyere* amongst the ceramic assemblage from Blansby Park, near Pickering (personal observation).

Pottery

Although it is clear that pottery was used in the area in the Late Iron Age we have little evidence relating to its production. Jeremy Evans has demonstrated a possible division between sites in the northern and southern part of the Moors with respect to preferred tempering agents (Evans 1995, 48-50, figure 5.2). Dolerite tempering dominates assemblages from the northern sites and calcite tempering those from the south. However there are exceptions: the Great Ayton Moor enclosure has produced some calcite tempered sherds (Willis 1997, 56), as has Roxby (Evans 1995, fig 5.2). In terms of craft this may mean little, the potters of the northern Moors utilising the readily available igneous material from the Cleveland Dyke, whilst their co-workers further south having easier access to calcite from the Wolds. However as Evans noted the distributions are not mutually exclusive, some dolerite tempered pottery appears in the Vales of Pickering and York, and calcite tempered wares dominate the assemblage at Roxby. It is possible that we are seeing the potters of the region choosing tempering agents suited to the eventual use of the vessels. Such attention to detail in the production of pottery could point to the existence of specialist producers, and herald the later development of large-scale potteries producing pottery broadly in the native tradition, for example Knapton ware in the third century and Huntcliff type pottery in the fourth century. However it could simply be indicative of a level of technical proficiency amongst the potters within the communities of the North York Moors.

Any greater understanding of the production of the native style pottery that was utilised by the inhabitants of the North York Moors during the late Iron Age and Roman period is hampered by the absence of recognised production sites. Except through the most fortuitous of circumstances this situation is unlikely to improve. The pottery was probably fired in surface clamps/bonfires, or possibly pit clamps, neither of which are likely to leave substantial traces in the archaeological record, although sites of both types have been recognised in southern England (Swan 1984, 53-55). The problem is perhaps compounded by the possibility that there was an extremely low level of ceramic usage, and therefore of production, in the late Iron Age and through the earlier part of the Roman period in the region (Wilson 1995, 73-74).

The appearance of Romanised ceramics in reasonable quantities on sites in the study area in the third and more particularly the fourth century possibly reflects their ubiquity in East Yorkshire as a whole. However, the discovery of a built kiln with a stone-lined flue producing Romanised ceramics at Cold Cam, Cockerdale Wood, Oldstead (Hayes 1963, 407-13) perhaps indicates a spread of ideas and adaptation by local producers. Equally the possibility of a kiln built by an itinerant specialist to cater to a local demand, or indeed a single order, is possible. In contrast with the limited range of vessel forms recorded from the Late Iron Age/Early Roman sites of the region this one 0.76m diameter kiln is recorded as producing `orange wares including hammer-head and flanged mortaria ... with ironstone grits and cream slip painted onto interior of rim and exterior of vessel, flanged bowls ..., grey gritty lid-seated jars ... also possibly grey dog-dishes with beaded rim and S-shaped bowls' (Swan 1984, microfiche 6.693).

Jet

Jet is one of the products most readily associated with the study area and it is clear from the apparently rapid expansion of the jet industry in Roman York (RCHME 1962, 141-44) in the third century (Allason-Jones 1996, 9) that jet was in demand. Recent analytical work has demonstrated that much of the material identified as jet on visual characteristics alone is in fact shale, cannel coal, or durain (Allason-Jones and Jones 1994). Jet-working is explored in detail by Allason-Jones elsewhere in this volume. However, allowing for the absence of reliable scientific analyses of the material from the study area, there is evidence for the utilisation of jet (or jet-like material) from the Bronze Age onwards (Smith 1994, 21-5, table 13). Allason-Jones (1996, 8) recognises a hiatus in the use of jet in the Iron Age that extends into the Roman period, but discoveries at Roxby provide evidence for the working of jet (or jet-like material) in the form of broken, unfinished artefacts from Houses 2 and 3 (Inman *et al* 1985, 199, 204). These discoveries appear to belong at the end of the Iron Age, or early in the Roman period given the apparent contemporaneity of the structures from which they derive and a radiocarbon date of 00± 150 ad (based on a date of 1950± 150BP (BM 2207 A) from House 3. The authors of the Roxby report derive further support for this date from the pottery. However, as suggested above, the possibly conservative nature of native-tradition ceramics in the region may mean that the date ranges offered have to reconsidered and extended. However recalibration of the radiocarbon date has given a date of 370 Cal BC - 420 Cal AD (at 95% confidence) (P Marshall *pers comm*). This suggests that House 3 cannot be assigned to the Late Iron Age, or Early Roman period with the confidence suggested by the authors of the report, but must be seen as belonging to the period between the Middle Iron Age and the end of the Roman period.

The question of the dating aside the finds of jet, or jet-like material, suggest small scale production of artefacts and this pattern is seen to continue on clearly Roman period moorland sites such as that to the north-west of Newbiggin Hall, Grosmont excavated by Hayes and dated to the fourth century (Hayes 1968, 121-3). The character of these Roman period finds is little different to those from Roxby and bears little resemblance to the highly sophisticated industry known in York (RCHME 1962; Allason-Jones 1996; this volume; Cool this volume).

As Allason-Jones (1996, 6) points out the origins of the material used are not certain. Most presumably originates in the Whitby area, but it could result from `beachcombing' for material washed out of the cliff, or the deliberate cutting back of the cliff to expose fresh faces. It is possible that the widespread inland deposits might have been exploited through pits and/or drifts where deposits outcropped in valley sides. The products of casual collection, such as the unworked jet found at Goldsborough Signal Station (Hornsby and Laverick 1932, 213; Allason-Jones 1996, 11), could easily account for the quantities apparently worked on sites in the study area. However collection and extraction to supply the jet industry of York and the lesser requirements of other sites such as Binchester, where possible jet-working waste has been noted (I Ferris and R F J Jones *pers comm*), can perhaps be regarded as a minor industry in its own right.

Glass

The late Iron Age/early Roman site at Roxby produced evidence of possible glassworking in the form of a length of blue-green glass rod from House 2. `The fragment had clearly been made by allowing to mix together molten blue glass and molten green glass' (Inman *et al* 1985, 199). No evidence of certain Roman period glass working is known from the region, but the evidence for small-scale working of glass would presumably leave little trace. (See Price, this volume.)

Invisible crafts

Many craft and industrial activities are virtually undetectable in the archaeological record, unless extremely favourable conditions pertain. Other activities may be more readily recognisable but are as yet not recorded from the study area, although their products may be.

Possibly the chief among these for the Roman period are those that relate to the provision of fuel for both domestic and craft uses. This by implication points to the management and exploitation of woodland. Charcoal and timber must have been in great demand, and woodland resources would have been both valuable and limited in extent, forming an essential element of the rural economy. As early as 1928 Corder recognised that the availability of fuel would have been a critical element in the successful operation of the Crambeck kilns complex to the south of the study area (Corder 1928, 134). Timber appears to have been the favoured fuel, with soft woods possibly being preferred as they yield more heat. The use of small diameter timbers, as has been noted for the Nene Valley, points to the practise of coppicing to secure a supply of fuel (Hartley 1960, 8), at least by the larger producers. However the operators of apparently isolated kilns such as those at Cold Cam may have relied on gathering what was available locally. Operating alongside charcoal burning and coppicing there may have been other woodland crafts, for example hurdlemaking.

Around the study area fuel would also have been of critical importance to the inhabitants of York and Malton/Norton, and any other settlements where any significant proportion of the population did not have control of woodland resources. This suggests that there was probably a significant element in the rural population servicing the needs of at least the larger population centres, and substantial production of wood for heating and cooking may have taken place alongside the charcoal burning suggested above.

The part played by coal is not clear, although coal is known from both Crossgates, Seamer just south of the Moors, and from Norton and in greater quantities from York. However as Dearne and Branigan (1995, 81 and fig 5) demonstrate there is a close association between the distribution of coal in East Yorkshire and the Roman road system and it is probable that the coal from York at least was supplied from the Yorkshire coalfield, while that from Norton might have been. Alternatively the Norton material could derive from a coastal trade originating in Northumberland and Durham, as could that from Crossgates. That having been said the possible exploitation of the deposits of low grade coal found in the Moorland block cannot be discounted. However the prevalence of pit mining in the modern period as described by Harrison (1989, 167-70), argues against casual exploitation, except where coal seams were exposed in valley sides. *The History of Helmsley* states that 'there is some slight evidence that moor coal was mined in Roman times' (McDonnell 1963, 180), although no possible Roman sites appear in the list of coal-pit sites compiled by Hayes (1963, 460).

In addition a range of domestic craft activities may be assumed including the production of textiles for clothing, the turning of wood for vessels and the tanning and working of leather for footwear and domestic vessels.

Discussion

From the above it is clear that there is little evidence of Roman industry from the North York Moors, other than processes that represent the direct exploitation of the geological resources of the area. The production of querns is the most obvious, by virtue of its products surviving to enter the archaeological record. Even so the scale of this production and indeed the extent that it represents an industry even on the specialist local production level is uncertain, and much must represent home production as is suggested above.

However, if our understanding of quern production can be obscured, at least in part by later reuse of rock outcrops, the position is so much worse for other extractive industries. Without detailed and extensive analysis of iron slags, and jet and jet-like materials we cannot hope to understand the patterns of exploitation of other natural resources. Such analyses might serve to identify sources that were exploited, but the scale of the exploitation will presumably remain unclear. However, with the possible exception of jet, the evidence we have points to home production as the norm in the area. The quantities of metal-working residues from Levisham Moor, Roxby and Crown End suggest production for local needs, and unless there was a significant movement of ore out of the study area, the levels of extraction were presumably equally low. This suggestion is unlikely to be overturned if future work suggests that many more of the local sites were smelting iron for their own needs and it should be emphasised that despite many Romano-British sites producing evidence of iron-working, in the majority of cases this is secondary-working (ironsmithing) not primary production (ironsmelting from ore).

If evidence of smithing can be expected to be found on more sites, so perhaps can evidence of bronze working and lead melting. Lead is not found in archaeological contexts prior to the Roman period and is known from sites on the Moors including Levisham Moor (Hayes 1983, 33-4) presumably indicating a Roman period origin for the objects concerned.

As suggested above collection and extraction of jet might well be regarded as 'industrial', or at least commercial production for export beyond the region and as such represents something more than local specialist production. However the collection of jet pebbles as evidenced at Goldsborough Signal Station might easily typify something nearer to a profitable sideline for many people. Certainly the evidence for the production of jet artefacts in the region, along with the possible evidence for glass-working at Roxby indicates small scale 'home' production of decorative personalia, probably for the community, or family group. Despite this the work may have been undertaken by specialists within the community as the possibility of glassworking suggests a reasonably high level of technical achievement (Guido 1978, 7-8), as does the employment of cylindrical, rather than biconical drilling in the production of jet artefacts. However, cylindrical, as well as biconical drilling, may be evidenced to the south of the study area at Staple Howe in the earlier Iron Age (Brewster 1963, 120) and cannot be regarded as a Roman period development.

In short within the study area Roman period crafts have only limited visibility and Roman period industry cannot readily be recognised on the North York Moors with any certainty. However the appearance of fragments of *opus signinum* on the site in Stonygate Newtondale provides an exception (Hayes 1988, 60) and presumably demonstrates the practice of a Roman-style craft. Otherwise there is little evidence of changes arising from the incorporation of the area into the Roman province other than the production of flat rotary querns, the Cold Cam kiln and perhaps the working of lead. However it is possible to argue for a continuation of pre-existing craft activity, typified by low levels of iron-working and pottery production for domestic use. The one area of production where a Roman period industry may appear is in the collection and/or extraction of jet presumably in the third century in response to expanded demand. Although freestone may have been quarried in the Iron Age to be used in the construction of huts, field clearance and other collection might have served; one introduction that may be assigned to the Roman army is the production of worked stone for building, evidenced by Lease Rigg and the coastal signal stations. Although the bulk of the local population did not take up masonry construction some civilian Roman period building projects may have involved the winning of stone from within the study area, for example the construction of the Stonygate building (Hayes 1988, 57-65), the Blansby Park building (Rahtz *et al* 2001) or perhaps the Beadlam villa (Neal 1996) located just south of the region.

Acknowledgements

Thanks are due to Peter Marshall of the English Heritage Centre for Archaeology for recalibrating the radiocarbon date from House 3 at Roxby. The drawings accompanying this paper were prepared by Eddie Lyons of the English Heritage Centre for Archaeology Graphics Studio. Figure 3 was kindly provided by Prof J Price.

Bibliography

Allason-Jones, L, 1996 *Roman Jet in the Yorkshire Museum*. York: Yorkshire Museum

----, and Jones, J M, 1994 Jet and other materials in Roman artefact studies. *Archaeologia Aeliana* 5th ser, 22, 22-32

Barker, G, 1985 *Prehistoric Farming in Europe*. Cambridge New Studies in Archaeology, Cambridge: Cambridge Univ Press

Brewster, T C M, 1963 *The Excavation of Staple Howe*. Wintringham: East Riding Archaeol Res Comm

Briggs, S, 1988 The lithology, origin and dispersal of some quernstones from Yorkshire. In Price and Wilson (1988), 289-311

Challis, A, and Harding, D W, 1975 *Later Prehistory from the Trent to the Tyne*. Oxford: Brit Archaeol Rep, Brit Ser, **20**

Cleere, H, 1976 Ironmaking. In Strong and Brown (1976) *Roman Crafts*. London: Duckworth, 127-41

---- , and Crossley, D, 1985 *The Iron Industry of the Weald*. Leicester: Leicester Univ Press

Corder, P, 1928 *The Roman Pottery at Crambeck, Castle Howard*. York: Roman Malton District Rep **1**.

Dearne, M J, and Branigan, K, 1995 The use of coal in Roman Britain. *Antiq J* **75**, 71-105

Evans, J, 1995 Later Iron Age and `native' pottery in the north-east. In Vyner (1995), 46-68

Guido, M, 1978 *The Glass Beads of the Prehistoric and Roman Periods in Britain and Ireland*. London: Rep Res Comm Soc Antiq London **35**

Harbord, N H, and Spratt, D A, 1975 Examination of Cinder from Early Iron Working at Crown End, North Yorkshire Moors. *J Hist Metallurgy* **9.1**, 3233

Harrison, J K, 1989 Landscapes of Industry. In Spratt and Harrison (1989), 159-83

Hartley, B R, 1960 *Notes on the Roman Pottery Industry in the Nene Valley*. Peterborough: Mus Soc Occ Pap **2**

----, 1978 Roman Fort at Lease Rigg. In Smith (1978), 30-32

---- , 1980 Roman fort at Lease Rigg. *Trans Scarborough Dist Archaeol Soc* **21**, 8-11.

----, 1993 Military Activity. In Spratt (1993), 160-65

Hayes, R H, 1963 Appendix A ii-xvi. In McDonnell (1963), 335-424

---- , 1968 A Romano-British site north-west of Newbiggin Hall, Grosmont, near Whitby. *Yorks Archaeol J* **42**, 120-25

----, 1974 Querns (a survey of handquerns found in East Cleveland, the North Yorkshire Moors and the Vale of Pickering). *Ryedale Hist* **7**, 2241

----, 1976 Further notes on North Yorkshire querns. *Ryedale Hist* **8**, 38

----, 1983 *Levisham Moor Archaeological Investigations. 1957-78*. Helmsley and Scarborough: North York Moors National Park and Scarborough Dist Archaeol Soc

---- , 1988 *North-East Yorkshire Studies: Archaeological Papers*. Leeds: Roman Antiqs Sec, Yorks Archaeol Soc

---- , Hemingway, J E, and Spratt, D A, 1980 The distribution and lithology of beehive querns in Northeast Yorkshire. *J Archaeol Sci* **7**, 297324

---- , and Rutter, J G, 1964 *Wade's Causeway. A Roman road in North-East Yorkshire*. Scarborough: Scarborough Dist Archaeol Soc Res Rep **4**

Heslop, D, 1988 The study of beehive querns. *Scottish Archaeol Rev*, **5**, 59-65

---- , *in prep The North Yorkshire beehive quern survey*

Hornsby, W, and Laverick, S D, 1932 The Roman Signal Station at Goldsborough, near Whitby. *Archaeol J* **89**, 202-19

Inman, R, Brown, D R, Goddard, R E, and Spratt, D A, 1985 Roxby Iron Age Settlement and the Iron Age in North-East Yorkshire. *Proc Prehist Soc* **51**, 181-213

Maxfield, V A, and Dobson, M J, (eds) 1991 *Roman Frontier Studies 1989. Proceedings of the XVth International Congress of Roman Frontier Studies*. Exeter: University of Exeter Press, 142-47

McDonnell, J G, (ed) 1963 *A History of Helmsley, Rievaulx and District*. York: Stonegate

Neal, D S, 1996 *Excavations on the Roman villa at Beadlam, Yorkshire*, Yorks Archaeol Rep **2**, Leeds: Roman Antiquities Section, Yorks Archaeol Soc

Peacock, D P S, 1987 Iron Age and Roman quern production at Lodsworth, West Sussex, *Antiq J*, **67**, 61-85

Price, J, and Wilson, P R, (eds) 1988 *Recent Research in Roman Yorkshire. Studies in Honour of Mary Kitson Clark*. Oxford: Brit Archaeol Rep, Brit Ser **193**

Rahtz, P A, Watts, L, and Jones, A, 2001 Blansby Park 2000 - interim report. *Yorks Archael Soc, Roman Antiqs Sec Bulletin* **18**, 3-12

RCHME 1962 *An Inventory of the Historical Monuments in the City of York. Vol I: Eburacum Roman York*. London: HMSO

Richmond, I A, 1925 The Roman camps at Cawthorn, near Pickering. First interim summary 1924, *Yorkshire Archaeol J* **28**, 332-39

Smith, D, (ed) 1978 An archaeological index for North-East Yorkshire. *Trans Scarborough Dist Archaeol Soc* 21, 30-38

Smith, M J B, 1994 *Excavated Bronze Age Burial Mounds of North-East Yorkshire*. Durham: Architect Archaeol Soc Durham Northumb Res Rep **3**

Spratt, D A, (ed) 1993 *Prehistoric and Roman Archaeology of North-East Yorkshire*. Rev edn. London: CBA and North York Moors National Park, CBA Res Rep **87**

---- , and Harrison B J D, 1989 *The North York Moors. Landscape Heritage*. Newton Abbot: David and Charles

Swan, V G, 1984 *The Pottery Kilns of Roman Britain*. London: HMSO, RCHME Supp Ser **5**

Vyner, B E, (ed) 1995 *Moorland Monuments: Studies in Honour of Raymond Hayes and Don Spratt*. York: CBA Research Report **101**

Whitaker, A H, 1967 The excavation of a Romano-British farmstead at Spaunton. *Ryedale Hist*, **3**, 12-25

Willis, S, 1997 The pottery of Iron Age tradition from the rectilinear enclosure site on Great Ayton Moor, North York Moors. *Durham Archaeol J*, **13**, 55-60

Wilson, P R, 1991 Aspects of the Yorkshire signal stations. In Maxfield and Dobson (1991), 142-47

----, 1995 North-east Yorkshire in the Roman Period: developments and prospects. In Vyner (1995), 69-78

---- , and Lee, G, 2000-2001 Cawthorn Camps: Trial excavations 1999. *Ryedale Hist*, **20**, 5-8

THE ROMAN POTTERY INDUSTRY AT HOLME-ON-SPALDING MOOR IN ITS LANDSCAPE

PETER HALKON

Background

The parish of Holme-on-Spalding Moor (NGR SE 810370) situated in the south-eastern Vale of York, 10km to the west of the Yorkshire Wolds and 20km north-west of the River Humber, contains one of the largest pottery industries in Roman East Yorkshire. Although the presence of this production centre has been known for a considerable time, its size and significance have only been recognised in recent years as a result of intensive survey and research aimed at examining it in the context of the landscape. This paper presents a summary of the section of the survey relating to the Romano-British kiln and settlement sites which has now been fully published (Halkon and Millett 1999).

The first reference to Roman pottery being found in Holme-on-Spalding Moor occurs in 1853 when the Rev R Whytehead gave a number of Roman `urns' from there to the museum of the Yorkshire Philosophical Society (YPSAR 1854, 21). Other early chance finds of pottery were made at Tollingham in 1873 (Clark 1935, 88) and Bursea Grange 1884/6 (Kingston-upon-Hull City Museum Accession number K.I.N.C.M. 1182, 1983), the latter being part of the Mortimer Collection (see also Hicks and Wilson 1975). A Roman lead coffin was found at Tollingham together with pottery in 1899 (*Yorkshire Weekly Post*, 14th January 1899), but was stolen from outside the vicarage, where it had been used as an ornamental trough (A Dowson *pers comm*). This was a high status item, making this area worthy of further investigation.

Romano-British pottery kilns were excavated at Throlam by Sheppard and Corder in 1930 (Corder 1930). After these excavations, the greyware pottery found in many parts of East Yorkshire became known as Throlam ware. As there are now around 34 known kiln sites producing similar pottery in the area, it is perhaps more correct to call the pottery Holme-on-Spalding Moor greyware.

In 1970 and 1971, further kilns were excavated at Hasholme Hall by the East Riding Archaeological Society (ERAS) (Hicks and Wilson 1975), but apart from this, archaeological work around Holme-on-Spalding Moor was sporadic until 1980, when the writer and members of ERAS began a systematic survey. The results of the work up to early 1982 appear elsewhere (Halkon 1983). In 1983 fieldwork was `formalised' as a basis for an MA thesis (Halkon 1987), the aim of which was to investigate Romano-British industry and settlement within an 8x8km square, with its corners at SE 7740, 7732, 8532 and 8540, examining any factors which may have been influential such as soils, drainage and environmental change.

In addition to fieldwork, a series of research and rescue excavations have been carried out as part of the larger project, conducted by Durham University Archaeology Department with the East Riding Archaeological Society (Millett and Halkon 1984; 1985; 1986; 1988); the most relevant to this paper were those carried out on the Romano-British kiln at Bursea House in 1983 and 1984. As a complement to fieldwalking and excavation, Mr J Pocock of Bradford University has undertaken extensive geophysical surveying, especially at Hasholme and Bursea (Pocock 1982; 1983; Halkon and Millett 1999). The setting of the industry in the wider landscape of the Foulness Valley has been reviewed elsewhere (Halkon 1998).

The first task was to suggest a reconstruction of the landscape at the time the kilns were in production, with special reference to geology, soils and communications.

The landscape - raw materials and communications

There is little variation in the solid geology of the landscape block, which comprises Triassic mudstones and Mercia mudstones (Kent 1980), though it does provide the area with its major feature, Church Hill, a Triassic inlier of marl capped with gravels of `Older Drift' (De Boer 1974). Gypsum occurs in the upper Mercia mudstone (Wilson 1948) which was exploited during the nineteenth century around Highgate (SE 828372). It is however the Drift rather than the solid geology which had the major bearing on the nature of the landscape around Holme-on-Spalding Moor (Selby sheet (71) of the Geological Survey). The glacio-lacustrine Devensian clays of Lake Humber, laid down *c* 11,000 years ago (Gaunt *et al* 1971; Catt 1990) underlie the majority of other drift deposits, and it was this almost stoneless blue clay which provided the Romano-British potters with their main raw material. Sand, `the most frequently used non-plastic constituent of pottery' (Swan 1984, 3) is also readily available, as the clays are overlain by ridges of aeolian sands and glacio-fluvial sands rising to above 8m 0D in some places, which form undulations in an otherwise flat landscape. The remainder of the Drift comprises alluvium and peat, which follow the original course of the River Foulness (Figures 1 and 2). Recent work by Jordan (1987) shows these deposits to have been formed by tidal salt marshes and a dendritic creek system. The River Foulness was particularly important, for study of aerial photographs, soil and geological maps and documentary research has shown that it once formed part of a dendritic creek system associated with a former tidal inlet of the Humber Estuary. It was not until the writer's work began that the full implications of the presence of this inlet were appreciated (Halkon 1989a). The Roman town of Brough on Humber and the extensive villa complex at Brantingham are near its eastern shore, and the Roman sites of Faxfleet A and

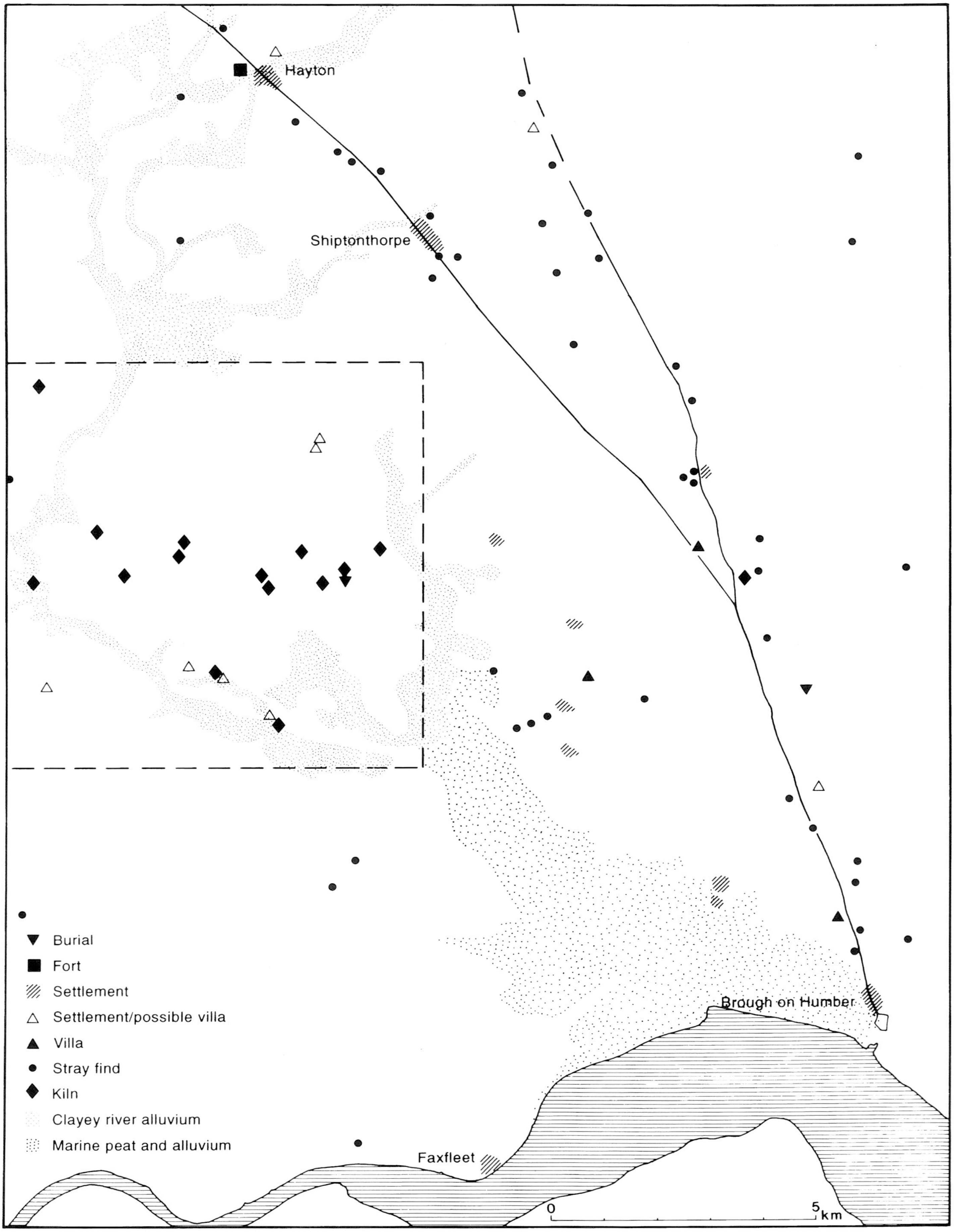

Figure 1. Romano-British sites and the Roman road against the alluvium of the Foulness Valley.

B (Bartlett 1968; Sitch 1989), lie at the south-western shore. Holme, therefore, with its ridges of Holme Moor aeolian sand (Furness and King 1978) provides the first dry land to the north, with its creek system allowing access to the Wolds foothills and the rest of the Humber drainage basin. The Roman roadside sites at Hayton and Shiptonthorpe may have had access to the estuary through this system of watercourses and the presence of the inlet also explains the route of the Roman road between Brough on Humber and York (Margary 1973, 2e; Millett and Halkon 1988). This feature is therefore crucial in our understanding of the development of Iron Age and Romano-British settlement in the whole of the region.

Beside clay, water and sand, the other raw material necessary in a furnace based industry is wood. Analysis of the pollen and other environmental evidence around the Hasholme boat (Turner 1987) shows the area near the creek system to have been wooded with oak-alder woodland with some hazel. The density of cropmark sites suggests that by the later Iron Age, most of the sand ridges were cleared for pastoral and arable agriculture. Palaeo-environmental investigation of deposits from Bursea House and Bursea Grange, where a large hut circle within a square ditched enclosure was partially excavated in 1986, (Halkon and Millett 1999, 67) shows that cereals had been cultivated at both locations in the Iron Age and Roman period. Evidence from the excavations of the at Bursea House pottery kilns demonstrates that alder and willow or poplar, were the most commonly used varieties of timber used for firing the kilns and it is likely that woodland management took place (Cartwright 1999).

The difference between today's well managed agricultural

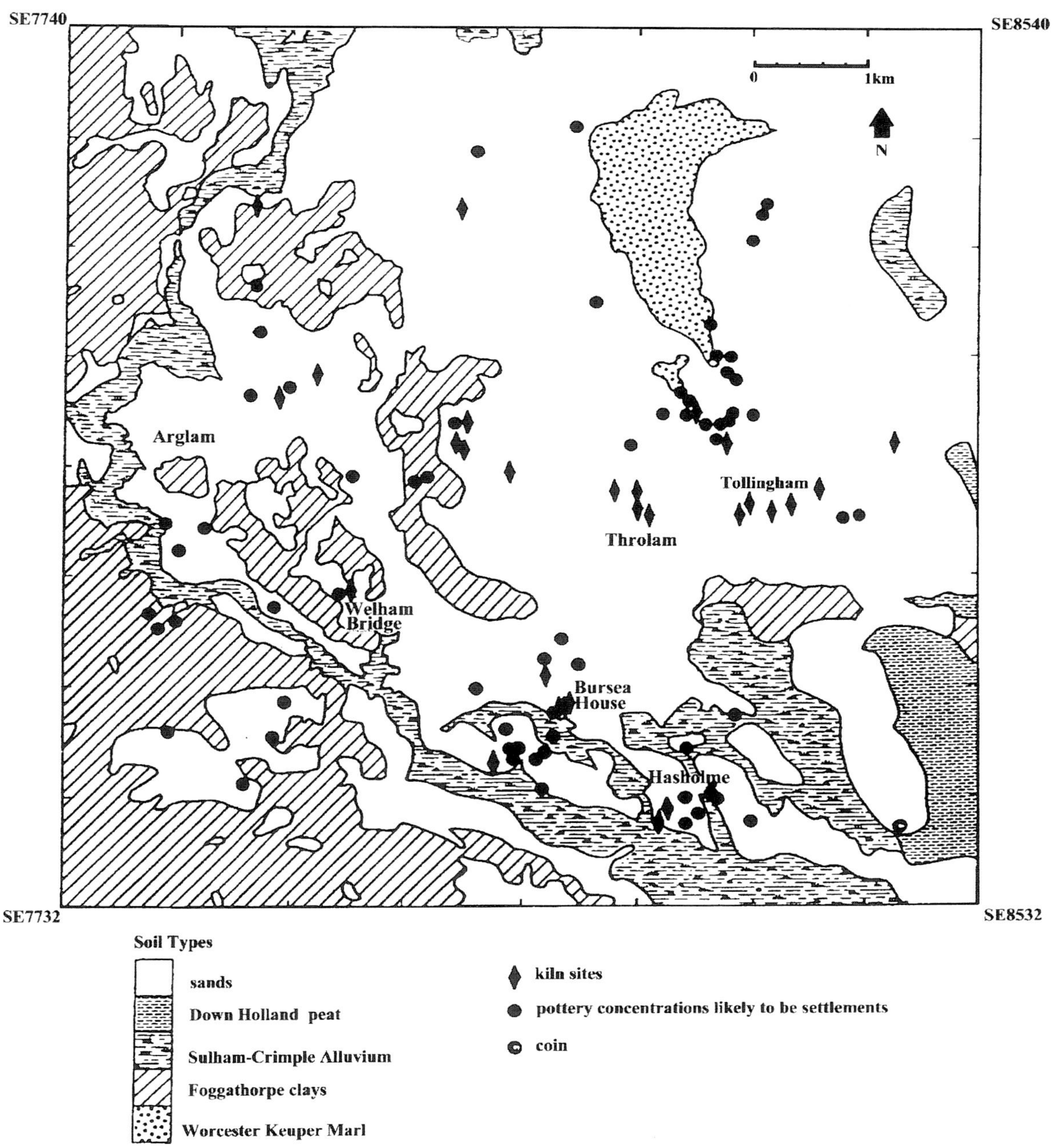

Figure 2. Romano-British sites in the 8x8km Holme Study Area against soil types.(Adapted from Soil Survey of England)

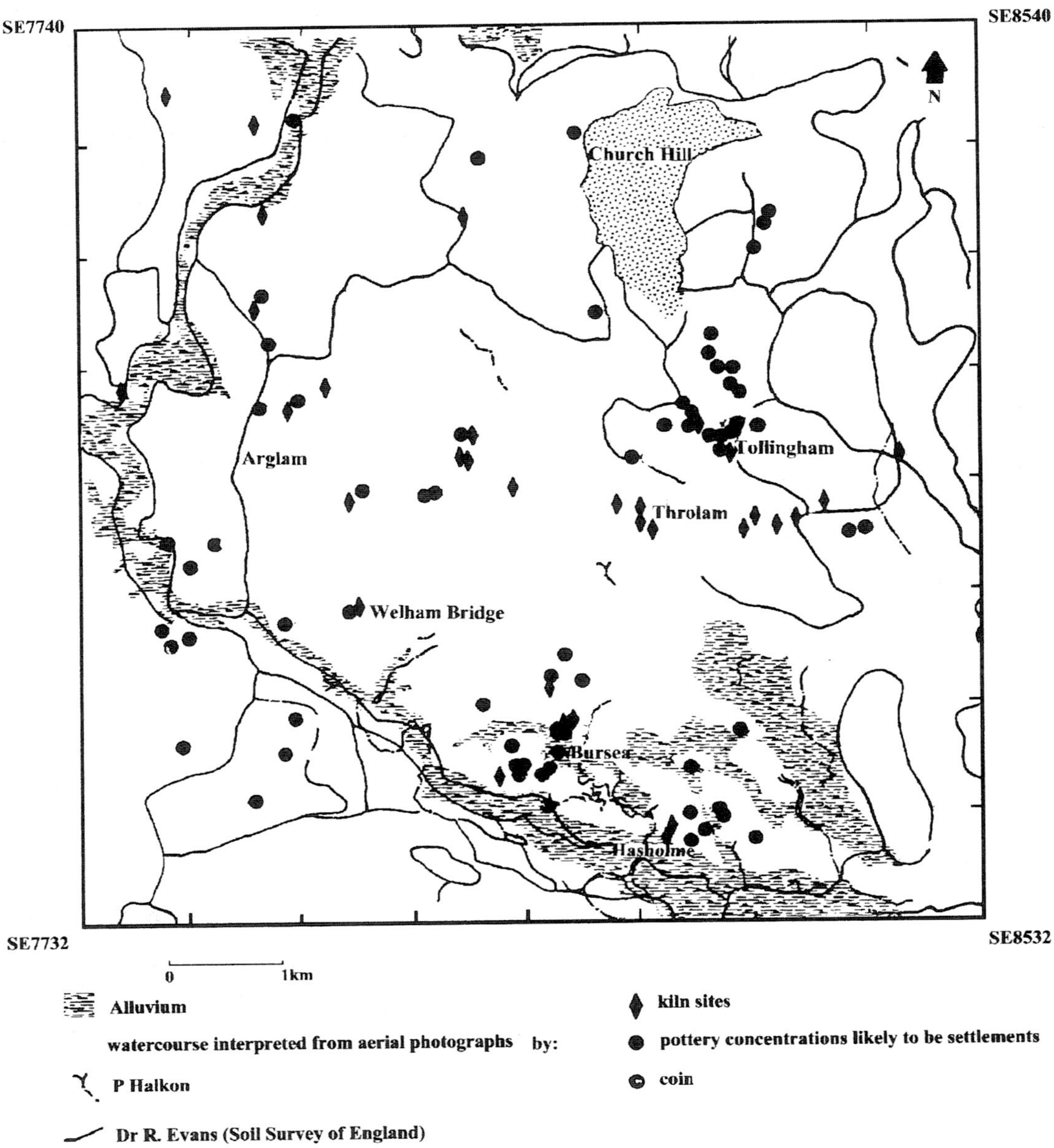

Figure 3. Romano-British sites in the 8x8km Holme Study Area against watercourses.

landscape and that of the past is therefore largely superficial, being the product of agricultural advances in the modern era, especially artificial drainage. In antiquity, the landscape consisted of a series of sandy ridges rising out of the marshland, and it was these that were the most suitable for agriculture and settlement. Although abundant water made conditions in some places difficult, it did provide a communications system and a resource for the Roman industries as it had done for the extensive Iron Age iron industry, represented by pieces of iron slag found during fieldwalking, which clustered around the creek system (Halkon 1987; 1990; 1997; Halkon and Millett 1999).

The methods and results of the survey.

Field walking, which until 1982 had been centred mainly around known kiln site areas at Hasholme (SE 820330) and Bursea (SE 815337) (Halkon 1983) took place on 55 farms. Finds previously discovered by farmers were also noted. The major proportion of fieldwork was carried out during the winters and springs of 1980-86. The fields selected were systematically walked by line and grid (Fasham *et al* 1980) and concentrations of pottery, burnt stone, kiln furniture and slag assumed to represent sites were recorded. Some geophysical surveying was also undertaken (Pocock 1982; 1983, 19 and 21). A complete gazetteer has been compiled (Halkon 1987; Halkon and Millett 1999, 49-61).

The survey resulted in the discovery of 22 hitherto unrecognised kiln sites bringing the total within the landscape block to 34 (Figure 1). Fifty-four concentrations of sherds and other material presumed to be settlement sites were found bringing the total of sites with Romano-British pottery within the landscape block (including kiln sites) to 106.

Subsequently a kiln site was recorded by S Harrison at Arglam in 1988, and an extensive Arglam settlement site was discovered in 1992 by R Lawson (site 134), which was field walked by ERAS members and pupils from Holme-on-Spalding Moor school in 1993 and 1994. Another kiln site was excavated by ERAS under the direction of R Mackey at Skiff Lane, to the north of the former airfield in August 1996.

When the position of kiln and settlement sites was plotted against the soil maps prepared by the Soil Survey of England (Figure 2), 89% of the total number of Romano-British settlements and kiln sites were found to lie on the better drained sandy soils of the Naburn soil series, Holme Moor soil series and Portington soil series, which form hillocks, ridges and low rises. Evans (1985, 96) noted the similarity between the landscape of the Holme area and that of the Fens and Iron Age settlement and Romano-British kiln distribution in the Fenland of Norfolk also provides a good comparison with Holme in showing preference for the Greensand, avoiding alluvium and peat (Gregory 1982, 351)

Seventy-six percent of all kiln sites were found to lie within 400m of a watercourse; 57% of all kiln sites were within 200m and the furthest distance of a kiln site from a watercourse was 875m. This fits well with Swan's hypothesis that:

'Few recorded Romano-British kilns appear to be sited farther than about 400m from a spring or other water source...those lying close to a river or large stream would have had the added advantage of ready access to water transport.' (Swan 1984, 6).

Eighty-six percent of all settlement sites in the Holme-on-Spalding Moor survey were within 400m of a watercourse and the furthest distance of a settlement site from a watercourse was 1225m (Figure 3). Most Romano-British sites within the Holme landscape block were therefore within easy reach of watercourses, some of which provided a means of access to the Humber estuary and beyond. This supports the relationship between sites and watercourses suggested by Hicks and Wilson (1975, 51), Ramm (1978, 110) and Evans (1985, 94). As with the Iron Age iron-smelting sites, the concentration of Romano-British kiln and settlement sites around the main creek system at Bursea and Hasholme is particularly noticeable. It is almost certain therefore, that this estuarine creek system provided the major form of transportation for the products of the Holme-on-Spalding Moor kilns.

The kilns *(Figure 4)*

Portable prefabricated kiln furniture in the form of baked clay bars, one elliptical and the other square in section were found at Bursea House (Halkon and Millett 1999, 51, site number 15) which resemble those illustrated by Swan (1984, fig IX, 69). This feature is thought to derive from a late pre-Roman Iron Age tradition. The earliest kilns excavated in the area were also at Bursea House, the stoke-hole of the later, smaller structure cutting through the back of the earlier kiln. Both kilns were made of clay blocks with internal lining. The earlier kiln was much larger being 2m in diameter, whereas the later was only a metre across. Their internal structures were also slightly different. The latter had parallel bars 0.30-0.40m long incorporated with the floor. They were set about 0.40m from each other and *c* 0.30m from the kiln wall, and were very similar to those of kiln 2 at Rossington Bridge pumping station, a possible surviving element of adjacent native technology (Swan 1984, 106). The earlier Bursea kiln had two much larger bars *c* 0.90m long and *c* 0.30m wide, with a cylindrical support between the bars attached to the kiln floor at one end, and is broadly in Swan's Linwood tradition. The kiln recently found at Skiff Lane bore a close resemblance to the earlier kiln from Bursea House.

Kiln 1 at Hasholme resembles kiln 1 at Linwood, dated typologically to the later third century whereas Hasholme kiln 2 is transitional between Linwood and Swanpool and should have a later date. Swan suggests that the Throlam kilns are the latest according to her typology, resembling those at Swanpool, which date from well into the fourth century.

Evidence for cross-Humber contact at an earlier date has been established at Bursea House in the final years BC, as the discovery of a Dragonby later Iron Age wheel thrown cordoned vessel illustrates (Halkon 1990). There are also typological links with other East Yorkshire kilns, as although not found *in situ*, banana shaped supports found at Hasholme, resembled those from Norton (Swan 1984, 110).

The pottery from the Holme survey.

The pottery from field walking was quantified and sherds sorted by fabric and form in an attempt to provide a chronological framework for the development of sites in the Holme area. The pottery reports of the excavations at Throlam (Corder 1930), Hasholme (Hicks and Wilson 1975) and Bursea House were used to create a typology. The work of Dr J Evans (Evans 1985) was drawn on in order to place the Holme Pottery industry in the wider context of northern Roman Britain and provide much needed external dating evidence.

Fabric

The present writer, following Peacock's guidelines (1977, 21-33) observed three broad local fabric types, whose main characteristics are outlined in Table 1.

It is possible that the first two fabric types may represent a deliberate attempt by the potters to produce different quality grades of product, as in all cases Holme-on-Spalding Moor fabric 1 vessels are superior to those of fabric 2 in firing and finish. Holme products fabrics 1 and 2 match Evans's fabrics 200 and 165 respectively (Evans 1985, 132-3). Creighton has more recently refined this writer's original fabric definitions based largely on excavated material from Bursea House (Creighton 1999). Thin section analysis on a sample from a Holme kiln site (Pryor 1997, 1031) showed that it was possible to isolate Holme-on-Spalding Moor fabrics using this method, though Monaghan (1997, 900) emphasises the difficulty of distinguishing one sand-tempered local grey ware from another.

Holme fabric 3 was almost exclusively reserved for forms of what was thought to have been a wheel-thrown copy in sandy-grey fabric of the "Dalesware" coarse shell-gritted handmade cooking pots, widespread in the North-east beginning *c* AD 200 (see Figure 5, J02a) (Loughlin 1977;

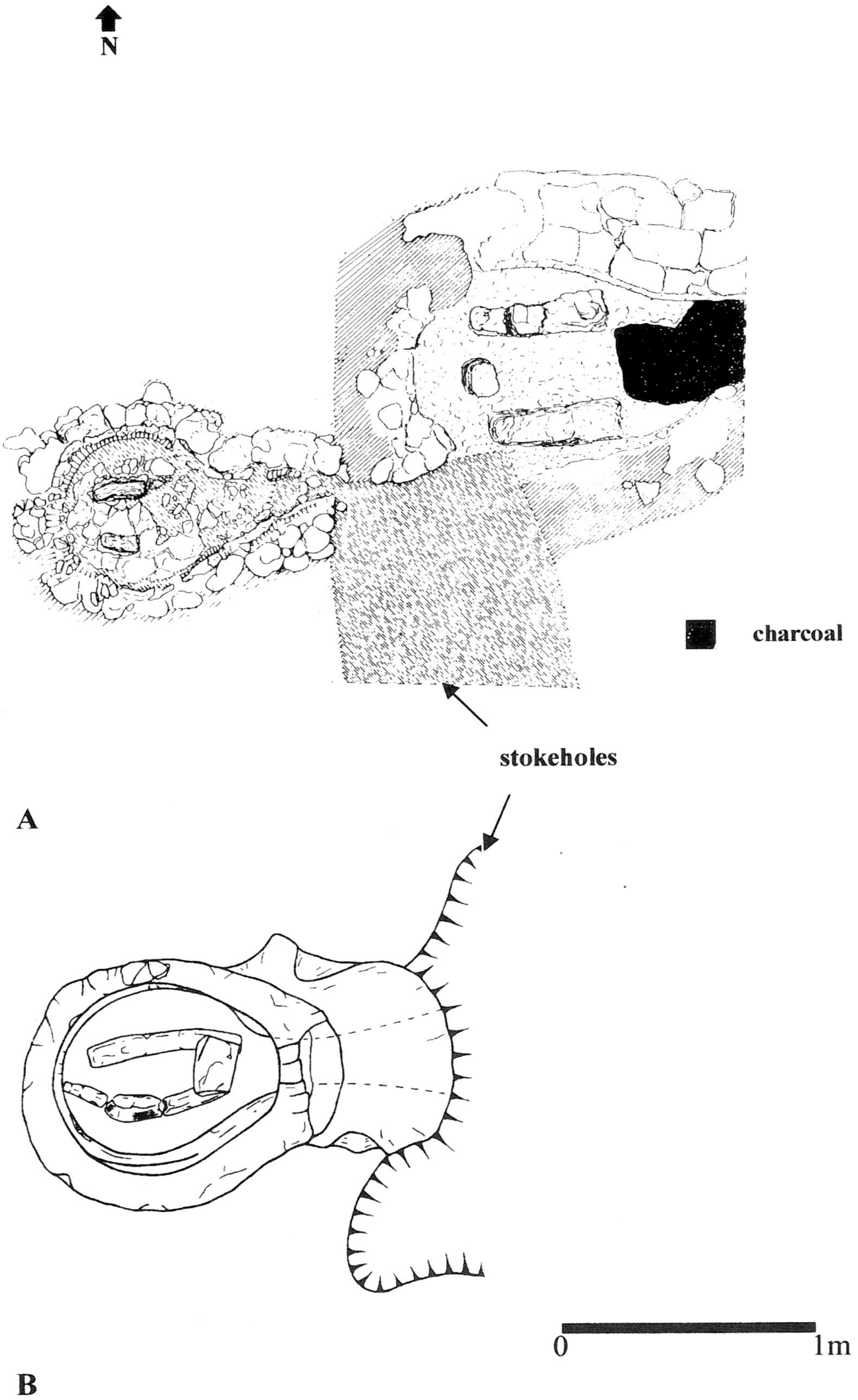

Roman Pottery Kilns at **A,** Hasholme Hall (after J. Wilson, Hicks and Wilson 1975) and **B,** Bursea House (after S. James Millett and Halkon 1984)

Figure 4. Roman Pottery Kilns at A) Hasholme Hall (after J Wilson, Hicks and Wilson 1975) and B) Bursea House (after S James, Millett and Halkon 1984).

Fabric	Colour	Hardness	Feel	Inclusions
1	dark grey	hard (wheel-thrown)	smooth some-times burnished	some sand and small flints
2	light grey to sandy brown	fairly hard (wheel-thrown)	slightly rough	frequent sand
3 (Dales-type)	mid-grey to sandy brown	fairly hard (wheel-thrown)	fairly rough	frequent, ill assorted 0.55-20mm angular flints, some black flecks, probably iron

Table 1: Definitions of the most common Holme fabrics

Evans 1985, 242) though Evans would date this from the mid third to mid fourth century (Evans 1985, 242) on the basis of the Trentholme Drive excavation at York (Gillam 1968, fig 23, nos 1-7; Evans 1985, 242; Swan 1980, 26). In a more recent paper, Swan has suggested that these wares were derived from a flat-topped, Ebor jar with a characteristic triangular sectioned rim. She argues that this form of vessel, brought to the York area by potters in the Sixth Legion, 'owed more to the north-west Mediterranean than to north-west Lincolnshire' (Swan 1992, 9; see also this volume). It does seem more logical that the wheel-thrown version should precede its browner, softer counterpart which is shell-gritted and handmade and has been attributed to the Lincolnshire kilns. However, the only example of the latter version of a Dalesware vessel discovered in the landscape block was found in association with a possible Lincolnshire product (J Evans *pers comm*) in a cream, hard fabric with grey and red flecking, caused by inclusions. Also present in quite large quantities was Huntcliff type pottery (Evans 1985, 369), though in nearly all cases the calcite grits had eroded away due to the acidic nature of most of the sandy soils of the study area.

Other fabrics found were relatively rare, the most common being Crambeck parchment ware (Corder and Birley 1937, 371-92) dated to the late fourth century by Evans (1985, 305). Samian ware (Hartley 1969, 235-51) was only found during fieldwalking and excavation on five sites and a very small amount of Nene Valley ware was also identified. The remainder of the pottery consisted of local handmade wares originating in the Iron Age and spanning the Roman period.

The three externally dateable pottery fabrics and fabric 3 were then used to create a broad temporal framework for local fabrics, thus providing a chronology for the sites where no external fabrics were found. The pottery from those sites with more than fifteen sherds was examined. Fabric 3 was present on 21 of the 24 sites with ready-dated fabrics and 19 of the 25 sites with more than 15 sherds. This would suggest a major period of settlement and industry in the mid third/early fourth centuries, agreeing with Evans (1985, 242, 275) dating of the Bursea House pottery industry. Creighton, although suggesting that this industry may begin in the early third century, argues that some of the differences in form and fabric, "do not relate to chronological shifts in forms, but in some degree of specialisation within the numerous kilns which made up the Holme-on-Spalding Moor industry" (Creighton 1999, 164).

In his recent survey of the pottery of Roman York, Monaghan (1997, 866) places Holme-on-Spalding Moor products in his Ceramic Period 3b (*c* AD 225-280) extending into Ceramic Period 4a (*c* AD 280–360). By Ceramic Period 4b (*c* AD 360–410), the Holme industry appears to have been eclipsed by Crambeck, to the extent that Crambeck pottery itself has been found in the Holme-on-Spalding Moor survey area. From the fieldwalked pottery it would appear that five sites were occupied either continually or sporadically until the late fourth century, others ending in the mid fourth. Five sites had no externally datable fabrics, and it was necessary to attempt to provide some idea of the date of settlement and industry by the comparison of non-local and local fabrics 1 and 2. This, though by no means conclusive, suggested that vessels with forms thought to be typologically later tended to have been made in fabric 1.

Forms

A typology of forms was devised by the writer based on the categories in Millett (1979). The published pottery drawings from the Hasholme excavations (Hicks and Wilson 1975, 61-66) and those carried out by Corder (1930, 23-39) at Throlam were arranged into functional groupings: jars, bowls, flagons and dishes. Material from the 1983 and 1984 Bursea House excavations (Millett and Halkon 1984; 1985) was also included in an updated typology by J Creighton and redrawn by B Sitch for the full publication (Halkon and Millett 1999).

There was a noticeable difference between the quality of the Hasholme wares and those at Throlam. Particularly striking at Hasholme were the large flagons with a collared, pie-crust decorated rim and strap handles running from the rim to the shoulder of the vessel (Figure 5, F02c), (Hicks and Wilson 1975, 64, fig 9.21). The Hasholme potters also specialised in sharply carinated bowls, which were presumably used for drinking. (Figure 5, B03a). There were also differences in decoration, with the Hasholme pottery displaying more variety than that from Throlam. The repertoire included open and narrow criss-cross and lattice decoration patterns,

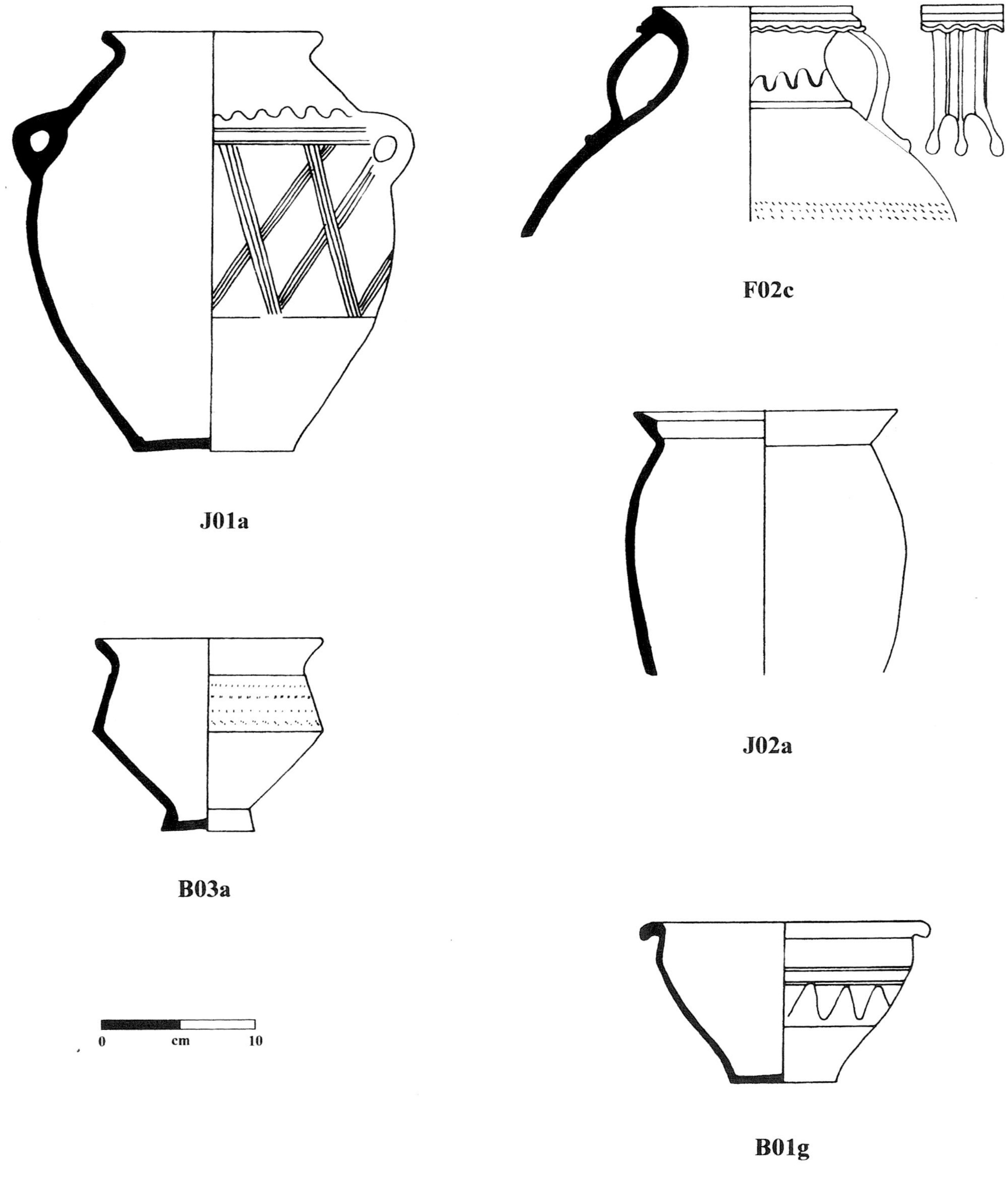

Figure 5. A selection of Romano-British greyware pottery forms from Hasholme Hall (after Hicks and Wilson 1975)

(largely on jars), wavy lines, (generally on wide-mouthed bowls), and rouletted patterns, occurring on the decorative flagons and small bowls with narrow bases, which were presumably for drinking. There was a striking similarity between the decoration on flanged bowls at Throlam and those at Crambeck (Evans 1989, 51), especially those with an internal wavy line. The form of this particular type of vessel was also close to those at Crambeck (see also Swan this volume).

The forms of the fabrics were then examined to see whether the dating of two local fabrics could be further clarified. Using samian, Crambeck, Huntcliff and local `Dales-type' fabrics, a comparison of sites was made to identify the relationship between the forms of vessels in Holme greyware fabrics 1 and 2 and the dated fabrics.

Chronology of sites based on the pottery evidence.

Using the conclusions so far reached for the chronology of the Holme pottery, the writer distinguished certain broad criteria characteristic of the third and fourth centuries, though this is not wholly supported by Creighton.

1. Characteristics of the second/third century.

Samian
Bowls greater than jars
Fabric 2 greater than 1
Dales types

2. Characteristics of the fourth century.

Flanged bowls
Flagons
Fabric 1 greater than 2
Crambeck parchment ware
Huntcliff type

These criteria were then applied to the sites with more than 15 sherds of pottery, and it was found that these sites could be placed into the following categories.

1. Sites with a 2nd to early 4th-century emphasis:-
(* = kiln site)

Bursea House *
Hasholme Hall *
Landing Farm *
Monk Farm*
Fir Tree Farm

2. Sites with either sporadic or continual occupation throughout the Roman period:

Beacon Farm
Bursea House (site 14)*
East Bursea Farm
Holme House*
Prospect Farm*
Welhambridge*
Woodlands Farm*

3. Sites with a later emphasis:-
Parlour Farm (Ladies Parlour)*
Park Farm
Rose Villa*
Sand Hill Farm *
Seaton Old Hall*
Skiff Farm*
Tollingham*
Throlam*

Markets for the pottery

It has already been noted that there was a dendritic tidal creek system in the landscape block during the Roman period (see Figure 1). This provided access to the Roman roadside settlements at Shiptonthorpe and Hayton. Fieldwork and excavation at Shiptonthorpe (Millett and Halkon 1986; Millett 1992) has already shown that much of the greyware pottery was being provided by the Holme kilns. Likewise field walking on the large civilian settlement at Hayton by B Sitch, and the present writer (Halkon, Millett and Taylor 1998) shows that much of the pottery is from the Holme-on-Spalding Moor kiln sites.

The identification during recent fieldwork of a Roman British settlement at South Cliffe, spreading over at *c* 4ha, almost directly across the tidal inlet from Hasholme, provides a further large consumer of Holme products (Halkon 1992).

Evans has identified the presence of Holme vessels on excavated sites at Brough in the third century (Evans 1985, 588), and at Brough, Malton, Beadlam, Castleford, Langton, Rudston and York in the fourth. Monaghan (1997) has also identified Holme products in third century York. Holme wares also predominated on "new" sites discovered during recent fieldwork by the writer including Ousethorpe near Pocklington (Halkon 1993); North Newbald; Hawling Road, Market Weighton (Halkon and Millett 1999), and Londesborough.

The dating of the period of production of the Holme kilns has already been referred to, and they seem to have dominated the markets in the later third and early to mid fourth century, before being eclipsed by Crambeck greyware products in the later fourth (Evans 1989).

The iron industry

Evidence for the working and smelting of iron has been recognised by local farmers for a considerable time, to the extent that slag and other metallic waste material are known as `Nosmun', a possible attribution of this to the Vikings. Although the *floruit* of this industry appears to have been in the Iron Age (Halkon 1997; Halkon and Millett 1999), some iron working and manufacture continued into the Roman period. An iron anvil, discovered near the kilns at Hasholme Hall 1970/1, was identified as being of typical Romano-British type (Manning 1975). This anvil (now displayed in the Hull and East Riding Museum) would have been set in a wooden block. It is probable that it was used in the working of blooms (Cleland 1981, 165).

A possible smithing hearth was also found during the 1970/1

excavations, and a geophysical survey undertaken by Mr J Pocock of Bradford University showed the presence of several possible furnaces or hearths in the surrounding area (Pocock 1983) and there was surface slag in the vicinity of these anomalies. Most of the iron had been completely oxidised and the slag was only weakly magnetic.

The field immediately adjacent to the kiln site contained a large concentration of slag and other debris, and should be associated with the kiln site as a field boundary is a modern division. Both sites are within 100m of the estuarine creek in which the Hasholme Boat sank. It is not surprising that a major regional pottery industry should follow such an established furnace based industry, the iron industry utilising bog ores, and both pottery and iron industries, the clay and managed woodland, and the creek system for transporting products and raw materials.

At North Cave (SE 880380), only 3km to the east of the Holme landscape block, smelting furnaces and a large amount of iron slag have been excavated in and around Iron Age round houses (Dent 1989). The material from North Cave is very similar to that found at the Holme smelting sites. Like several of the Holme sites with evidence for iron-smelting, this site continued into the Roman period. Swan (1984, 49) writes that there is clear evidence from numerous kiln sites that other industries were associated with potting, iron smithing being the most common adjunct. Wakerley in Northamptonshire (Jackson and Ambrose 1978) is a good example of an industrial site producing pottery and iron. Here iron working seems to have been carried out first and the potting does not seem to have been as long-lived as in Holme-on-Spalding Moor.

Glass manufacture?

Glass bangles were excavated on the kiln sites at Bursea House and Hasholme Hall (Halkon 1983; Millett and Halkon 1984; Hicks and Wilson 1975). Henderson would date the Bursea House example to the mid first to second century AD (Henderson 1983). A further bangle was found by D McLeod at Hasholme Grange in 1999, during a reconnaissance visit by the English Heritage Aerial Photographic Unit and the writer.

With the discovery of a glass slag fragment at Bursea House during fieldwalking, the possibility of glass manufacture has to be considered. More slag was uncovered in the garden of the house by Mr A S Johnson. During the excavation of 1983, some blue glass waste, similar to the bangle was found, although it should be noted that none of this material was derived from certain Roman-period contexts. Prof J Henderson has suggested (1999, 165-67 and *pers comm*) that it represents different phases in glass manufacture and a post sixteenth-century date of production. Prof J Price (*pers comm*) and Prof J Henderson do not accept a Roman attribution due to the chemical composition of the slag. It remains possible that the Bursea House bangles were produced on the site, but do not relate to the glass making waste found there. P Crew (*pers comm*) would argue from his own work and Nepalese ethnographic parallels, that no large scale industry was needed for the production of such bangles (but see also Price, this volume). Crew (*pers comm*) has also remarked on the possibility of a distinctive regional grouping in East Yorkshire.

Conclusions concerning industrial sites in the Holme landscape block.

Evidence within the Holme landscape block and beyond shows that the Holme-on-Spalding Moor pottery industry grew to be the main provider of coarse wares in the East Yorkshire area until well into the fourth century (Evans 1989, 93; Swan this volume), out-competing the Crambeck potteries and indeed some of the earlier forms of Crambeck pottery (types 4 and 14) may be inspired by Holme products (Evans 1989, 52).

It has been possible to suggest a broad chronology for Romano-British industrial and settlement sites, though this should be confirmed by further excavation. The areas nearest the main channel of the River Foulness, close to the estuarine creek system appear to have been exploited first, with the major centre of pottery production moving to the Throlam/Tollingham area in the fourth century. The concentration of sites discovered during fieldwalking and cropmark features (see Figure 6) implies that settlement was more intensive here. There are hints in the form of kilns (Swan 1984) of links with North Lincolnshire and it is likely that the estuarine creek system provided a convenient method of communication. Further research on both banks of the Humber is needed to clarify this relationship.

The evidence from aerial photography and fieldwalking shows clustering of sites around the hamlets of Bursea, Hasholme, Arglam, Welhambridge, Tollingham, Throlam and Skiff. Due to the lack of evidence it is not known whether there is any continuity (compare Taylor 1974, 12) between the Romano-British settlements and those hamlets listed above. Although it is almost certain that the pottery industry collapsed prior to the arrival of any Anglian settlement (Brown 1974, 19). The coincidence between `site' and `hamlet' may simply be caused by the same settlement constraints being present in both periods, the sand ridges upon which most sites occurred, remaining the most suitable locations for exploitation. The presence of late Roman pottery does show that the Holme landscape block continued to be occupied, and was not subjected to any total marine inundation as has been previously suggested. In such a marginal area, however, even a minor change in climate and/or sea level could have made a significant difference to settlement potential.

Many of the factors which enabled the Roman pottery industry in Holme-on-Spalding Moor to be so successful re-emerged later, in the High Middle Ages, as the location became a major centre for the production of the Humberware, which supplied Medieval Hull, Beverley and York, as well as numerous smaller settlements with pottery.

Endnote

This article was originally submitted in 1994, but contains some updating in the light of more recent work.

BIBLIOGRAPHY

Bartlett, J, 1968 A pig of lead and other finds from Broomfleet, East Yorkshire. In J Bartlett (ed) (1968) *Kingston upon Hull Museums Bulletin* **1**, 2-6

Brown, D, 1974. Problems of continuity. In Rowley (1974), 16-19

Catt, J A, 1990 Geology and Relief. In Ellis and Crowther (1990),

Figure 6. The Pot Hill field, Throlam, from the air. (Photo P Halkon, July 1995)

13-29

Cartwright, C, 1999 Charcoal identifications. In Halkon and Millett (1999), 126

Clark, M K, 1935 *A Gazetteer of Roman remains in East Yorkshire*. Leeds: Roman Malton and District Report **5**

Cleland, J, 1981 Roasting, blooming or smithing. *Current Archaeology* **77**, 165-7

Corder, P, 1930 *The Roman pottery at Throlam, Holme-on-Spalding Moor, East Yorkshire*. Hull: Roman Malton and District Rep **3**

----, and Birley, M, 1937 A pair of fourth-century pottery kilns near Crambeck. With a note on the distribution of Crambeck Ware. *Antiq J* **17**, 392-413

Creighton, J, 1999 The pottery. In Halkon and Millett (1999), 144

De Boer, G, 1974 Physiographic Evolution. In D H Rayner and J E Hemingway (eds) (1974) *The Geology and Mineral Resources of Yorkshire*. Leeds: Yorkshire Geological Society, 287

Dent, J S 1989 Settlements at North Cave and Brantingham. In Halkon (1989b), 26-32

Ellis, S, and Crowther, D R, 1990 (eds) *Humber Perspectives. A Region through the Ages*. Hull: Hull Univ Press

Evans, J, 1985 *Aspects of Later Roman pottery Assemblages in Northern England*. Unpublished Bradford Univ PhD thesis

----, 1989 Crambeck; the development of a major northern pottery industry. In P R Wilson (ed) (1989) *The Crambeck Roman Pottery Industry*. Leeds: Roman Antiqs Sect, Yorks Archaeol Soc, 43-91

Fasham, P J, Schadla-Hall, R T, Shennan, S T, and Bates, P J, 1980 *Fieldwalking for Archaeologists*. Winchester: Hampshire Field Club Archaeol Soc

Furness, R R, and King, S J, 1978 *Soils in North Yorkshire IV. Sheet SE63/73 (Selby)*. Harpenden: Soil Survey Record **56**

Gaunt, G D, Jarvis, R A, and Matthews, B, 1971 The Late Weichselian sequence in the Vale of York. *Proc Yorks Geol Soc* **38**, 281-4

Gillam, J P, 1968 The coarse pottery. In L P Wenham (1968) *The Romano-British cemetery at Trentholme Drive, York*. London: HMSO, Ministry Public Build Works Res Rep **5**

Gregory, T, 1982 Romano-British settlement in west Norfolk and the Norfolk fen edge. In D Miles (ed) (1982) *The Romano-British Countryside*. Oxford: Brit Archaeol Rep, Brit Ser **103** ii, 351-76

Halkon, P, 1983 Investigations into the Romano-British industries of Holme-on-Spalding Moor, East Yorkshire. *East Riding Archaeol* **7**, 15-24

----, 1987 *Aspects Of The Romano-British Landscape around Holme-on-Spalding Moor, East Yorkshire*. Univ Durham Unpubl MA thesis

----, 1989a Iron Age and Romano-British industry around Holme-on-Spalding Moor. In Halkon (1989b), 15-22

----, (ed) 1989b *New Light on the Parisi*. Hull: East Riding Archaeol Soc and Dept Adult and Continuing Education, Univ Hull

----, 1990 The Holme-on-Spalding Moor landscape. In Ellis and Crowther (1990), 347-56

----, 1992 Fieldwork in the southern Vale of York and southern Yorkshire Wolds. *Yorks Archaeol Soc Roman Antiqs Sect Bull* **9**, 14-15

----, 1993 From elephants to hypocausts - more fieldwork in the south-eastern Vale of York. *CBA Forum 1993*, 12-14

----, 1997 Fieldwork on early iron working sites in East Yorkshire. *Hist Metallurgy* **31.1**, 12-16

----, 1998 The Foulness Valley in the Iron Age and Roman Period. In P Halkon (ed) (1998) *Further Light on the Parisi - Recent Research in Iron Age and Roman East Yorkshire*. Hull: East Riding Archaeol Soc, Univ Hull Dept Hist and East Riding Archaeol Res Trust, 14-23

-----, and Millett, M, (eds) 1999 *Rural Settlement and Industry: Studies in the Iron Age and Roman Archaeology of Lowland East Yorkshire*. Leeds: Yorks Archaeol Soc, Yorks Archaeol Rep **4**

----, ----, and Taylor, J, 1998 *The Landscape Archaeology of Hayton. Interim Report 2*. Hull: Univ Hull

Hartley, B R, 1969 Samian Ware or Terra Sigillata. In R G Collingwood and I A Richmond (1969) *The Archaeology of Roman Britain*. London: Methuen, 235-51

Henderson, J, 1983 An interim report on two glass bangle fragments from Holme-on-Spalding Moor. *East Riding Archaeologist* **7**, 23-24

Hicks, J D, and Wilson, J, 1975 The Romano-British kilns at Hasholme. *East Riding Archaeologist* **2**, 49-70

Jackson, D A, and Ambrose, T M, 1978 Excavations at Wakerley, Northants, 1972-5. *Britannia* **9**, 115-242

Jordan, D, 1987 A minerogenic deposit. In Millett and McGrail (1987), 90.

Kent, P, 1980 *Eastern England from the Tees to the Wash*. London: HMSO, British Regional Geology. 2nd edn.

Loughlin, N, 1977 Dalesware: a contribution to the study of Roman coarse pottery. In D P S Peacock (ed) (1977) *Pottery and Early Commerce: Characterisation and Trade in Roman and Later Ceramics*. London: Academic Press, 85-146

Manning, W H, 1975 An iron anvil from Hasholme, Yorkshire. In Hicks and Wilson (1975), 49-70

Margary, I V, 1973 *Roman Roads in Britain*, (3rd edn). London: Baker

Millett, M, 1979 An approach to the functional interpretation of pottery. In M Millett (ed) (1979) *Pottery and the Archaeologist*. London: Inst Archaeol Occ Pub **4**, 35-48

----, 1992 Excavations and survey at Shiptonthorpe, East Yorkshire 1991. *Univ Durham Newcastle Archaeol Rep 1991* **15**, 29-33

----, and Halkon, P, 1984 Excavations at Bursea House, Holme-on-Spalding Moor, 1983. *Univ Durham Newcastle Archaeol Rep 1983*, **7**, 45-48

----, and ----, 1985 The Romano-British landscape in Holme-on-Spalding Moor: Interim report on fieldwork and excavation in 1984, *Univ Durham Newcastle Archaeol Rep 1984*, **8**, 36-38

----, and ----, 1986 Excavations at Shiptonthorpe, Welhambridge Farm and East Bursea Farm, East Yorkshire 1985, *Univ Durham Newcastle Archaeol Rep 1985*, **10**, 40-43

----, and ----, 1988 Landscape and economy: recent fieldwork and excavation around Holme-on-Spalding Moor. In J Price and P R Wilson (eds) (1988) *Recent Research in Roman Yorkshire*. Oxford: Brit Archaeol Rep, Brit Ser **193**, 37-47

----, and McGrail, S, 1987 The Archaeology of the Hasholme boat. *Archaeol J* **144**, 69-156

Monaghan, J, 1997 *Roman Pottery from York*. Archaeol York. The Pottery **16/8**. York: CBA

Peacock, D P S, 1977 Ceramics in Roman and Medieval

Archaeology. In D P S Peacock (ed) (1977) *Pottery and Early Commerce: cCharacterisation and Early Trade in Roman and Later Ceramics*. London: Longman, 21-33

Pocock, J A, 1982 *A Magnetometer Survey at Bursea House*. Bradford: Bradford Univ School Archaeol Sci

---- , 1983 A note on the magnetometer survey at Hasholme Hall, Holme-on-Spalding Moor. In Halkon (1983), 22

Pryor, R, 1997 Thin section analysis. In Monaghan (1997), 1031-37

Ramm, H, 1978 *The Parisi*. London: Duckworth

Rowley, T (ed) 1974 *Anglo-Saxon Settlement and Landscape*. Oxford: Brit Archaeol Rep, Brit Ser, **6**

Sitch, B, 1989 A small Roman port at Faxfleet. In Halkon (1989b), 10-14

Swan, V, 1980 *Pottery in Roman Britain*. Princes Risborough: Shire

---- , 1984 *The Pottery Kilns of Roman Britain*. Royal Commission on Historic Monuments Supp Ser **5**. London: HMSO

---- , 1992 *Legio* VI and its men: African Legionaries in Britain. *J Roman Pottery Stud* **5**, 1-33

Taylor, C, 1974 The Anglo-Saxon countryside. In Rowley (1974), 5-15

Turner, J, 1987 The pollen analysis. In Millett and McGrail (1987), 85

Wilson, V, 1948 *East Yorkshire and Lincolnshire*. London: HMSO, British Regional Geology

THE ROMAN POTTERY OF YORKSHIRE IN ITS WIDER HISTORICAL CONTEXT

Vivien G Swan

The Background

When the Roman army invaded the territories of the tribes of Yorkshire, the Brigantes and Parisi, it found that very little of the local pottery was appropriate to its wide ranging and quite specialised needs. At best, some of the relatively coarsely tempered, crude jars, handmade in East Yorkshire (nos 46, 235-7), the Tees Valley, and sporadically elsewhere, were just good enough for basic cooking (Evans 1995). However, to step up production, improve its quality and diversify suddenly, in order to cater for an army of many thousands, was clearly beyond the capabilities of the native potters. Hitherto, these had been geared to serving small, relatively local markets, and many, if not most producers were probably working on a semi-domestic or very localised basis. The Roman army was thus faced with several choices: either to import its pottery from southern Britain or from the Continent, or to set up its own military *figlinae* (workshops making a variety of fired clay products) using soldier-potters and slaves or contractors, or to encourage the natives to upgrade and expand their production, the last inevitably a much more long-term proposition. That it chose all three options is a testimony to Roman pragmatism and adaptability.

Flavian to early Hadrianic military workshops

From a detailed study of types of Roman pottery vessels in Northern Britain for the period from *c* AD 70 to the early 120s, it seems likely that many (perhaps most) of the forts and fortresses from South Yorkshire to the Tay in Scotland would have had their own pottery workshops. Most of these northern *figlinae* were probably manned either by army personnel or occasionally by civilian potters working under direct military control; many would have made tiles as well, but only at periods when military building-work was taking place. In Yorkshire, as elsewhere, relatively few actual kiln sites of the pre-Hadrianic period have been recorded. Best known is Grimescar Wood, near Huddersfield, where, in the Trajanic period and possibly earlier, *Cohors IV Breucorum* made stamped tiles (*RIB* II, 2470) and pottery vessels for the auxiliary fort of Slack, almost 5km to the west (Purdy and Manby 1973; Swan 1984, MF 6.720); and also certainly tiles for the fort at Castleford (Betts 1998, 226, 231). The orange/red-yellow cooking wares, all typical of late first-century military contexts, included carinated or hemispherical bowls with a reeded rim (no.1) and lids to match, high-shouldered cooking-jars with a short everted rim (nos. 3-4), cheese wrings (no. 5), flagons with a pulley-wheel-type rim (no. 4), and possibly mortaria and rusticated cooking-jars.

At Templeborough (Rotherham), during the First World War, Thomas May, whose standards were well in advance of his time, rescue-excavated the auxiliary fort of *Cohors IV Gallorum.* Among his published pottery, were several mid/late first-century vessels, some of them wasters, which he thought were locally made (May 1922: forms 222-7, 173, 178, 235-7). These included ring-necked flagons (no. 6), rusticated cooking-jars (no. 9), imitations of samian forms Drag. 27 (no. 7) and 29/37, segmental flanged bowls (no. 10), and plain-rim dishes perhaps imitating Gallo-Belgic or Pompeian red ware forms (no. 11). All were in ordinary sandy Romanised, utilitarian fabrics (mostly oxidised), such as would have been made in a military works-depot.

A dump of ash and pottery, including wasters, recorded in a pipe trench at Aldborough, immediately east of the later *civitas* capital, near the junction of the York-Milby-Catterick roads suggests a workshop of Trajanic and later date, presumably originally established to serve a fort in the vicinity (Jones 1971, 53-60, figures 13 and 18). The latter is currently unlocated but was probably sited north of and on the northern fringes of the later town (replacing the Flavian military establishment at Roecliffe, just over a mile to the west). Again the products included several utilitarian forms typical of military sites, ring-necked flagons, cooking-jars with a short everted rim (no. 16), and carinated bowls with a reeded rim and lid to match (nos. 13, 15), as well as plain-rim dishes (no. 17), and less standard campanulate bowls; also made in the vicinity were stamped mortaria in a white fabric (nos. 14, 18). These last, and several other coarse ware forms may suggest a continuation of production until at least 140 in civilian, or possibly in veteran hands, at the period when the town was being established as the *civitas* capital of the Brigantes, following the evacuation of the fort (Bishop 1996, 1-3). In fact one of the mortarium potters, VIATOR also seems to have had a workshop further south at the former fort of Castleford (Hartley 2000, 183-87)

Just over the county boundary at Piercebridge, Co. Durham, a Trajanic-early Hadrianic kiln was excavated in 1980 by the late Peter Scott (Swan 1984, MF 1.265; Swan *forthcoming*). It lay just outside the site of the third- and fourth-century fort, but probably related to a much earlier military establishment, in whose *vicus* the workshop had probably been sited. Its products also included ring-necked flagons, cooking-pots with linear rustication in relatively low relief, flat-rim carinated bowls, lids, small everted-rim jars, rouletted bowls and cups imitating samian forms, and mortaria (L Hird and A Fitzpatrick *pers comm*). Almost all typify military supply of the early second century.

York: legionary production and supply in the pre-Hadrianic period

At York (*Eboracum*), the fortress of *Legio VIIII Hispana*,

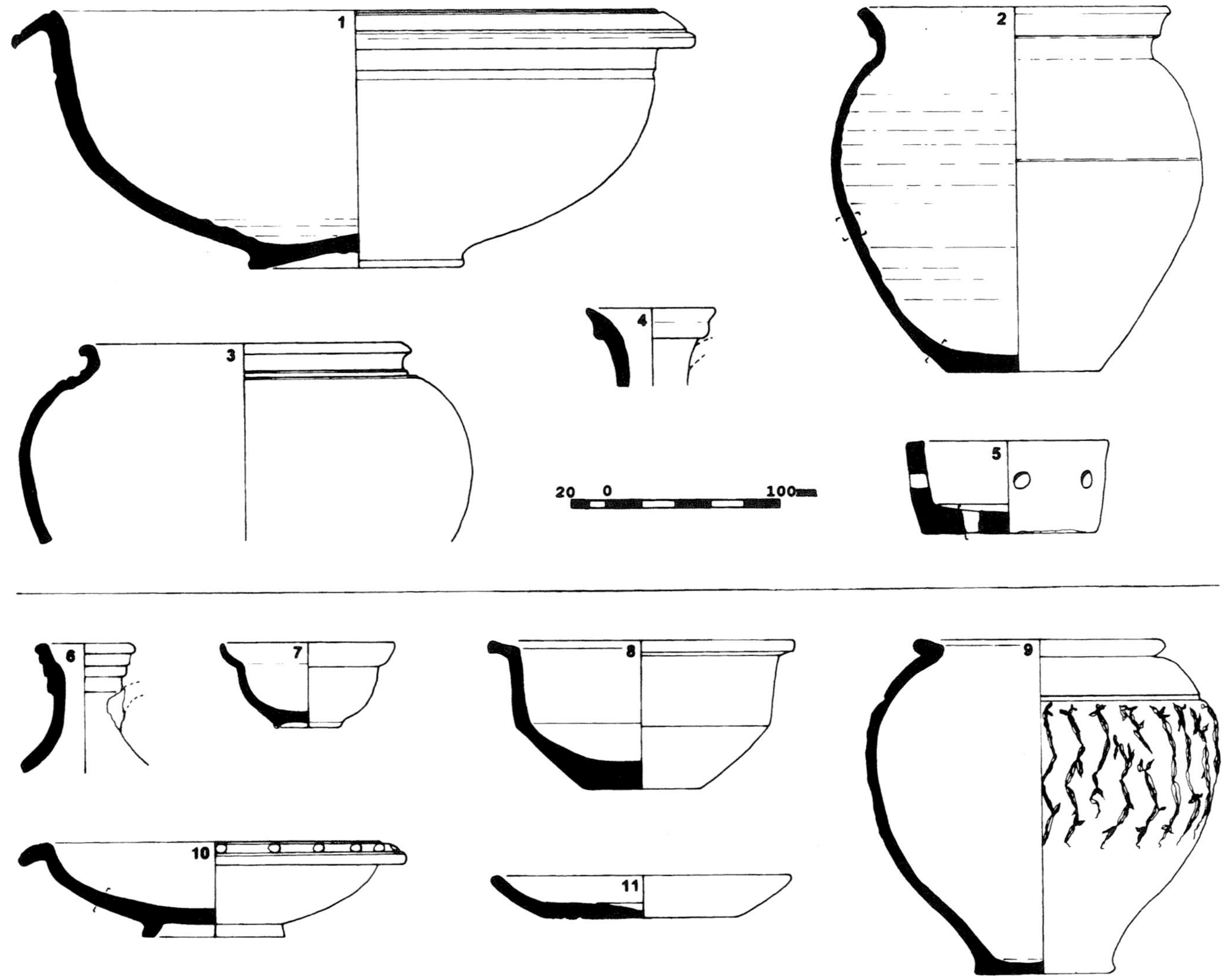

Figure 1. Workshops serving early military sites: Slack and Templeborough.
Nos. 1-5: products of Grimescar Wood Flavio-Trajanic tilery and pottery (after Purdy and Manby 1973, with revisions); nos. 6-11: Flavian and Trajanic vessels from Templeborough fort, probably produced in its vicinity (after May 1922, with revisions). Scale 1:4.

1 *Hemispherical bowl with reeded rim.*
2 *Cooking-pot.*
3 *Jar with everted rim; cf. no. 27, a closely similar form in Ebor ware.*
4 *Flagon with pulley-wheel rim ('Hofheim-type'), a form out of production before the end of the first century; cf. no. 19.*
5 *Cheese wring.*
6 *Ring-necked flagon with even rings; cf. no. 22.*
7 *Cup imitating samian form Drag. 27.*
8 *Carinated bowl with simple flattened rim; cf. contemporary examples of same general type from Aldborough, York and Malton (nos. 15, 28, 36-7).*
9 *Cooking-jar with applied rustication; cf. similar contemporary vessel (no. 26) and much later version (no. 151).*
10 *Segmental flanged bowl with white painted dots.*
11 *Plain-rimmed convex-walled dish with under-grooving, probably imitating a Pompeian red ware form (cf. Peacock 1977, figure 3).*

evidence from the Blake Street barracks indicates that at least one Flavio-Trajanic military *figlina* making orange and grey wares (nos. 19-33), must have existed somewhere in the vicinity, commencing immediately on the foundation of the fortress, and supplying at least 70% of its needs (Monaghan 1993, 684-90, figure 273). These earliest products appear to find their closest parallels in the Lower Rhineland (see below). Even so, at York, only the probable site of the tilery of that legion is known for the pre-Hadrianic period. It was evidently sited about 200m east-south-east of the east corner of the fortress, on or near the area now occupied by St. Cuthbert's church, Peaseholme Green (Hargrove 1818, 346-47; 1838, 60-61). It is important to emphasise, however, that the dumps of fired clay and pottery rejects, found in the next-door garden of the Borthwick Institute, and originally published as probable Ninth Legion kiln waste (King 1974), are now thought to represent debris from later production by *Legio VI Victrix* (see below), which became mixed with earlier tile wasters (some stamped) from the adjacent Flavio-Trajanic production site of *Legio VIIII* (Swan 1992, note 3). Nevertheless, the Ninth Legion's pottery workshops may not have lain very far away. The rescue excavations in 1996 outside the King's Pool Building, Peaseholme Green (opposite St Cuthbert's Church), recovered a number of sherds of pre-Hadrianic reeded rim bowls and mortaria (including possible wasters), which had become incorporated in medieval deposits sealing Hadrianic and Severan kiln debris (Swan and McBride *forthcoming*). Moreover, two vessels from the Borthwick Institute excavations, a large collared storage-jar and a fragment of a face-pot, would, on typological grounds, be more appropriate in a pre-Hadrianic context (King 1974, figure 9.6 & 12). Clearly, more extensive excavation is needed to test whether the Flavio-Trajanic ceramic workshops had occupied the same general zone of the fortress *canabae* as the known Hadrianic and Severan potteries.

In northern Britain in general, much more needs to be done to seek out the sites of early *figlinae*, particularly in the environs of army installations, since disproportionate attention has hitherto been concentrated on the defences and internal buildings of forts. Geophysical prospection and systematic fieldwork near military establishments, in areas with the appropriate raw materials, might well prove informative. In the absence of known kiln sites, the recognition of local production in a military tradition is generally dependent on the repetitive occurrence in a fort (or series of forts) of stamped tiles (e.g. those of *IIII Gallorum* at Castleford; Betts 1998, 226, 231-32), or of a range of distinctive and well-known basic vessel-types in the same fabric (perhaps the essential mess-items ordered by the military quartermasters), though with the precise detailing tending to differ slightly from locality to locality. Such vessels are more often oxidised orange or buff, but are sometimes reduced grey (usually the rusticated cooking-jars). In addition to the more standard vessel types, such military assemblages often include unusual, or exotic forms in the same ware. The latter could imply that many early workshops were probably manned by the army's own soldier-potters, each with his own individual ceramic traditions. At least some of such men are likely to have been classed as *immunes*, that is, artisans, who had joined the army as ordinary soldiers with recognised technical skills, which they used in the service of the regiment (Breeze 1977, 136-37; cf. *Digest* L, 6.7[6]). In return, they gained exemption from some of the heavier fatigues, though they were not absolved from actual fighting. At the auxiliary fort of *Vindolanda*, on the Stanegate (the Tyne-Solway military zone), one of the writing tablets (*Tab Vindol* II, 155), a military report of the early 90s AD, lists the number of men involved in, or being sent to, various activities connected with the building of the fort baths and ?the hospital,

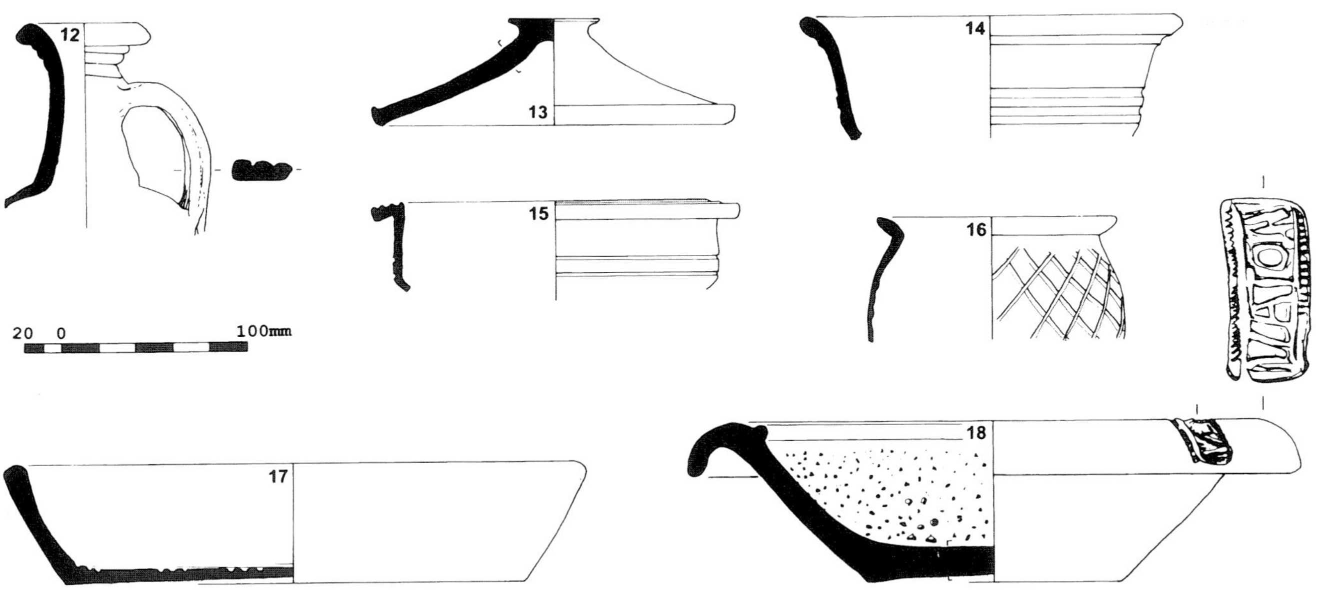

Figure 2. Workshops serving early military sites and their aftermath: Aldborough. Vessels from the Trajanic-Hadrianic kiln waste-dumps (after Jones 1971, with revisions). Scale 1:4.

12 Ring-necked flagon with splayed neck and expanded top ring, a typologically later development of such forms as nos. 6 and 22.
13 Lid with squared-off rim, a common form in contemporary military contexts.
14 Cordoned campanulate bowl.
15 Carinated bowl with reeded rim.
16 Jar with everted rim; the burnished lattice may have been influenced by that on the (then) newly-introduced BB1 cooking-pots.
17 Plain-rimmed convex-walled dish with internal basal grooving, a form probably imitating Pompeian red ware (cf. Peacock 1977, figure 3).
18 Flanged mortarium with stamp of the potter NATOR.

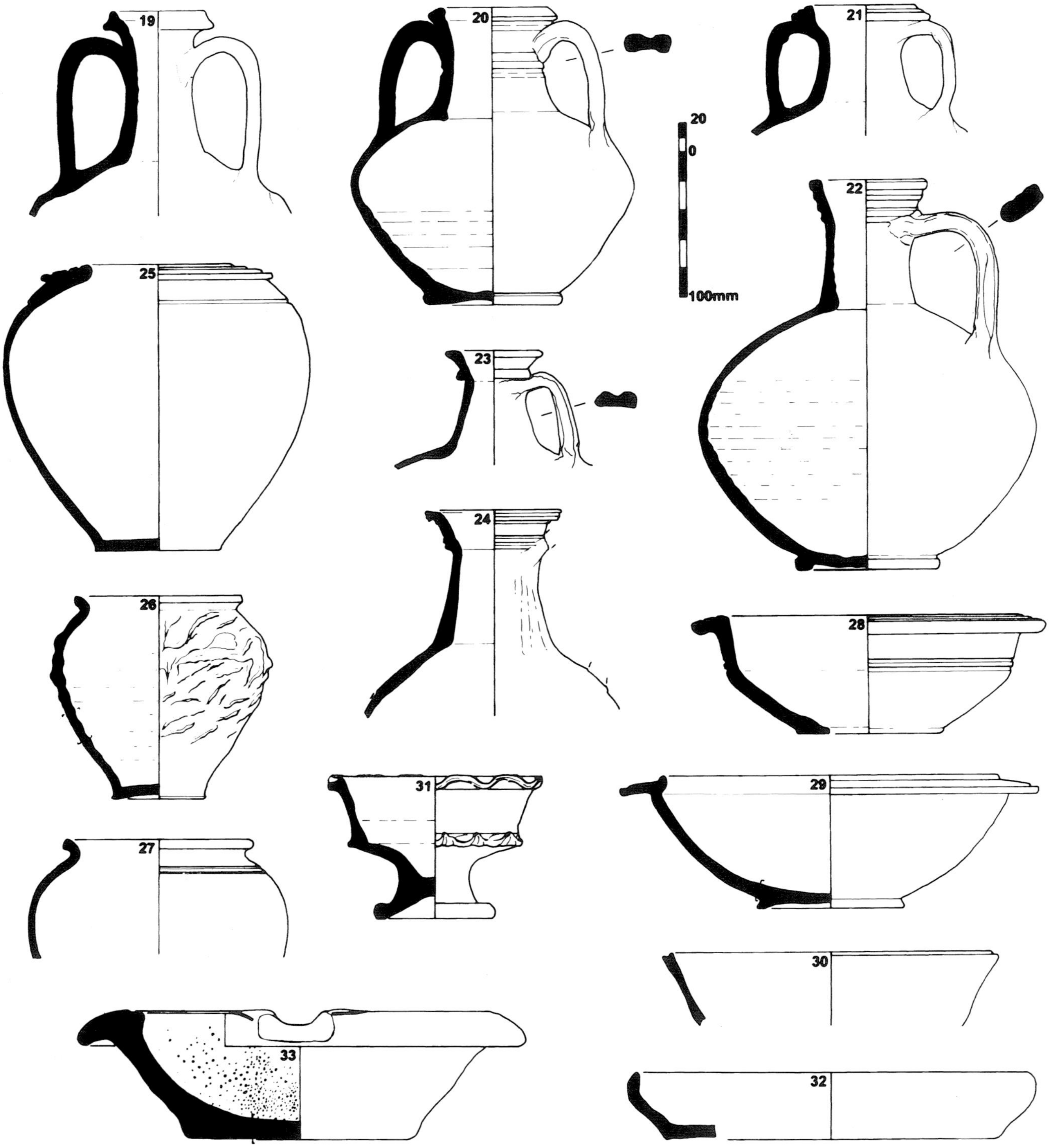

Figure 3. Workshops serving early military sites: York. (Captions continued on the opposite page)
Flavio-Trajanic vessels in the utilitarian Eboracum Oxidised ware (Ebor Fabrics 1 & 2); unless otherwise stated, all probably produced in workshops operated by the Ninth Legion. Scale 1:4.

19 Hofheim-type' flagon (after Perrin 1995); cf. no. 4.

20 Double-handled flagon with sub-triangular rim (after Monaghan 1993); cf. the typologically later version made at Appletree Farm, York (no. 69).

21 Double-handled flagon with rilled lid-seated rim (after Perrin 1995).

22 Ring-necked flagon with even rings (after Monaghan 1993); cf. the typologically later versions made at Appletree Farm, York (nos. 70-71).

23-4. Collared flagons (after Monaghan 1993 and 1997).

25 Storage-jar with reeded collar (after RCHME 1962).

Figure 3. Workshops serving early military sites: York. (Captions continued)
26 *Rusticated jar in a reduced Ebor fabric (after R Perrin, York Minster Excavations Archive); cf. the detailing of a contemporary Templeborough vessel (no. 9).*
27 *Everted-rim jar (after R Perrin, York Minster Excavations Archive); cf. a closely similar Grimescar vessel (no. 3).*
28 *Carinated bowl with reeded rim (after R Perrin, York Minster Excavations Archive); cf. nos. 15 and 72.*
29 *Flanged bowl, an unusual form intrinsic to Flavian York (after R Perrin, York Minster Excavations Archive).*
30 *Straight-sided bowl or dish with external lid-seating (after Monaghan 1993).*
31 *Frilled tazza or 'incense cup' (after Monaghan 1993).*
32 *Dish with concave wall, perhaps imitating a Pompeian red ware form (after Perrin 1995); cf. nos. 11, 17.*
33 *Flanged mortarium (after Monaghan 1993).*

and may well be referring to the manufacture of tile (and by association probably of pottery as well).
Though fragmentary, it reads,:

"April 25th, in the workshops, 343 men.
of these: shoemakers, 12
builders to the bathhouse, 18
for lead ...
for ... wagons(?) ...
... hospital ...
to the kilns ...
for clay ..."

In this instance, the men were probably being sent to the kilns and clay-pits of the Brampton tilery (between *Vindolanda* and Carlisle), whence Trajanic *Vindolanda* seems to have drawn a significant portion of its supplies of mortaria and cooking-vessels (Swan *in prep*). This general phenomenon is further illuminated by the analyses of the ceramic materials found at the Flavian legionary fortress of Inchtuthil, Scotland (Williams 1985), and from an unpublished study of the petrology of Roman tile produced in York (Betts 1985; I Betts *pers comm*). These indicate that the same basic clay source often served both the pottery and the tile workshops of early military establishments, as well as providing daub for cladding the wattle-work of timber buildings.

The recognition of probable military *figlinae*

The case for the probable existence of military *figlinae* at many late-first and early second-century installations is well illustrated by the Flavio-Trajanic pottery from York (the so-called *Eboracum* or Ebor ware), and that from Trajanic to early Hadrianic Malton, sites only 34 km apart (Figures 3, 4). Though in its earliest phase Malton seems, from Corder's (1930a) evidence, to have been drawing considerable amounts of its pottery from the *Eboracum* workshop and from north-west Lincolnshire (nos. 54, 57), by the Trajanic period local products can be discerned (possibly suggestive of a change in garrison and/or supply policy). To the naked eye, some of the orange, and occasionally grey, fabrics associated with the establishments at York and Malton are very similar to one another, sometimes virtually indistinguishable, yet a substantial number of the forms and the decorative techniques are completely different, as Bidwell and Croom noted (1997, 100-01). Even where vessels of the same basic type occur at both establishments, each still has its own subtle individual characteristics, presumably intrinsic to the craftsmen or workshops involved in their production. As the late John Gillam pointed out (1973, 53) of the pottery of this period at various military dispositions along the Stanegate road, it was as if the potters on all these early sites were trying to do much the same thing, but were only succeeding up to a point. Such inter-site differences are often most evident in the detailing of the rims, or in simple decorative features. For example, in the earlier second century, the Malton potters were fond of rouletting (nos. 39, 41) and the rims of their carinated bowls had a pronounced undercutting (Figure 4.36-7 & 42), but they also made local campanulate and hemispherical bowls imitating samian forms Drag. 29 and 37 in the same fabrics (nos. 38-9, 41; Bidwell and Croom 1997, nos. 29-30, 47-8, 73, 79, 134-5, 230, 277, 421). In contrast, at York, rouletting was virtually absent on Flavio-Trajanic Ebor ware; the rims of the carinated bowls were rectangular, not sharply undercut (no. 28); the early everted-rim jars had more than one shoulder-groove (no. 27) and a distinctive type of bowl with a grooved flange is also characteristic (no. 29; Perrin 1995, figure 130, nos. 6, 8, 10-11; Monaghan 1993, nos. 2813, 2824, 2838, 2879, 2881, 2908). For elsewhere in Yorkshire, the identification of early military production awaits the publication of large assemblages, though there are now pointers in the recently published ceramic sequences from the fort at Castleford, West Yorkshire (Rush 2000, 148-49).

The origins of potters

Though the more widespread basic forms evidently relate to general military traditions (for example, ring-necked flagons, collared storage-jars, high-shouldered everted-rim jars, rusticated cooking-pots, carinated bowls with a flat or reeded rim, and plain-rimmed platters copying the Gallo-Belgic form, Cam. 16 or Pompeian red rare), it is their detailing as well as some of the more unusual, but often utilitarian forms that best reflect the differing origins and backgrounds of the local soldier-potters and sometimes their consumers (see below). Bidwell and Croom have suggested (1997, 100-01) that some of the Malton craftsmen may have originated in the burgeoning potteries of the Verulamium region (Herts./Middlesex), which had already been supplying military dispositions since the Neronian period. The late Flavian-Trajanic period was a time when other potters from that industry seem to have moved in the general direction of military markets (Swan 1984, 98-99; unpub. info. on pottery from the Scottish fort of Elginhaugh, K Hartley *pers comm*), and the Malton craftsmen seem more likely to have been civilians, though they could have been auxiliary recruits. In contrast, one or more of the Flavian potters at York seem likely to have originated in the Lower Rhineland, possibly from the legionary base at Neuss (*Novaesium*), where local

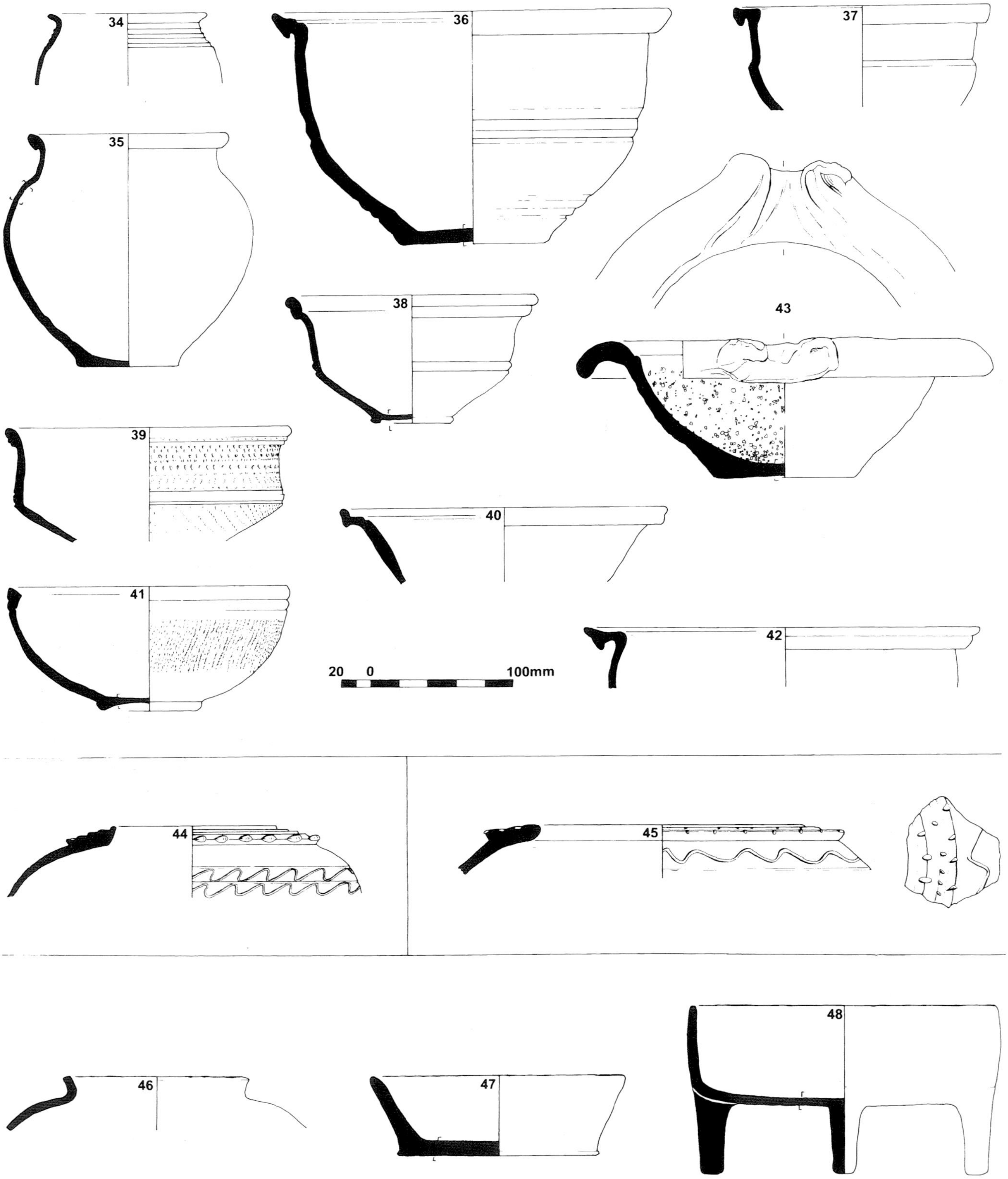

Figure 4. Workshops serving early military sites (Captions opposite)

kilns were producing a number of closely similar forms (Filtzinger 1972); these seem more likely to have been military recruits or transfers. An exploration of the whole subject of military recruitment and transfers, and the origins of potters at early forts in northern Britain would require more space than is available here. Suffice it to say that though much more research and publication is needed, the very varied and sometimes unusual or exotic character of much pottery in pre-Hadrianic military assemblages in the north Midlands and northern Britain, tends to accord with the conclusions of epigraphers such as John Mann and Brian Dobson, namely that at this early period, the garrisons of *Britannia* were drawn from a greater diversity of provinces of the Empire than at any other time during its history (Mann 1983, 23, 89,

Figure 4. Workshops serving early military sites: Malton.
Flavian-early Hadrianic pottery probably made in the vicinity of the fort (except no. 45): 34-43: grey or orange Romanised forms and fabrics; 46-8: coarse calcite-gritted ware made for the military by native potters. All pots come from Malton fort or vicus *unless otherwise indicated. Scale 1:4.*

34 *Jar with rilled shoulder (after Bidwell and Croom 1997).*
35 *Necked jar (after Corder 1930a).*
36 *Carinated bowl with undercut reeded rim (after Bidwell and Croom 1997).*
37 *Carinated bowl with undercut flattened rim (after Mitchelson 1964).*
38 *Campanulate bowl with cordoned rim and shoulder reminiscent of samian form Drag. 29 (Malton Vicus Excavation Archive, after E King).*
39 *Rouletted campanulate bowl with cordoned rim and shoulder (Malton Vicus Excavation Archive, after E King).*
40 *Campanulate bowl with undercut lid-seated rim (Malton Vicus Excavation Archive, after E King).*
41 *Rouletted hemispherical bowl with cordoned rim, probably an imprecise imitation of samian form Drag. 37 (Malton Vicus Excavation Archive, after E King).*
42 *Bowl with undercut lid-seated rim (after Mitchelson 1964).*
43 *Flanged mortarium (after Bidwell and Croom 1997).*
44 *Storage-jar with reeded collar and stabbed and incised wavy-line decoration, found in Malton vicus and possibly but not certainly a local product (after Mitchelson 1964). The stylistic details are precisely paralleled in Gaul (no. 45).*
45 *Storage-jar closely similar to previous, but from a kiln site producing such forms at Beaumont-sur-Oise, l'Ile-de-France, France (after Vermeersch et al 1995).*
46 *Handmade jar with neat everted rim, an upgrading of an indigenous form presumably for military supply (after Bidwell and Croom 1997).*
47 *Handmade plain-rimmed dish or bowl; a native imitation of a Roman prototype (after Bidwell and Croom 1997).*
48 *Handmade bowl with tripod feet, probably made by an indigenous potter specifically for a Gallic element in the garrison (after Corder 1930a, revised).*

table 9; Dobson and Mann 1973, 198-99, 202-03).

Native potting and pre-Hadrianic supply

Despite the inevitable early emphasis on production by, or under, the military in Yorkshire, the indigenous potters were not totally ignored. The simple, handmade, jars native to East Yorkshire and north Lincolnshire, tempered with mineral calcite, sometimes with the addition of shell, chalk and other gritty particles (nos. 235-7), can be seen in the earliest levels of the forts at Hayton (Johnson 1978, figure 24, nos. 42-5), Malton (Corder 1930a, figure 11, nos. 11, 14, 19, figure 17, no. 20; Mitchelson 1964, figure 7, no. 5, figure 8, no. 26; Bidwell and Croom 1997, 98-100, figure 22, no. 7), Castleford (Rush 2000, figure 46, no. 70, figure 48, no. 103, figure 49, nos. 107-11, figure 52, no. 172), and Brough-on-Humber (Corder 1935, figure 6, A5; Corder and Romans 1937, figure 10, nos. 18-19, figure 13, nos. 75-8; ibid 1938, figure 14, nos. 3, 7, 18; Wacher 1969, figure 53, nos. 1, 2, 4, 6; figure 54, nos. 28, 30, 32). In East and North Yorkshire and beyond, such wares continued to be the mainstay of most of the rural settlements away from roads, well into the third century. Nevertheless, these native products were rare in the early levels at York. Bidwell and Croom have suggested (1997, 100) that the needs of the legionary fortress were so large, that in the early stages of Roman occupation, the native potters were unable to expand their production and marketing enough to meet its demands, and preferred to serve the more local markets represented by the forts. Moreover, there are hints that individual forts may themselves have each played a direct role in the purchase of at least some of their pottery (through local and non-local contacts), rather than depending wholly on the more unified military supply systems, from which they could apparently obtain samian and other products.

At Malton, the specialised tastes of certain ethnic elements in the garrison seem to have stimulated the indigenous potters to fashion some vessel-types which were different or totally alien to their own traditions. High-shouldered jars (no. 46), plain-rimmed dishes (no. 47) and tripod bowls in coarse, handmade, calcite-gritted fabrics constitute cases in point (no. 48, and Bidwell and Croom 1997, no. 237). Tripod vessels had not been made anywhere in Britain in the late Iron Age, but related to a cooking style which originated in Gaul. There such vessels were a regular feature of the Centre (as far south as Aquitaine) and the North (excluding the extreme northern fringes of France and modern Belgium), and were particularly popular in the Ile-de-France. Presumably these coarse Malton tripod bowls represent the response of the indigenous East Yorkshire potters to requests or commissions from Gallic soldiers based in the fort. Indeed, the probable reality of such a Gallic presence at the fort may be further emphasised by the occurrence in the *vicus* of a distinctive decorated collared storage-jar (no. 44; Mitchelson 1964, figure 8, no. 35), the form of which can be precisely paralleled among the products of kilns at Beaumont-sur-Oise, Ile-de-France (no. 45; Vermeersch et al 1993, pl. 83, nos. 20-21; Vermeersch 1995, pl. 3, no. 9). This vessel (or traditional Gallic regional style) must have been brought to Malton from there, perhaps by an individual soldier.

Institutional and personal imports

In this early period, there were numerous imports, mainly tablewares and specialist vessels (more often from the Gaulish provinces), which satisfied specialist demands; for example, samian from southern, and eventually central Gaul, and pale-coloured mortaria and flagons from workshops in the region of Noyon (Oise), and other mortaria from Aoste, Isère and possibly from Central Gaul, the Rhone Valley and Bavay (K Hartley 1998, 202-09 with references). These lie outside the orbit of this paper. However, it is appropriate to mention in passing, that quite ordinary cooking-vessels were sometimes carried into the region by the troops themselves, either as personal possessions, or as baggage items of official 'kit', probably issued to a century or *contubernium* for the basic preparation and serving of food (see also p 47). The latter mechanism probably accounts for the presence at York, in areas of barracks and early rubbish dumping, of a small amount of Exeter Fortress Ware Fabric B (nos. 59-60; cf. nos. 61-2; Swan and Bidwell 1998, n 3). One or more centuries, perhaps a vexillation, from *Legio II Augusta* may have been sent North for action with *Legio VIIII* and ultimately transferred into the York-based legion in *c* AD 75, at the time of the Second Legion's move from its Exeter base to establish the fortress at Caerleon. Excavations at Caerleon have indicated that no barracks were constructed on the plots designated for Barrack XII and possibly Barrack XI, probably until the Antonine period, so that legion was evidently under strength at the time when it arrived there (Casey and Hoffman 1995).

Pottery manufacture in the Humber Estuary

For Yorkshire, however, it seems that the existence of well-established, and good quality potting traditions on the southern side of the Humber Estuary, between the Rivers Trent and Ancholme, and the low cost and relative ease of water transport, resulted in some early pottery (and influences, as will be seen later) being drawn from that region (Figure 5). Kilns such as those excavated or suspected at Dragonby and its vicinity, near the mouth of the Trent were geared, at least in part, to the needs of the army's supply-systems, since many of their wares were not in use locally (nos. 49-53; Stead 1976, 96, kiln 3; Rigby and Stead 1976, 136-39; May *et al* 1996, 103-6, 574-83, kiln 4 and pit F2567). Some of their products evidently travelled up the River Don to the Doncaster fort (Buckland and Magilton 1986, figure 39, no. 218, figure 40, no. 295), up the River Derwent to Stamford Bridge (Croom *et al forthcoming*) and Malton (no. 54; Bidwell and Croom 1997, nos. 125, 409), and in small quantities up the Ouse to York (no. 58; Monaghan

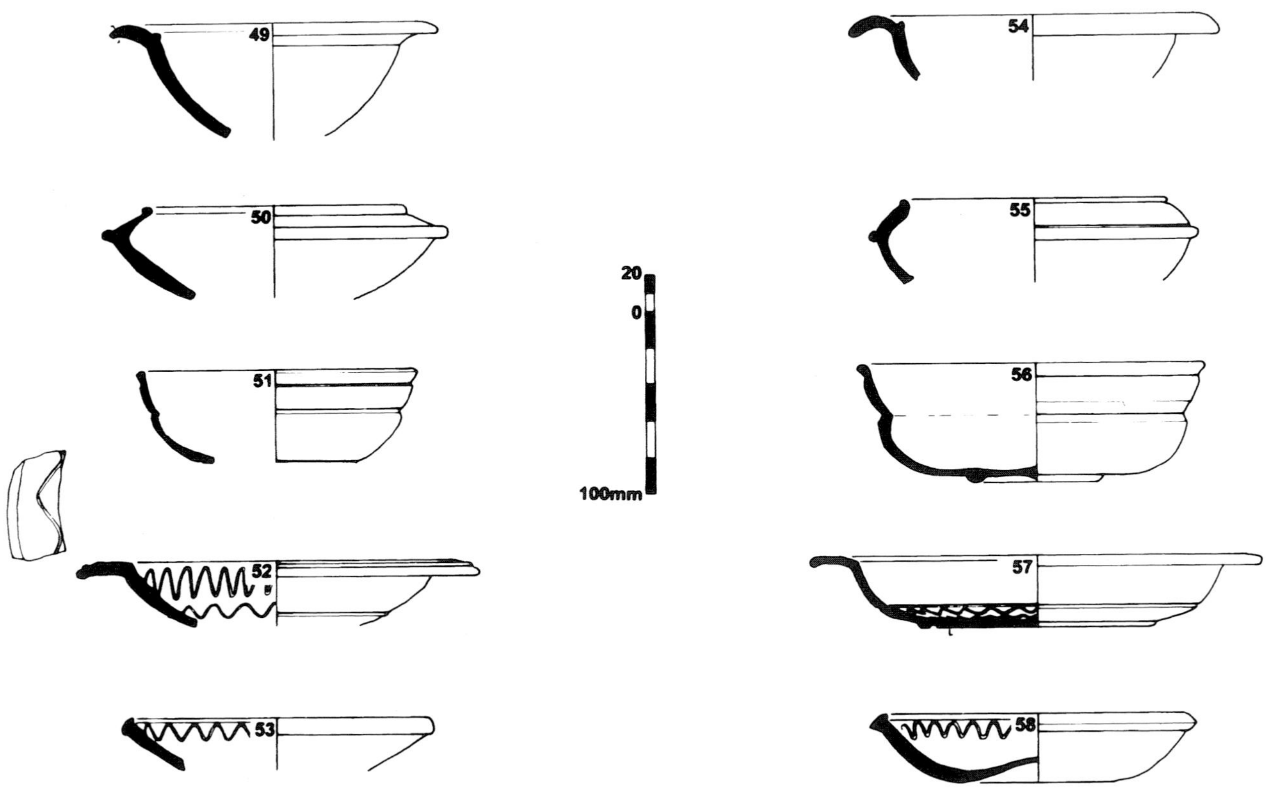

Figure 5. Workshops involved in military supply in the late first and early second century: Dragonby and its neighbourhood (South Humberside). Romanised grey or orange products of kiln(s) at Dragonby, Scunthorpe (left) and similar vessels (probably originating in the same area) found on military sites in Yorkshire (right). All decoration is burnished. Scale 1:4.

49, 54: Segmental bowls with hooked flanges from Dragonby and Malton fort (after May et al 1996; after Corder 1930a).

50, 55: Collared segmental bowls from Dragonby and Brough-on-Humber fort (after May et al 1996; after Corder and Romans 1937).

51, 56: Cups or small bowls, from Dragonby and Brough-on-Humber fort; these imitate, in a rather angular style, the samian form Drag. 27 (after May et al 1996; after Corder and Romans 1937).

52, 57: Segmental dishes with flattened lips and grooved walls from Dragonby and Malton fort (after May et al 1996; after Corder 1930a).

53, 58: Convex-walled dishes with faceted and internally beaded rims, from Dragonby and York (after May et al 1996; after Monaghan 1997).

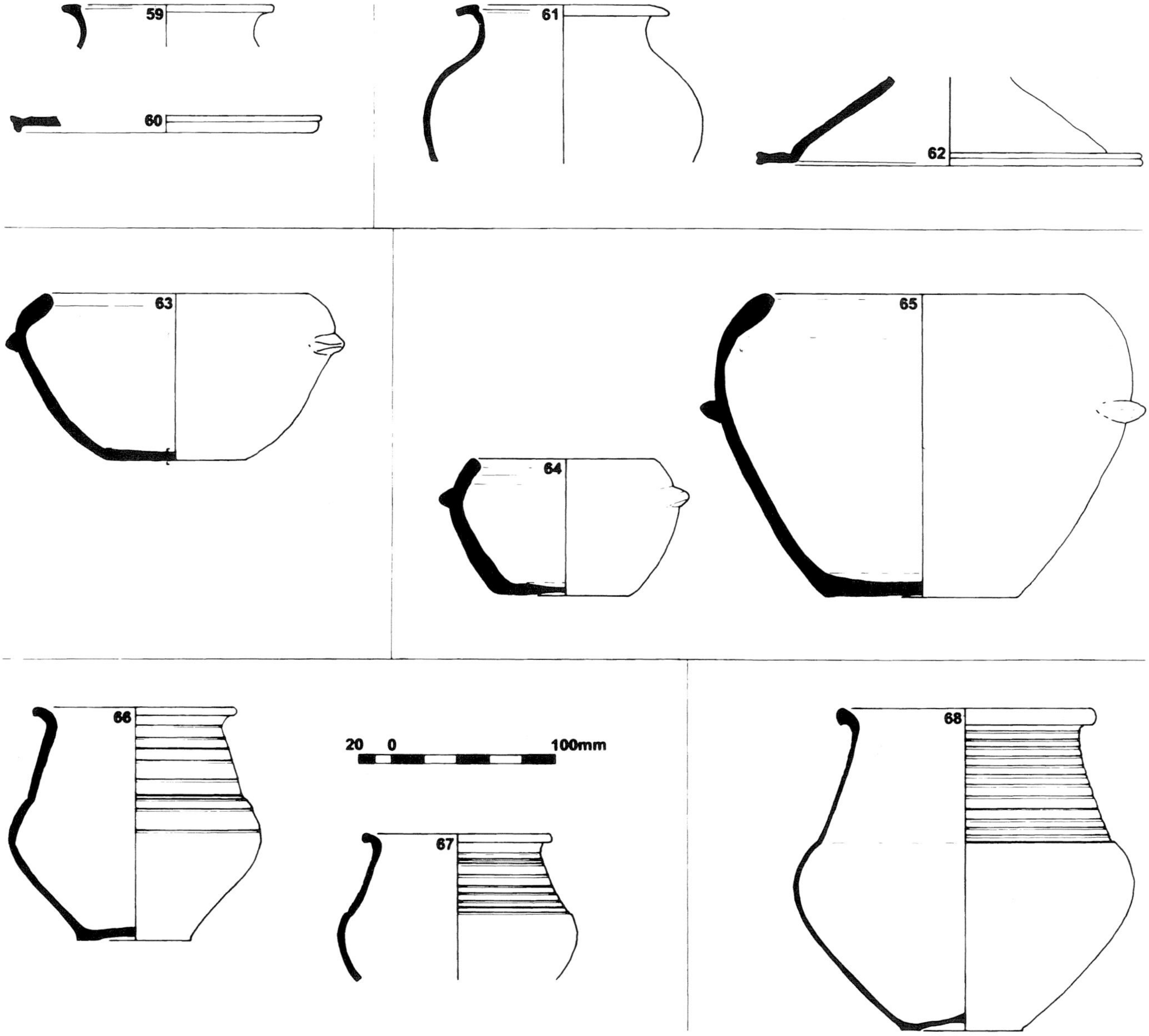

Figure 6. York: Flavio-Trajanic non-local utilitarian vessels: from Exeter (59-62), from the Nijmegen region, Netherlands (63-5), and from the Nord or Pas-de-Calais, France (66-8). Scale 1:4.

59, 61 Exeter Fortress ware lid-seated jars with flattened rim, respectively from York and Exeter (after Monaghan 1993, after Holbrook and Bidwell 1991).

60, 62. Exeter Fortress ware lids with a bi-lobed rim from York and Exeter (after Monaghan 1997; after Holbrook and Bidwell 1991); the last was published as a bowl but Bidwell (pers comm) now considers this to be a lid.

63-5: Nijmegen region deep bowls with lug handles in handmade or hand-finished ware, from York (63, after Monaghan 1993) and from Nijmegen (64 after Haalebos 1992; 65 after Holwerda 1941).

66-8: North Gaulish grey ware 'vases troconiques' with characteristic horizontal burnishing on the neck, from York (66, after RCHME 1962; 67, after Monaghan 1993) and from Baralle, Pas-de-Calais, France (68, after Tuffreau-Libre 1995).

1993, no. 2845; 1997, nos. 4003, 4071, 4073 and possibly 3870), up the Aire to Castleford (Rush 2000), figure 45, no. 58, figure 74, no. 496), and up the Trent or Ancholme to the Lincoln fortress and colonia. Many more were shipped across the Humber to the military establishment at Brough-on-Humber (nos. 55-6; Swan 1996, 575-6, 579-82). The author has suggested that some of these workshops may have been operated by the military (in tandem with local ironworking); indeed, one or two of Dragonby's Trajanic products seem best paralleled in Pannonia.

Early workshops and Hadrianic change

Though small amounts of first and early second-century pottery of types used by the military may have reached the native settlements north and west of the Humber (particularly those near roads), by various *ad hoc* mechanisms, in general, the indigenous peoples of Yorkshire had not yet wholeheartedly adopted such sophisticated commodities. Thus when many of the forts in Yorkshire had their garrisons

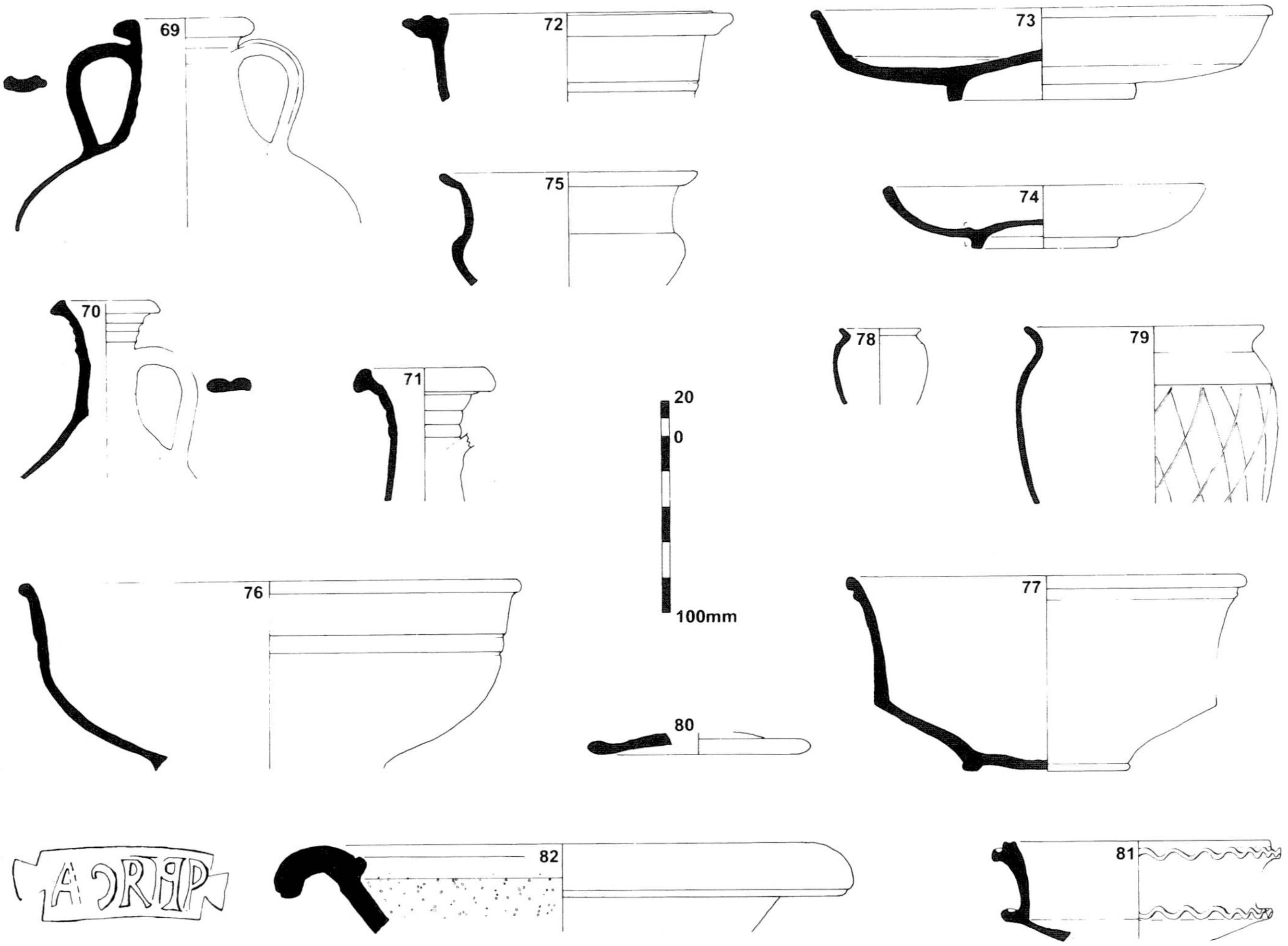

Figure 7. York: second-century workshops: the Apple Tree Farm kiln site, Heworth. Trajanic to mid Antonine; all the products are in Eboracum Oxidised ware (Ebor 1 or 2), and the scale is at 1:4, unless otherwise indicated. Apart from no. 77 (a form recorded at Apple Tree Farm), all illustrated vessels come from the Heworth kiln sites.

69 Double-handled flagon with elongated rim (after Lawton 1989), a development of the earlier Ebor form, no. 20; probably Trajanic-early Hadrianic.

70 Ring-necked flagon with flaring neck, expanded top ring and with lower relatively shallow rings, a developed version of no. 22, Trajanic-Hadrianic (after Wenham 1968).

71 Ring-necked flagon, as previous but with internal seating for a stopper, probably Hadrianic-early Antonine (after Lawton 1989).

72 Carinated bowl with gently undulating reeded rim, a slightly later version of no. 28, Trajanic-early Hadrianic (after Lawton 1989).

73 Dish copying samian form Drag. 18 or 18/31, Trajanic-early Hadrianic (after Lawton 1989).

74 Concave-walled dish, probably imitating Gallo-Belgic form, Cam. 16, Trajanic (after Wenham 1968 with reconstruction derived from author's examination of original assemblage).

75 Campanulate bowl in grey reduced Ebor ware (after Lawton 1989), Hadrianic-early Antonine, perhaps an imprecise copy of a Hadrianic introduction to York (cf. nos. 122, 124).

76 Hemispherical bowl (after Lawton 1989). The outward-sloping rim, proportionately wide upper border, and raised cordon below, suggest a form, derived, not from the Gaulish samian Drag. 37 bowl, but from a late Italian sigillata type, Conspectus 37.4.3. This form (introduced to York with Legio VI) was also made in striped ware at Apple Tree Farm (no. 145).

77 Carinated campanulate bowl with double-beaded rim (after Perrin 1990); the origin of the form is uncertain, but it may ultimately have developed from Gallo-Belgic terra nigra bowls (cf. Holwerda 1941, pl. xii, nos. 633-4); a probable Hadrianic introduction to York.

78 Beaker in grey reduced Ebor ware (after Wenham 1968); possibly a copy of a BB1 form (Gillam Type 64/65), though the vessel might have affinities with earlier everted rim beakers (Hadrianic-Antonine).

79 Copy of a BB1 or BB2 cooking-pot in grey reduced Ebor ware with burnished lattice (after Lawton 1989); Hadrianic-Antonine.

80 Lid of lagena or amphorette (author unpublished); second century.

81 Tazza with frilled rim in Eboracum White-slipped ware (author unpublished); second century.

82 Mortarium in Ebor White-slipped ware, with stamp of the potter AGRIPPA in an ansate frame (after Dickinson and Hartley 1971); the impression (scale 1:1) shows striations suggestive of wood-graining on the original die (Hadrianic-mid Antonine).

moved northwards at the time of the construction of Hadrian's Wall early in the 120s, there was insufficient local demand to sustain the continuance of their potteries as civilian suppliers. With the exception of York and possibly Aldborough, most of these early military-inspired workshops appear to have ceased production. For a while, the new frontier garrisons utilised the works-depots and potteries that had served the Trajanic Stanegate forts (e.g. Brampton, Scalesceugh and Corbridge). They also probably opened a few additional workshops, one perhaps in the lower Tyne Valley, serving South Shields and Wallsend (P Bidwell *pers comm*), and another at Catterick, but at the same time the forts on the frontier and its hinterland also switched over to a partial dependence on supplies from civilian potteries much further south, such as the Dorset producers of (BB1) Black-burnished ware (Gillam 1973, 54). For most of Yorkshire, however, it was as though the old military workshops had never been, for even by the reign of Hadrian, their ceramic traditions seem to have made little or no lasting impact on native pottery production and consumption in that region.

York: Trajanic into Hadrianic: garrison changes and ceramic chronology

York, being a legionary fortress, had a slightly different ceramic history. At some time, most probably in the early Hadrianic period, there was a drastic change in the character of the military pottery production there. It is highly likely that this should be linked to a major change in the occupancy of the fortress, that is, the departure of *Legio VIIII Hispana* and its replacement by *Legio VI Victrix*. However, the lack of well-dated, published sequences of stratified deposits within the fortress, for the period from 100/110 to 140/160 means that the details of the transition are obscure, and there have been suggestions (implausible, as will be demonstrated), that some of the observable ceramic innovations (discussed below) might belong to the late Trajanic period (Monaghan 1997, 863-4, 870-1). A proper understanding of the historical background and its problems is therefore essential to the interpretation of the pottery.

It is not known precisely when *Legio VIIII Hispana* was withdrawn from *Britannia*. The samian evidence from York suggests that there had already been some reduction in the legionary presence from the late Flavian period onwards (B Hartley 1971, 59). Some detachments were probably at Carlisle, and at the nearby works-depot of Scalesceugh, where the legion was evidently making (and stamping) tiles (*RIB* II, 2462.2-4 and 2462.16); precise dating is, however, lacking. There were, indeed, significant building activities in the Carlisle fort in AD 93/5 and again in *c* 105+ (Caruana 1992, 106-07; McCarthy 1991, 5), and a mixed legionary and auxiliary garrison (attested there at other times) seems quite probable at this period too. The latest epigraphic record of the Ninth from York is the fine building dedication (*RIB* I, 665) from near the *porta principalis sinistra* (the south-east gate of the fortress), dated to between 10 December 107 and 9 December 108. On the other hand, tiles stamped LEG VIIII [(presumably implying the whole legion, not just a vexillation), and a mortarium stamped] G VIIII HIS have been found respectively at the fortress of Nijmegen (Netherlands), and at its nearby works-depot, de Holdeurn. These are thought to date to after AD 104/5 and before *c* 120/122 (Bogaers 1967, 63-65; Bogaers and Haalebos 1977, 107). Nijmegen's previous garrison, *Legio X Gemina*, moved to *Aquincum* (Budapest) in 104/5, and in *c* 122, the Lower Rhine military dispositions were reorganised on the transfer of *Legio VI* from Xanten to York. In fact, other tiles from Nijmegen, assigned to AD 90 onwards, record the participation in building activities of a *vexillatio Britannica* (Saxer 1967, nos. 42 and 255; Brunsting and Steures 1997, 323). A detachment from the Ninth Legion might well have constituted part of one such group. At York, the continuous decline in the quantities of samian which had commenced in the late Flavian period, accelerated in *c* AD 100, with a drastic drop from *c* 110 onwards (the lowest point being in *c* 110-120), and a significant rise from about AD 130 (B Hartley 1971, 59-61; Dickinson and Hartley 1971, 128-32). A general reduction in samian supply in the late Trajanic period is a phenomenon recorded on many sites in Britain (Marsh 1981), and cannot automatically be taken as an indicator of reduced levels of activity. However, a comparison of the relative quantities at York of the products of Les Martres de Veyre (the main source of samian in Britain between 110 and 125), with the relative quantities of that found on other British sites, including military establishments such as Carlisle, Chester, Caerleon, Holt and Wilderspool (Dickinson and Hartley 1971, 128-9; Monaghan 1997, 948; Ward 1998, 54-5), indicates that York is quite exceptionally deficient in this material. This suggests that there was, indeed, a very significant reduction in the density of occupation in the fortress between 110 and 125.

It is not impossible that a detachment from *Legio VIIII* (perhaps as part of the *vexillatio Britannica*) was transferred to Nijmegen in about 105, a time when there were widespread troop movements triggered by Trajan's Second Dacian War. In Britain, these included the abandonment of the last forts held in Scotland, and the dispatch, in the spring of AD 105, of *Cohors IX Batavorum* from *Vindolanda*, ultimately to the Lower Danube (A Birley 1997a, 273-74). However, Keppie (2000, 94) has plausibly suggested that the whole of the Ninth might well have transferred to Nijmegen by *c* 114, the time of preparations for Trajan's Parthian campaigns of 114-7. This was another occasion when many units in frontier provinces across the Empire were 'shunted' slightly eastwards. The presumption that *IX Hispana* went to, or was still in Nijmegen in 120 makes no sense (cf. Ottaway 1993, 45; Brunsting and Steures 1997, 329). As Keppie (1989, 248-50) has argued cogently, the troubles in Britain late in Trajan's reign and on Hadrian's accession (*SHA Hadr* 5) were most likely caused by the absence of the Ninth (Britain's most northerly legion), and this ultimately resulted in the transfer to Britain of *VI Victrix*, the nearest available legion in Lower Germany, then at Xanten (*Vetera*). As he points out, 'If the Ninth had then been at Nijmegen, it is inconceivable that *VI*

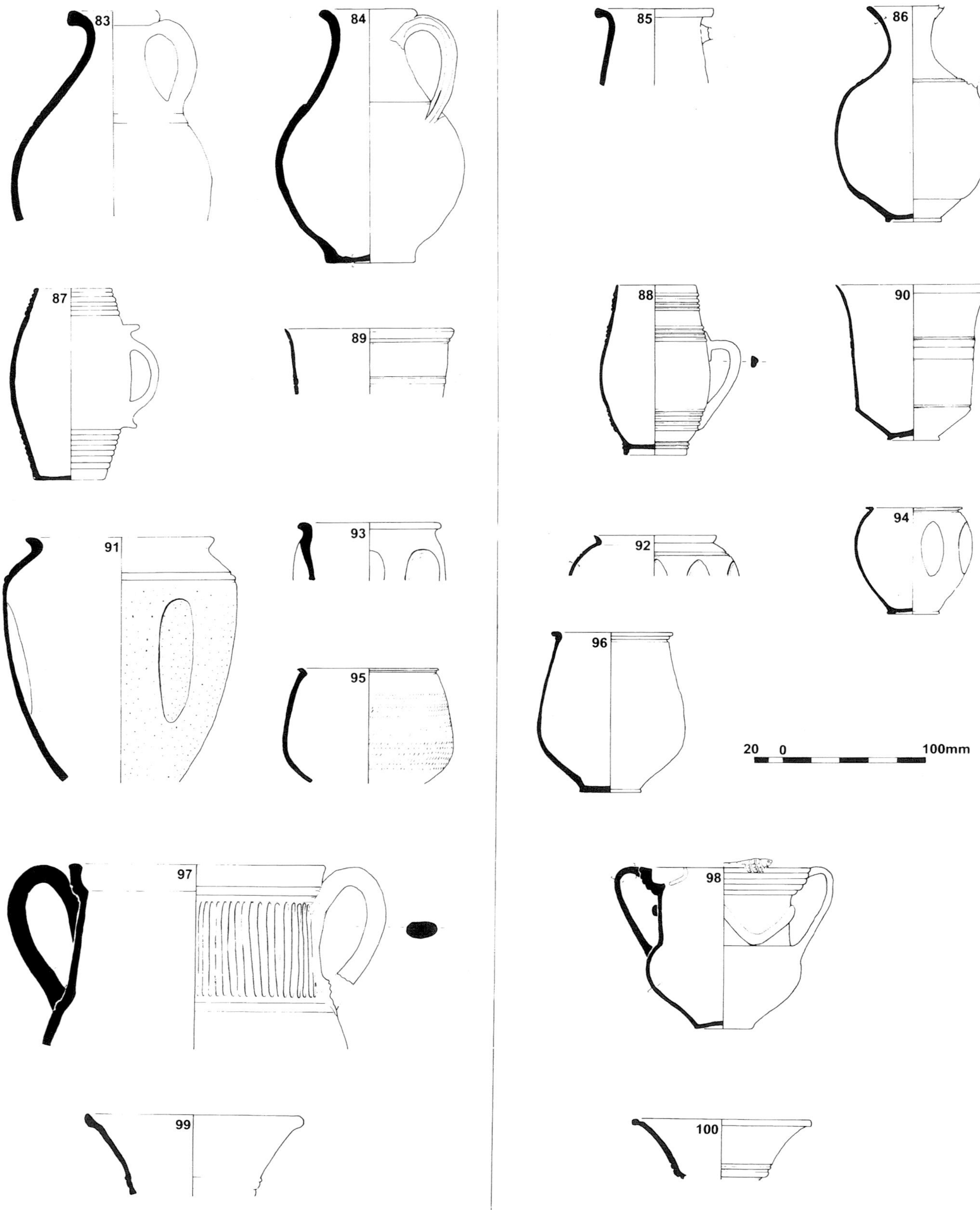

Figure 8. York (Captions opposite)

Victrix would have been moved to Britain in preference.' *Legio VIIII* must have already completed its time at Nijmegen and been moved on elsewhere well before the transfer of the Sixth to Britain in *c* 122. It presumably went further east, since there is epigraphic evidence of its continuing existence after that date (Keppie 1989, 249-52; 2000, 93-94). It must therefore have moved to Nijmegen some years earlier. Whether legionaries in the *vexillatio Britannica* had been sent back briefly, before the ultimate departure of *IX Hispana* to Nijmegen, to their parent unit(s) in York (and elsewhere), cannot be determined for certain, though significantly, the samian levels at York provide no hint of this. The occurrence

Figure 8. York: pottery probably made in the military workshops just outside the fortress in the Hadrianic-early Antonine period: (left) jugs, tankards, beakers and cups from York; (right) comparable Wetterau ware vessels (after Rupp 1988). The York products are in the utilitarian Eboracum Oxidised wares (Ebor 1 & 2), unless otherwise stated. All were found in York. Scale 1:4.

83-4 Pear-shaped jugs with cordoned shoulder and flaring rim, in reduced (grey) Ebor ware (after Perrin 1981; Perrin 1990); cf. nos.85-6 in Wetterau ware; generally similar forms occur in the late Italian (Tardo-Padana) sigillata repertoire (Conspectus Form K).

87 Barrel-shaped tankard in probable fine oxidised Ebor 3 ware (after May 1912); cf. no. 88 in Wetterau ware.

89 Tall cordoned conical beaker in fine oxidised Ebor 3 fabric (Borthwick Excavation Archive, after E King); the form ultimately reflects details found in late Italian sigillata (Conspectus 50.1.2/50.4); cf. no. 90 in Wetterau ware.

91 Rough-cast indented beaker with double shoulder grooves, a widespread form at this period (after Monaghan 1997); cf. no. 92 in Wetterau ware.

93 Plain indented beaker (after King 1974); cf. no. 94 in Wetterau ware.

95 Bag-beaker in fine (Ebor 3) with cornice rim and coarse rouletting ware (Swan and McBride forthcoming); cf. no. 96 in Wetterau ware.

97 Lid-seated cantharus with cordons and burnishing (Swan and McBride forthcoming); cf. Wetterau parallel with applied snake, no. 98.

99. Lid-seated conical cup with concave wall and cordoned carination (after Perrin 1995), a type was also made in fine (Ebor 3) ware, and at the Apple Tree Farm workshops. The ultimate ancestry of this form might be seen in Tardo-Padana sigillata cups (Conspectus 49, or 52.2.1), but some of the fine detailing has been lost in the course of copying; cf. Wetterau parallel, no. 100.

at the York fortress, in an area of barracks, of several coarse lug-handled bowls in a corky grog-tempered fabric such as occur in Flavio-Trajanic assemblages at Nijmegen, might represent items (most probably comestibles supplied in such containers) brought back by soldiers who had been there (no. 63, cf. nos. 64-5; Monaghan 1993, nos. 2849 and 3096; cf. Holwerda 1941, nos. 1393, 1398; Haalebos 1992, figure 4, no 11). However, the Trajanic drain in which the only well-stratified example occurred, was back-filled substantially later than its construction, and this and the other examples could equally well have been associated with the presence of newcomers from Lower Germany in the Hadrianic period. In sum, because the evidence of the York samian suggests that the late Trajanic garrison was significantly depleted or absent, it does not make good sense to associate major changes in the nature of the legion's own pottery with that period, as Monaghan has suggested (1997, 870, 876), or with the possible return of a vexillation from Nijmegen for a short period. Indeed, the ceramic innovations discussed below are much more likely to stem from the installation of a new legion in the York fortress, a decade or so later, a proposal further supported by the Hadrianic dating of similar pottery at other sites in Britain and the Continent (see below).

Local potting at York: Trajanic into Hadrianic

In the period following the departure of *Legio VIIII*, there must still have been some minor military presence at York, simply for care and maintenance of the fortress. Perhaps an auxiliary detachment fulfilled this role. It is just possible that the Nijmegen-type corky lug-handled bowls (already mentioned) and the sudden and apparently quite brief influx of North Gaulish grey ware beakers (*vases troconiques*; nos 66-8), recorded at about this period (Monaghan 1997, 863, 889), often in barrack areas within the fortress (and in contemporary extra-mural rubbish deposits), should be linked to the origins or affinities of caretaker troops (?from North Gaul or the Lower Rhineland, near their production areas). Of relevance is the fact that *vases troconiques* are also quite well-represented at *Vindolanda*, in the Trajanic assemblages relating to the Batavian and Tungrian cohorts, which seem to have contained men from those regions (Swan *in prep*). Such units might have brought with them supplies of their traditional (and idiosyncratic) drinking-vessels (?for beer) at the time of their transfers to Britain (cf. the influx of North African olive oil in York when troops from that region arrived with Severus; Williams and Carreras 1995, 237-38). Alternatively such units may have provided the contacts or stimulus for trade with that part of Gaul (cf. the commencement of the supply of BB1 to the northern frontier zone when *Legio II Augusta*, which already used the ware, was sent north to work on the construction of Hadrian's Wall).

At York, however, after 40 or so years of legionary occupation, by a garrison of five and a half thousand Roman citizens, there would also have been a sizeable nucleus of Romanised legionary veterans and their dependants. These would have wanted the sort of pottery to which they had been accustomed. The pottery workshop at Apple Tree Farm, Heworth (Wenham 1968; Swan 1984, MF 6.692; Lawton 1993), most likely founded at about this period to fill the gap left by the departure of the legion, may well have initially catered for just such a Romanised civilian population. It lay barely 3 km north-east of the fortress. Could it perhaps have been located on land allocated to, or leased by a veteran? The old tilery of *Legio VIIII* had been sited very close to the fortress, presumably within an annexe or military compound, a location which, no doubt, would have rendered it inappropriate for civilian exploitation. A link between the Heworth workshop and the first-century garrison seems to be implied by the profiles of some of the earliest Apple Tree products. These included developed versions of forms produced by workshops of the Ninth Legion, for example, ring-necked flagons (nos. 70-71), double-handled flagons

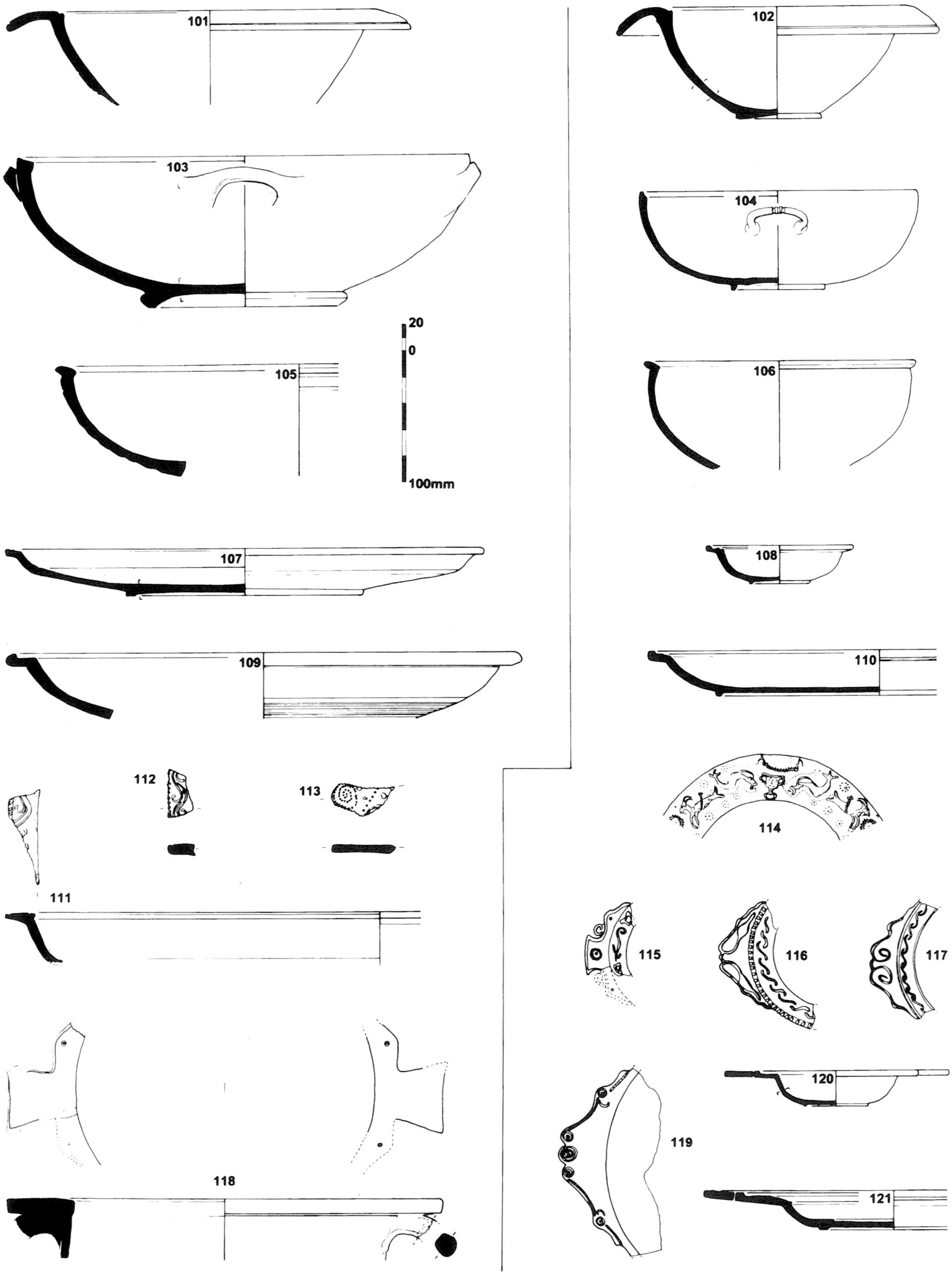

Figure 9. York (Captions opposite)

Figure 9. York: pottery probably made in the military workshops just outside the fortress in the Hadrianic-early Antonine period: (left) bowls, dishes and platters from York; (right) comparable Wetterau ware vessels (after Rupp 1988). The York products are in utilitarian Eboracum Oxidised wares (Ebor 1 & 2), unless otherwise stated. All were found in York. Scale 1:4.

101 Segmental bowl in fine oxidised Ebor 3 ware with grooved flange (Swan and McBride forthcoming); cf. 102 in Wetterau ware.

103 Lug-handled segmental bowl with rim expanded and flattened internally (after King 1974); cf. 104 in Wetterau ware.

105 Segmental bowl with flattened everted rim (after Monaghan 1997); cf. 106 in Wetterau ware.

107 Carinated dish in fine oxidised Ebor 3 ware with flattened and gently grooved rim. The general profile may ultimately have been influenced by late Italian sigillata vessels, such as Conspectus 42, but no precise parallel is evident (after Swan and McBride forthcoming); cf. 108 in Wetterau ware.

109 Shallow segmental bowl or dish in fine oxidised Ebor 3 ware, with a flattened rim lobed at the tip, a profile probably ultimately derived from the late Italian sigillata form, cf. Conspectus 41 (Swan and McBride forthcoming); for a Wetterau parallel see no. 110.

111 Dish or platter in fine oxidised Ebor 3 fabric; the flattened rim, grooved on the upper edges, has white barbotine scrolls bordered with dots. The profile ultimately derives from a late Italian sigillata form, Conspectus 40, also with scroll and dot decoration (Swan and McBride forthcoming); cf. 116 for a comparable arrangement of dots and scrolls on a Wetterau vessel.

113 Fragment of a fine oxidised Ebor 3 grooved rim with white barbotine decoration, a dish form similar to no. 111 (Swan and McBride forthcoming).

114 Fragment of a rounded lug, an expansion of the rim, probably of a dish or platter, with white barbotine dots, in fine oxidised Ebor 3 ware (Swan and McBride forthcoming); cf. no. 114 for similar decoration and nos. 117-9 for flat lug-like rim extensions with rounded edges in Wetterau ware.

115 Bowl in fine oxidised Ebor 3 fabric (Swan and McBride forthcoming); the edge of the rim is cut in facets and has a flat ansate-shaped expansion (lug-handle) impressed with concentric circles, probably ultimately imitating metal prototypes; cf. Wetterau platter no. 116 for ansate rim-expansions, and no. 119 for similar decorative motifs.

(no. 69), imitations of Gallo-Belgic and samian dishes (nos. 73-4), collared storage-jars (cf. no. 25 and Monaghan 1997, no. 2884), rusticated cooking-pots and distinctively similar versions of the carinated bowl with reeded rim (no. 72; Lawton 1989; I Lawton *pers comm*). Morphologically, none of these earlier Apple Tree products shows any relationship to the vessels apparently associated with the arrival of the Sixth Legion.

York: ceramic innovation and the military in the Hadrianic period

Legio VI Victrix was brought to *Britannia* from *Germania Inferior* in about 122 by the new governor Platorius Nepos, but spent a substantial portion of its earliest years in the province (certainly in the building season) assisting in the construction of Hadrian's Wall (A Birley 1971, 82-83; 1997b, 123-24, 130). In fact, it may not have been present at York full-time until at least AD 128. Either then or perhaps earlier, in its first winter season in the fortress, the legion seems to have opened a works-depot in the *canabae*, just 180-200 metres south-east of the east corner of the fortress defences and close to the old tilery of *Legio VIIII*, in the Peaseholme Green area. The material recovered from waster dumps on the site (the Borthwick Institute: King 1974, figures 9-10; King's Pool, Peaseholme Green: Swan and McBride forthcoming), included a range of utilitarian pottery (Ebor ware Fabrics 1 and 2) and a small number of unusual, and very fine tableware forms (Ebor Fabric 3; Perrin 1977, figure 7.1, nos. 13-21). Though the assemblages of second-century reject vessels on the site had been subsequently contaminated by later tiles and pottery wasters discarded from the nearby Severan kilns (when the abandoned legionary kiln dumps of both periods were back-filled into disused clay-pits in the mid/late third century), it is possible to separate most ceramic groups typologically with some confidence. Virtually all the Hadrianic products, both coarse and fine, marked a clear break in the legionary potting traditions of York; almost every class of vessel in use by the previous garrison had been replaced by a new form (Figures 8-11). There was a notable absence of ring-necked flagons, (ubiquitous in Britain in military and military-orientated potteries of the period, though much more rare at corresponding sites on the German *limes*); the carinated bowls were much deeper and simpler (no. 132); large segmental bowls with lug handles were prominent (no 103), and unusual cups, beakers and jugs appeared (Figures 8-10). Some vessels were coated with a white slip. The sophisticated and beautifully crafted orange tablewares (Ebor 3) were not only different in form, but finer and much more highly polished than any earlier Ebor products with burnishing. Some had a red slip and there were also limited attempts at simple stamped motifs (no. 118), facets in imitation of cut glass (King 1974, figure 10, no. 14), and white barbotine decoration (nos. 111-13).

New products also appeared at the Apple Tree Farm workshops, possibly at about the same time or certainly within a couple of years, since several of the forms new to the fortress workshops also seem to have been manufactured there, though sometimes with different finishes (nos. 75-7). Among these Heworth vessels, were small quantities of bowls in a variety of Ebor oxidised ware decorated with vertical red-painted stripes (Figure 11); a number of their profiles resembled those used for the fine and utilitarian Ebor wares made at the Peaseholme Green *figlina* (e.g. no. 76;

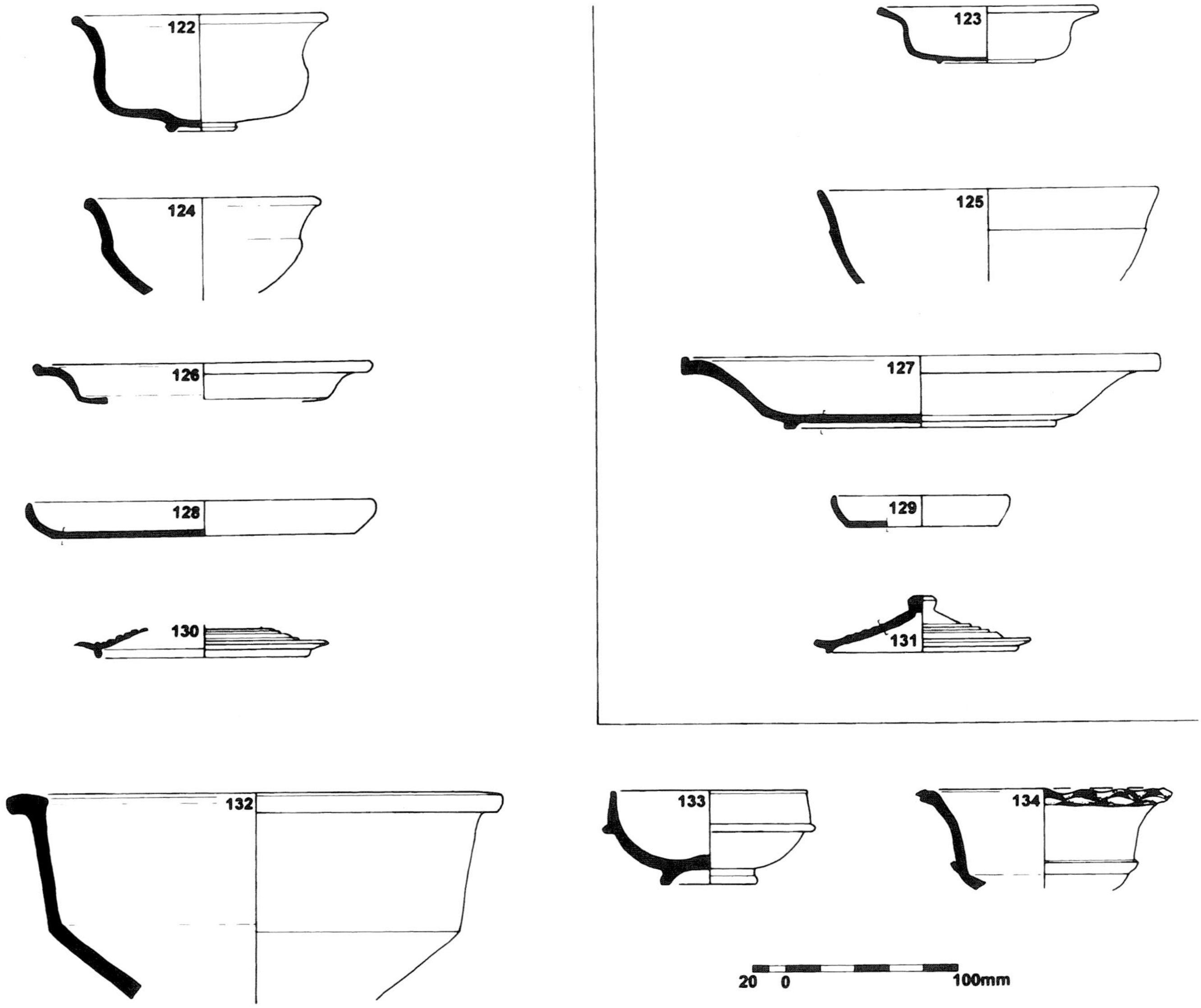

Figure 10. York: pottery probably made in the military workshops just outside the fortress in the Hadrianic-early Antonine period: (left) cups, bowls, dishes, lids and tazze from York; (right) comparable Wetterau ware vessels (after Rupp 1988).The York products are in the utilitarian Eboracum Oxidised wares (Ebor 1 & 2), unless otherwise stated. All were found in York. Scale 1:4.

122 Fine oxidised Ebor 3 campanulate cup (after Perrin 1977); cf. 123 in Wetterau ware.

124 Campanulate cup (after Perrin 1990); cf. no. 125 in Wetterau ware.

126 Fine oxidised Ebor 3 dish with concave wall (Swan and McBride forthcoming). No exact parallels occur in the Tardo-Padana sigillata repertoire, but the profile may ultimately have been influenced by those of the contemporary dishes, Conspectus 47 or 48; cf. no. 127 in Wetterau ware.

128 Fine oxidised Ebor 3 plain-rimmed dish reminiscent of a Pompeian red ware form (unpub Borthwick Excavation Archive, after E King); cf. Wetterau dish no. 129.

130 Fine oxidised Ebor 3 lid; cf. Italian sigillata form, Conspectus 54.3.1. (unpub Borthwick Excavation Archive, after E King); cf. Wetterau parallel no. 131.

132 Carinated bowl with plain lip (Swan and McBride forthcoming); contrast the profile of the earlier Ebor carinated bowl, no. 28; this later form is much deeper and simpler, and clearly represents the re-introduction of a new version of this basic culinary type.

133 Flanged cup (after Perrin 1995). On chronological grounds, this cup is unlikely to be an imitation of the pre-Flavian South Gaulish samian form Drag. 24/25, and the profile almost certainly derives from late Italian sigillata models (Conspectus 34.2.1, and 33). The same form also occurs at York, with red painted stripes (no. 147), and in Turret 26a on Hadrian's Wall, both confirming its Hadrianic date.

134 Frilled tazza (Swan and McBride forthcoming).

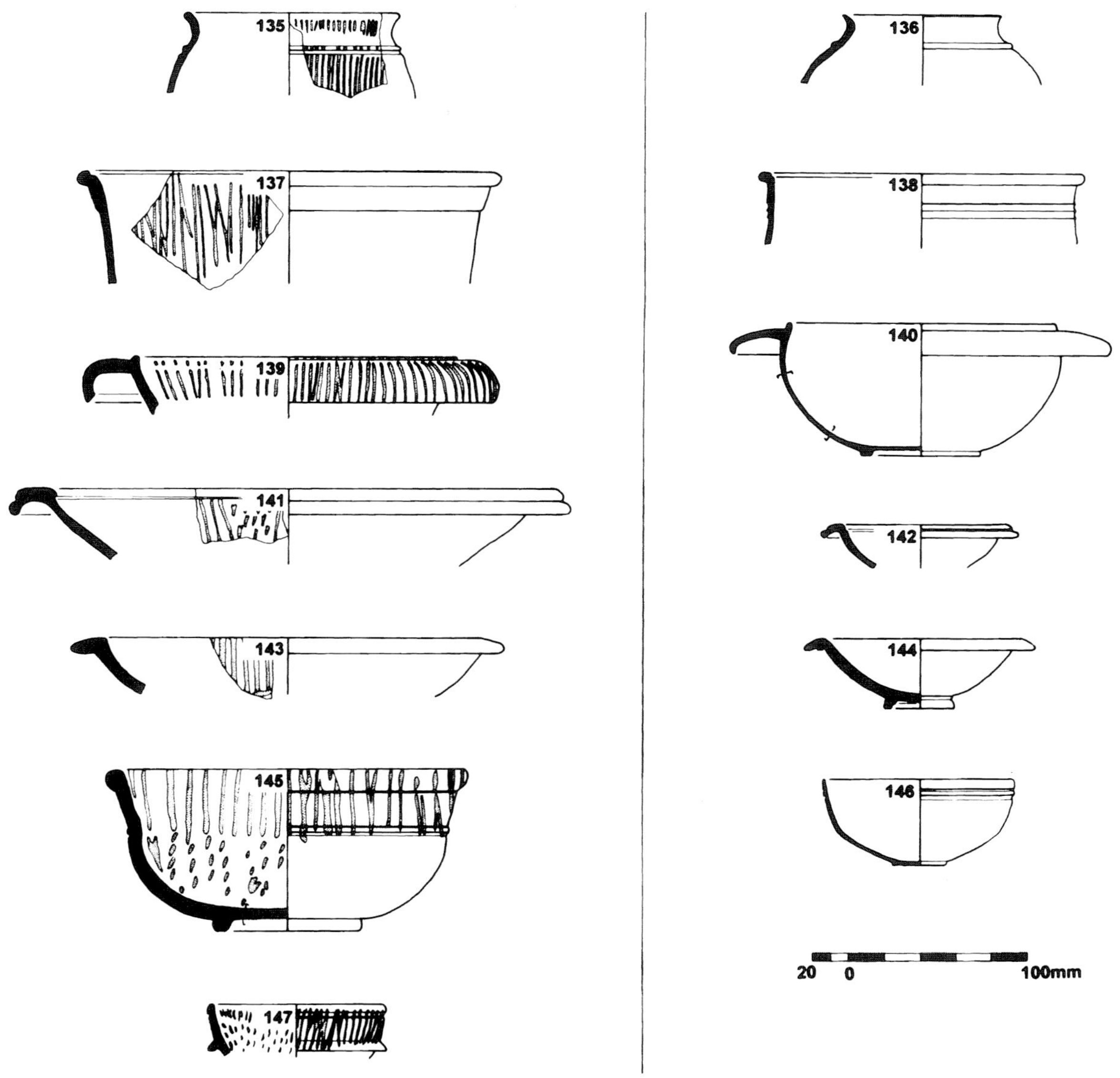

Figure 11. York: Eboracum Oxidised tableware with red-brown painted decoration (Hadrianic-early Antonine), probably made at Apple Tree Farm, Heworth, (left) beaker, cups, bowls and dishes; (right) comparable forms made in the Wetterau workshops (after Rupp 1988). Unless otherwise stated, all the Ebor ware vessels were found in York. Scale 1:4.

135 Necked beaker (unprov. in the Yorkshire Mus., but probably from York area; after Keene 1970); cf. Wetterau beaker, no. 136.

137 Lid-seated bowl (after Keene 1970). The ancestry of this type is unclear, but ultimately, it may owe its inspiration to the much earlier forms, Conspectus 52 or R 99. Cf. Wetterau bowl no. 138.

139 Flanged bowl (after Keene 1970); cf. no. 140 in Wetterau ware.

141 Bowl or dish with grooved lip (from Elmswell, East Yorks; after Keene 1970). The ultimate inspiration for this profile may be the Tardo-Padana forms, Conspectus 39.1.3 and 140; cf. Wetterau cup no. 142.

143 Dish (after Keene 1970), the form probably has late Italian sigillata ancestry, eg. Conspectus 44.2.1; cf. the small Wetterau bowl no. 144.

145 Hemispherical bowl (from Piercebridge; after L Hird in P Scott unpub Piercebridge Excavation Archive). For affinities, see no. 76; the miniature Wetterau bowl, no. 146 might be comparable.

147 Cup (after Keene 1970); for affinities see no. 133.

Lawton 1993; I Lawton *pers comm*). Whether the Apple Tree Farm centre now passed back into the hands or supervision of the military is a moot point. It would certainly have lain within the *prata legionum*, that is the land controlled by the legion, and could theoretically have been repossessed by it. Several of the Hadrianic-early Antonine white-slipped Ebor ware mortaria made in these workshops have name-stamps with ansate borders (Dickinson and Hartley 1971, 135-9, figure 19); these include the products of a potter named Agrippa, who also stamped tiles, some also with ansate frames (no. 82; *RIB* II.5, 2489.2-3). Tiles with ansate-bordered name-impressions which are presumed to emanate from private tileries (*RIB* II.4 and II.5), were apparently made only at Gloucester and York, both sites of legionary fortresses. There is thus a strong possibility that such tiles were the products of either serving soldiers, veterans, or other officials; so too, by implication, may be the Apple Tree Farm mortaria with ansate stamps. Clearly there were very close links between the York legion and the Apple Tree Farm workshops.

Probable military origins of the York Hadrianic potters

The possible origins of these new Hadrianic ceramic styles at York merit some discussion. Many vessels are reminiscent of contemporary ranges of pottery from military sites on the Middle to Upper Rhine and Upper Danube limes and at Nijmegen, some of which seem to have been legionary, or military products. At this period in Britain, generally (but not precisely) similar vessel-types and finishes seem to have been adopted quite suddenly by the workshops of several other military establishments, where they (particularly the fine wares) have often been referred to as 'legionary ware'. The phenomenon, mainly but not exclusively legionary (Greene 1977), is evidenced at the fortresses of Caerleon and Chester (and its works-depot at Holt), at Wilderspool industrial settlement (perhaps linked to the activities of *Legio XX;* Swan 1997), at the short-lived Hadrianic fort at Hardknott (probably supplied by the Muncaster military kilns; Bidwell *et al* 1999, 94-6), and at Brough-on-Humber (discussed below). The character and contexts of this distinctive group of products almost certainly indicate the presence in *Britannia* of potters from Upper Germany, *Noricum* and *Raetia* (whence similar traditions had probably also been transmitted by the military to Nijmegen). The 120s was a period when there were several epigraphically attested transfers of legionary and auxiliary units from the Continent to *Britannia* (others unrecorded are not improbable). Soldier-craftsmen from such provinces may have arrived with them on a number of separate occasions. The possible circumstances of the transmission of such ceramic traditions to York may be tentatively suggested from a close scrutiny of the forms, finishes and decoration of the vessels and their frequency at some of the Continental production sources (Swan 1997; Swan and McBride *forthcoming*). A significant number of the details of many of the York utilitarian and fine ware forms (most of the latter platters, dishes, and cups derived ultimately from Italian sigillata prototypes), together with their finishes, such as the beautifully polished orange surfaces, the use of red slips, red-painted stripes and interrupted lines (Figure 11), and their decoration, for example white barbotine scrolls with dotted borders (nos. 111-3) , and simple impressed stamps (no. 118; King 1974, figure 10, no. 17) can be paralleled quite closely in the Hadrianic ranges of a distinctive type of pottery, known collectively as 'Wetterauer ware' (Figures 8-11 *right*); the name refers to the region of Upper Germany in which it occurs most commonly, that is, an area east of Mainz, between the rivers Rhine and the Main. A study of the industry by a German scholar, Vera Rupp (1988, 35), has shown that *Legio XXII Primigenia*, the legion based at Mainz, was almost certainly involved in the manufacture of this pottery to supply legionary and auxiliary units based in that region. It therefore seems it seems highly likely that there was at least one potter in the York legionary workshops, who had first-hand familiarity with the products of *Legio XXII Primigenia* (the well-known range of striped vessels and orange polished 'legionary' wares made near Nijmegen provides fewer and less precise parallels; Haalebos and Koster 1981; Holwerda 1944).

In the 120s, York's new legion would have consisted largely of men who been recruited in the province in which it had been recently based, that is, at Xanten in the Lower Rhineland (*Germania Inferior*). However, we know, too, from the tombstone of a certain T. Pontius Sabinus, at Ferentinum, in Italy (*CIL* X, 5829 = *ILS*, 2726), that under Hadrian, probably either in *c* 122 or in *c* 125/6, at a time of crisis in the province (Jarrett 1976, 146-47; Casey 1987; Frere 2000), the army of *Britannia* had been augmented by an *expeditio Britannica*, led by Sabinus and comprising three thousand men, drawn from the two legions of *Germania Superior* (*XXII Primigenia* and *VIII Augusta*), and from the *VII Gemina* (the legion based in *Hispania Tarraconensis*). Vexillations of this size were normally large enough to include men with a wide range of basic artisan skills (eg potters and tilers), for they often operated as independent units (*cf.* the *vexillatio Britannica*: *op cit*). It is more than likely that on their arrival or at the end of the building or campaigning season, the men in Sabinus' force would have been transferred into the three British legions, to bring them up to strength. Among those drafted into *Legio VI Victrix* may have been soldier-craftsmen originally drawn from *Legio XXII Primigenia*, who thus became responsible for the introduction and production of the new pottery styles at York. Alternatively, legionaries may have already been transferred into *VI Victrix* from *XXII Gemina* (the nearest legion) at the time of its departure from Xanten, to raise its complement, or even to provide surplus for transferring on into other units in Britain. The epigraphic evidence hints at some considerable shuffling of men between units in Britain at this period. One soldier, whose *origo* was *Traiana* (Xanten) is recorded in *Legio II Augusta* (*RIB* I, 357) and a second from there was apparently in *Legio XX Valeria Victrix* (*RIB* I, 506); both are most likely to have come to Britain initially with *VI Victrix*. Whatever the precise circumstances of the arrival of the York military potters, it is quite probable, that at the same time, there would have been transfers into that legion of other men of similar origin or background; these soldiers would very likely have provided a significant demand within the York fortress for the distinctive ranges of vessels with which they had been familiar.

The distribution of York fine wares

The exotic Hadrianic fine wares with unusual profiles and a polished, painted or red-slipped finish made little impact beyond York itself; trade further afield was evidently extremely limited. They are even relatively uncommon finds in the fortress and colonia. The few vessels recorded beyond York mostly occur at military establishments, or sites with probable military associations, and the majority of such vessels may well have been carried there as the personal possessions of individual soldiers or veterans. Examples occur at Doncaster (Buckland and Magilton 1986, figure 39, no. 245), at Malton (Keene 1970, 88, note 11, figure 1, no. 13;

Bidwell and Croom 1997, nos. 136, 311, 411, and 354); at Aldborough (Myres *et al* 1959, figure 25, no. 35; Jones 1971, figure 10, no. 92; and unpub. vessels in site museum, P Bidwell *pers comm*), at a possible port site on the River Ouse at Selby (unpub vessels identified by the author from excavations by MAP Archaeological Consultancy), at Scoreby, Stamford Bridge, on the site of the probable fort and on the opposite side of the River Derwent (Lawton 1993, 8; Croom *et al* forthcoming), at Castleford (Rush 2000, figure 68, no. 412, figure 78, no. 538), at Brough-on-Humber (Corder and Romans 1938, figure 15, no. 25; Wacher 1969, figure 62, nos. 225, 242, figure 78, no. 683), at Piercebridge fort (no. 145; Hird forthcoming), at South Shields fort (Dore 1983, figure 66, no. 1010 and P Bidwell *pers comm*), at Wallsend fort (P Bidwell *pers comm*), and in Turret 26a on Hadrian's Wall (confirming a Hadrianic date for the striped products; Woodfield 1965, T.26a, no. 27). At the last five sites, there is evidence of military activity under Hadrian, including building, perhaps involving the legion. The site at South Elmswell, near Driffield, where striped bowls also occur (no. 141; Congreve 1938, figure 5A, nos. 1-2; Corder 1940, figure 10, nos. 42-47), need not be out of place in this list; crop-marks of a military-type enclosure (D N Riley unpub air photo), and a range of other early Roman artefacts of types which are unusual in civilian contexts in the region, may hint at some military connection. The transport of cooking-equipment in the baggage of groups of men from the Sixth Legion, engaged in building and other activities on the Antonine Wall, will also have been a factor in the occurrence there of isolated examples of Apple Tree Farm products, for

Figure 12 (page 54). The South Yorkshire potteries (Doncaster area).All vessels are in reduced grey ware, and the decoration is burnished, unless otherwise stated. Scale 1:4 except where indicated.

- *148 Flavio-Trajanic campanulate bowl with bifid grooved rim (Gillam Type 301), possibly from same workshops as no. 150 (after Buckland et al 1986).*
- *149-50 Flavio-Trajanic dish with domed base and inturned carinated lip (Gillam Type 337; after Rigby and Stead 1976), impressed in centre interior of base with the name-stamp of the potter, CEN (scale 1:1; after Rigby 1998); cf. Dragonby product, no. 53.*
- *151 'Late' rusticated cooking-pot from Doncaster vicus (after Buckland et al 1986). Probably dating to within a period from about 135 to the late second century, the longer curved rim distinguishes these forms from first and early second-century rusticated jars often associated with military sites (nos. 9, 26).*
- *152 Rossington Bridge Black-burnished ware 1 (BB1) cooking-pot, c135-late second century (after Buckland et al 1980).*
- *153 Probable Rossington Bridge Black-burnished ware 1 (BB1) dish; not a type found in the Dorset BB1 repertoire and probably made in the Rossington Bridge potteries. From South Shields supply-base (after Bidwell 1994); Antonine.*
- *154 Beaker or tankard in fine reduced 'parisian ware', probably imitating a BB1 type (Gillam Type 65), from Blaxton kiln site (after Buckland and Dolby 1980); mid-late second century.*
- *155 Mortarium with hooked flange and internal bead rim in an oxidised fabric with a white slip and ironstone trituration grits. The form is similar to those made by the potter-entrepreneur, SARRIUS (stamp above at 1:1, not from the same vessel); from Cantley kiln site (after Annable 1960); Antonine.*
- *156 Mortarium in Cantley white-slipped ware (orange-brown fabric with ironstone or slag trituration grits), from Cantley, (after Buckland et al 1980); probably late third-fourth century. The reeded collar, decorated with grouped stripes in red-brown paint is similar in style to Mancetter-Hartshill products; the profile and fabric together resemble Catterick late mortaria (no. 280);.*
- *157 Jar with looped handles from Branton kiln site (after Buckland 1976); perhaps late third century. The type is local to a wide region, and related to the East Yorkshire series of jars with looped and ultimately counter-sunk handles (nos. 200-01, 219, 243-52).*
- *158 Lid-seated jar, a regional type characteristic of the South Yorkshire industry; from Doncaster vicus (after Buckland et al 1986); most probably late second to mid third century.*
- *159 Flanged lid-seated jar with shallow combing on the body, from Blaxton (after Buckland and Dolby 1980); mid/late second to first quarter of the third century.*
- *160 Jar with lip grooved externally and internal reeded lid-seating and lightly combed body, from Cantley (after Annable 1960). Probably dating to within the period c 135 to the late second century; examples of the form reached York (cf. Monaghan 1993, 3022).*
- *161 'Shouldered' (S-shaped) bowl, from Blaxton (after Buckland and Dolby 1980). Produced for most of the life of the South Yorkshire industry, this basic type was common generally in the North-East from the mid-second century onwards (cf. nos. 206, 227, 248, 258).*
- *162 Large bowl of truncated sub-conical form without a shoulder, from Doncaster vicus (Buckland et al 1986). This type, particularly characteristic of the South Yorkshire potteries was long-lived, being produced from the mid second to the fourth century.*
- *163-4 Grey ware wheel-thrown copies of BB2 forms (Gillam Types 224, 319), from Blaxton (after Buckland and Dolby 1980); end of second to early/mid third century; such forms were also imitated by several East Yorkshire potteries (nos. 207, 209-10*

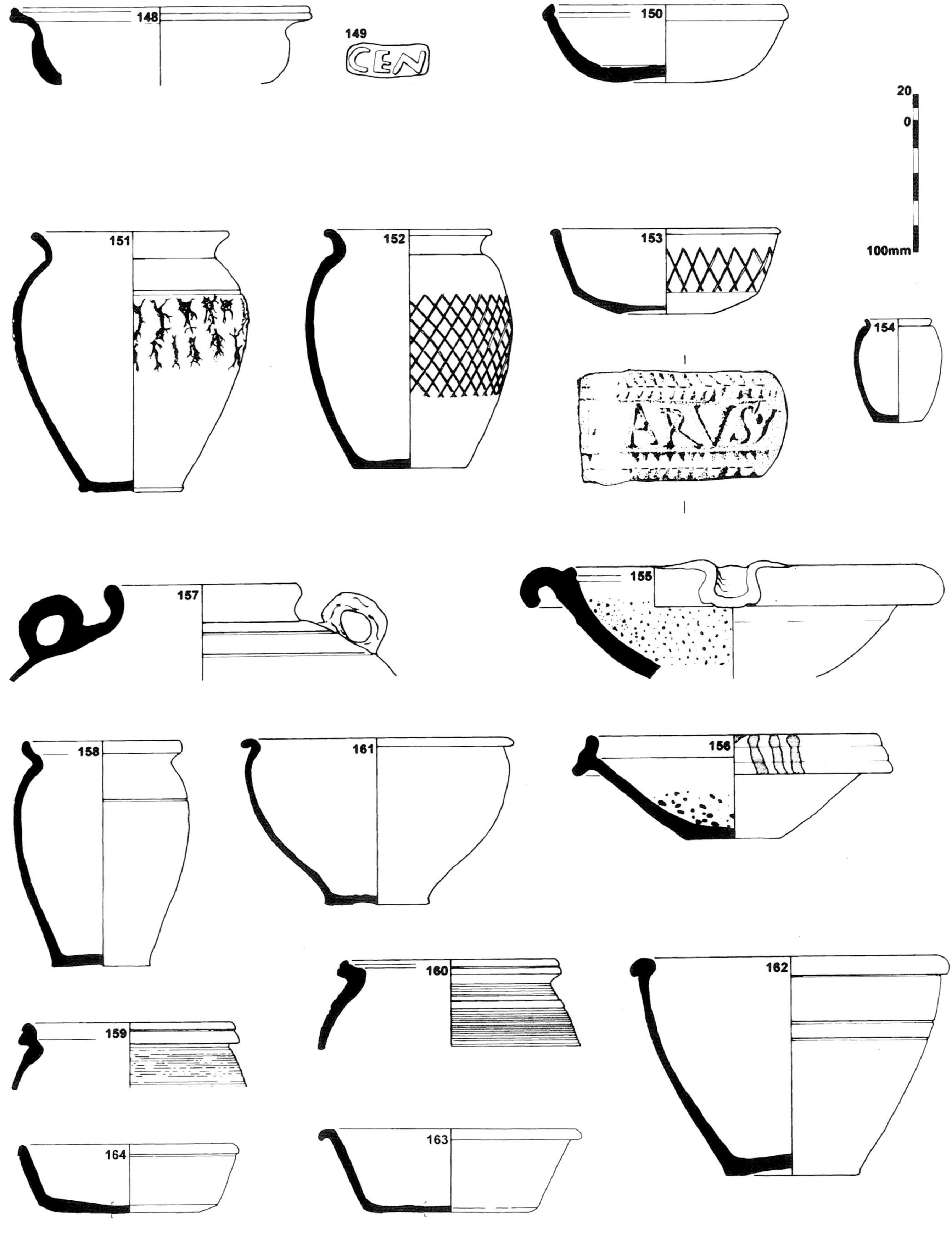

Figure 12. The South Yorkshire potteries

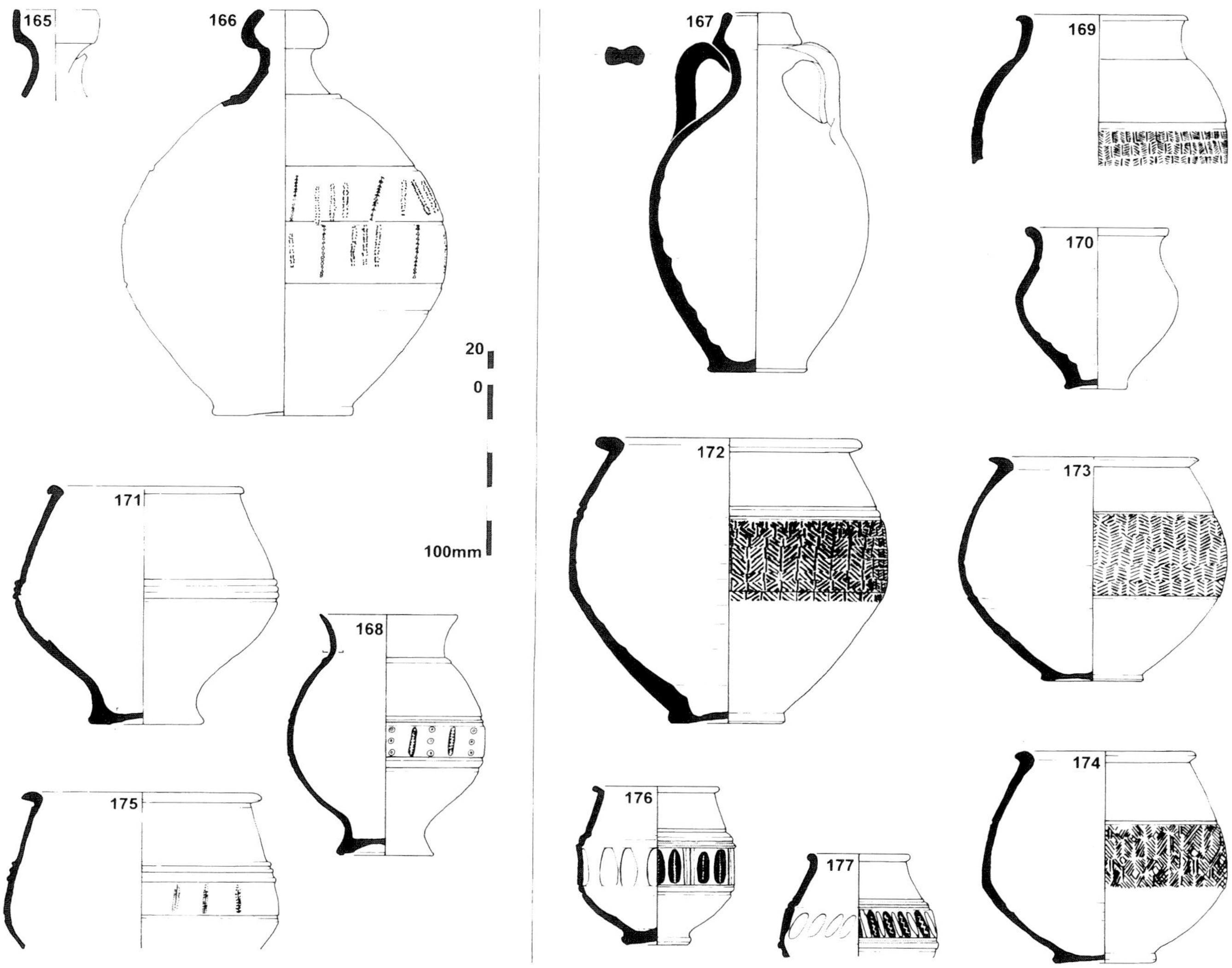

Figure 13. 'Parisian ware': (left) Rossington Bridge fine reduced ware and similar fine grey ware products from other South Yorkshire or north-west Lincolnshire workshops (mid second to very early third century); (right) products of the Saintonge, France, in similar style (after Santrot & Santrot 1979). Scale 1:4.

165 Flask with cupped mouth from Blaxton (after Buckland and Dolby 1980). Such vessels were also made with handles; cf. no. 167 from Gaul.

166 Flask with cupped mouth and impressed frond motifs from Rossington Bridge (after Buckland et al 1980).

168 Cordoned globular beaker of 'poppy-head' profile with impressed decoration (after Buckland et al 1980); cf. possibly related Saintonge forms (nos. 169-70), and similar stamped motifs (no. 176).

171, 175 Beakers with angular profile, multiple shoulder grooves and small sharply everted curved rim; cf. generally similar angular or semi-angular cordoned beakers from the Saintonge, with cordoned zones of heavily impressed individual stamps or roller-stamping (nos. 172-4, 176-7).

example the mortarium from Bar Hill fort with a stamp of the potter MVCO (K Hartley 1975, 142, figure 49, no. 1).

Brough-on-Humber (Petuaria): immigrant potters

At about the same time as the production of exotic vessels at York, there seems to have been a short, but remarkable, pottery-making episode at Brough-on-Humber. These products too were unusual for Britain at this period. They included colour-coated beakers, mortaria and other vessels, a number of which have clear affinities with pottery occurring in the forts of the Upper Rhine *limes*, particularly in *Raetia* or *Pannonia*, and suggest the presence of potters from somewhere in that general region (Darling forthcoming; M Darling *pers comm*). The precise circumstances of the establishment of such a workshop at *Petuaria* are uncertain, but it is probably best envisaged in a military context. For most of its existence, the fortified area at Brough-on-Humber seems to have had some military function, and the epigraphically attested *civitas* is now presumed to have been situated elsewhere in the general vicinity (Wacher 1974, 393-97). A brief change in the occupation of the fort, involving significant building activities, is thought to have taken place in about AD 125 (Wacher 1969, 3). This might have constituted the occasion for short-lived pottery manufacture of this nature. As already mentioned, the early

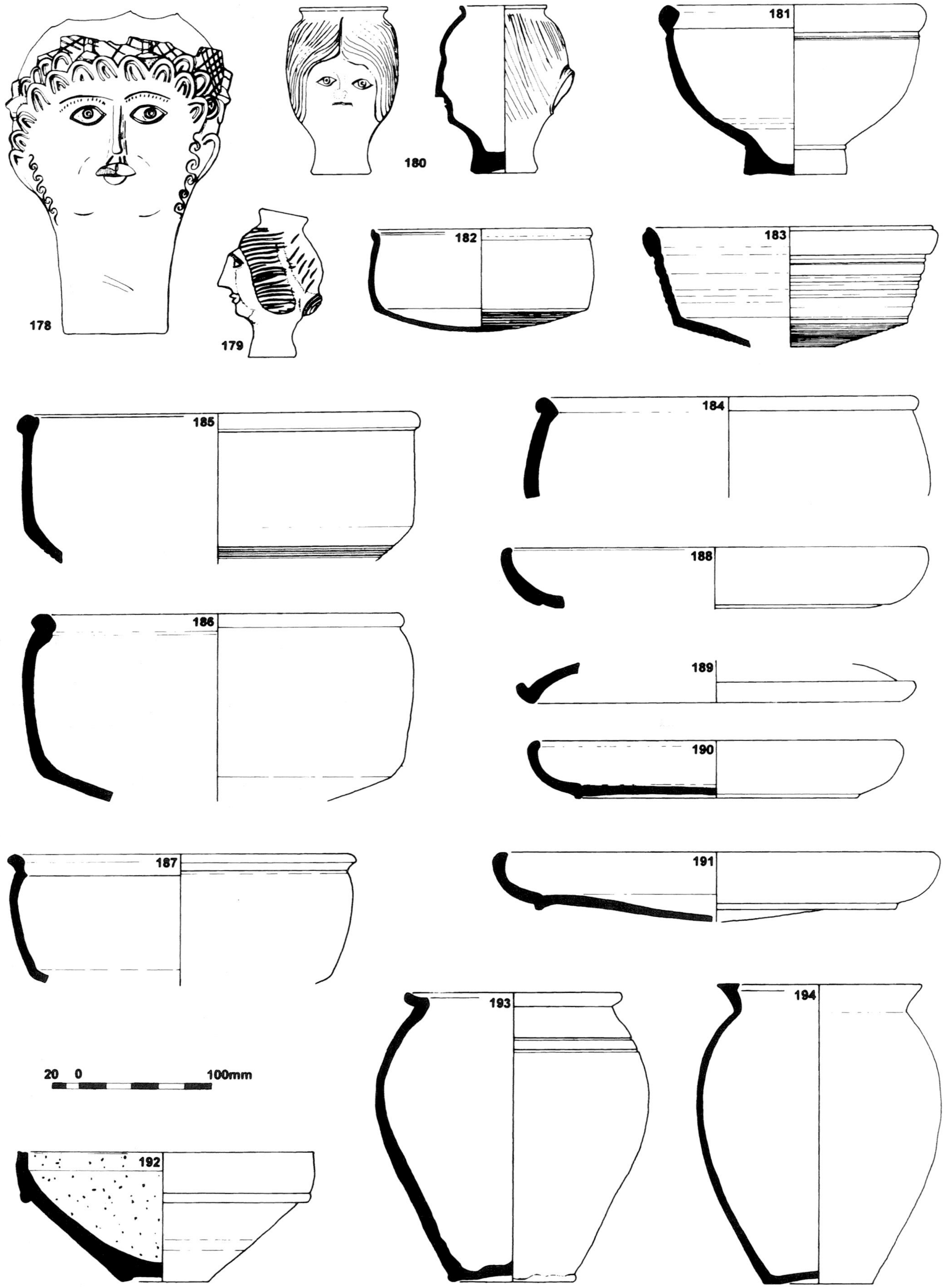

Figure 14. York (Captions opposite)

Figure 14. York: vessels with non-local affinities made locally, by or for the military, in the first decades of the third century, probably from 208 onwards.All are in utilitarian Eboracum oxidised wares (Ebor 1/2), and from York itself, unless otherwise stated. Nos. 178-80 and 182-91 are North African in style. Scale 1:4.

178 *Head-pot in the form of a probable portrait of the young Caracalla in c 209 (after Swan and Monaghan 1993).*
179 *Head-pot, probably intended to portray the Empress, Julia Domna, in 209-211/2, the period when the royal family was in York (after May 1911).*
180 *Head-pot, possibly another representation of Julia Domna during her time in Britain (after Gillam 1968).*
181 *Deep lid-seated bowl with an interned 'club-shaped' rim and internal lid-location groove; probably reduced Ebor ware (after Monaghan 1997); parallels can be found in the Rhineland (Swan and McBride 2002).*
182 *Casserole with lid-locator and combed rounded base (after Monaghan 1993).*
183 *Casserole with almond-shaped rim with internal and external lid-location recesses (after Monaghan 1997).*
184 *Casserole with undercut beaded rim and bulging girth (after Perrin 1981).*
185 *Casserole with rilled rounded base; the bead-rim is grooved on top for seating a lid (after Perrin 1981).*
186 *Bead-rimmed round-based casserole with internal lid-groove (after Swan 1992).*
187 *Casserole with internal lid-seating and gently rounded base (Swan and McBride forthcoming).*
188 *Round-bottomed platter with external notch at the wall-base junction (after Perrin 1990).*
189 *Domed lid with upturned rounded rim (after Swan 1992).*
190 *Platter with an internal groove and external flange/foot-ring at the wall/base junction (after May 1911).*
191 *Round-bottomed platter with an internal groove and external flange/foot-ring at the wall/base junction an internal groove and small external flange at the wall/base junction (after Swan 1992).*
192 *Wall-sided mortarium in Eboracum white-slipped ware (Swan and McBride forthcoming); the rim profile and use of a groove on the collar and top of the rim at this period is generally associated with products of potteries in the Middle/Lower Rhine and Moselle, such as Soller, Kreis Dren (Haupt 1984, Taf. 172; Richardson 1986, 109-12).*
193 *Grey Ebor ware lid-seated butt-jar (after Gillam 1968).*
194 *Grey probable Ebor ware Dales-type jar (after Gillam 1968); the form may have been introduced to York from Gallia Narbonensis.*

Hadrianic period was a time when there had been a number of transfers to Britain of troops from outside the province, and the Brough potters might have arrived with an auxiliary unit, or with a vexillation from *Legio VI Victrix.*

The decline of legionary production at York

The manufacture of the exotic fine wares at York also seems to have been relatively short-lived. They are not common there, and most occur as rubbish survival in later contexts. Presumably, when the soldier-potters who made them reached retirement age, craftsmen with a Romano-British background were put to work in the legion's pottery. It is also likely that, as new recruits from other sources gradually replaced long-serving men in the legion, so the fashion for pottery of this nature changed and demand waned, though utilitarian vessels with more ordinary profiles, less finely burnished surfaces and white slips continued to be produced locally. Moreover, for the greater part of the 140s, much of *Legio VI* was absent in Scotland (certainly in the building season), and vexillations drawn from York may well have been on detached garrisoning duties there in the late 140s and 150s (Swan 1999, 434). It is therefore conceivable that the military kilns just outside the fortress had closed down before the middle of the second century. A similar contemporary run-down seems to have occurred at the legionary workshops at Holt, when much of *Legio XX* was on detached duties, and a number of the barracks in the Chester fortress were empty or derelict (Ward 1998; Swan and Philpott 2000, 63-65). By this period, most Ebor ware was probably manufactured by civilian potters (?veterans) working on the Apple Tree Farm site. The forms of many of their products were reasonably normal by Romano-British standards, though some still seem to have been devolved versions of exotic Hadrianic introductions (nos. 74-5, 77). Moreover, in addition to the orange oxidised (often white-slipped) wares, this workshop had begun to make grey reduced imitations of Black-burnished ware (BB1) forms (nos. 78-9), thus presaging a ceramic trend that was to become universal in the following century.

The South Yorkshire potteries: the early background

By the Antonine period, other major potting centres had began to develop in Yorkshire. To what extent these were linked to a local military presence, or alternatively, to demand from troop concentrations further north, evidently varies from site to site. On the whole, rural markets were not yet a very significant factor in the siting of kilns in Yorkshire. The South Yorkshire potteries, an industry of some significance for northern Britain in the Antonine period, consisted of a number of centres just south-east of Doncaster, of which the best known comprise the workshops at Cantley and Rossington Bridge (Buckland *et al* 1980; Buckland *et al* 2001). Their development in the second century could conceivably represent a revival of production, which had initially been established in the late first-century at one or more related workshop(s) (probably sited somewhere in the general vicinity but as yet unlocated), presumably to serve the Flavio-Trajanic fort at Doncaster, but possibly with an eye on military markets further afield. This early output, found mainly on military sites near the Humber-Mersey axis, but occasionally further north (Rigby 1998, 192-3, 197), had included convex-walled dishes with a concave base and a sharply inturned rim, impressed with the name-stamps of the potters CEN, REDITAS and SACE (nos. 149-50; Gillam Type 337); also in a similar light grey fabric with darker surfaces, were campanulate bowls with a bi-lobed rim (no. 148; Gillam Type 301; Buckland and Magilton 1986, no. 14). The underlying reason for a late Hadrianic-Antonine resurgence at this locality is, nevertheless, obscure, since between the mid 120s and the late 150s or early 160s, the fort at Doncaster seems to have been held on a care and maintenance basis, or was only partially garrisoned (Buckland 1986, 13). Not far to the north-east, however, the adjacent area of north-west Lincolnshire (between the Rivers Trent and Ancholme) had ceramic traditions which were quite exceptional for the North-East. This region had been regularly using and producing reasonable quality pottery since well before the Roman conquest (May 1996, 437-8), a phenomenon which had apparently contributed to the relatively rapid Romanisation of ceramic techniques and forms in the post-conquest period (when they were exploited by the military), and to the spread of their traditions to surrounding areas, and indeed, across the Humber in the course of the second and early third centuries. The late Iron Age and early Roman styles of North Lincolnshire are, indeed, reflected in some of the grey kitchen wares made by the South Yorkshire potteries in the second century and later: for example, in the late rusticated cooking-pots (no. 151), the S-shaped bowls (nos. 161-2), and the cordoned jars with heavy looped handles (no. 157; the forerunner of the sequence of East Yorkshire jars which gradually developed countersunk handles (nos. 201, 219, 249-51). Similarities in the kilns of both areas are also clearly apparent, for example, Antonine kilns with raised oven floors of prefabricated bars over-plastered with clay, occur at Cantley (Kilns 20 and 40) and Rossington Bridge Farm (Swan 1984, MF 6.706, 6.717, 6.721), and also at Dragonby Kiln 5 (May *et al* 1996, 135).

The South Yorkshire potteries: the origins of some potters

In addition to relatively local influences, the South Yorkshire industry also drew several craftsmen from outside the immediate region of Doncaster at this period. One of these, a certain Sarrius, a mortarium manufacturer, who had kilns in the Midlands at the important centre of Mancetter/Hartshill, near Nuneaton (Swan 1984: MF 5.647-8, and 5.654), also set up workshops with his associates at Rossington Bridge (no. 155; Swan 1984: MF 6.703-4; Buckland *et al* 1980: 146, 154, figure 3; Buckland *et al* 2001). He was evidently more than just an artisan-potter; he seems to have had tremendous business drive and entrepreneurial flair, particularly where there were profitable military markets to be exploited. Recent research by K Hartley indicates that Sarrius had established several other, more or less contemporary, mortarium workshops, one near the Antonine Wall fort of Bearsden, and another somewhere in the North-East (K Hartley forthcoming; K Hartley *pers comm*). Indeed, it may well have been Sarrius himself who attracted to his Rossington Bridge factory-compound another potter (or potters) from Dorset, skilled in the traditional handmade techniques of Black-burnished ware (BB1; nos. 152-3). He could hardly have failed to notice that this remarkable type of handmade pottery, native to the tribal territory of the Durotriges, had already scored an enormous success with the army of Hadrian's Wall, Wales and the North.

Again, the lure of potentially lucrative northern military markets may have been the underlying reason for the introduction to Rossington Bridge (and probably nearby Cantley and Blaxton), of the manufacture of a distinctive tableware, traditionally (but erroneously) known as 'parisian ware', after the tribal area of the Parisi in East Yorkshire, in which it was first recognised (Figure 13; Corder 1958; Elsdon 1982). It was characterised by a fine polished fabric, apparently in a 'Gallo-Belgic' tradition, with a sandwich fracture (dark grey surface and lighter grey core). The limited range of forms, probably mainly drinking-sets, included cup-mouthed flasks (nos. 165-6), simple beakers with a very short everted rim, probably influenced by the profile of a BB1 type (no. 154; Gillam 1976, nos. 20-21), beakers of poppy-head shape with a flaring rim (no. 168), sometimes indented, and, most distinctive and unusual of all, squat, almost biconical, beakers with small everted rim and girth and shoulder grooves (nos. 171-175). Some of the vessels were decorated with zones of rouletting, roller-stamping, or impressed motifs including vertical frond-like or infilled sausage-shaped stamps (nos. 168, 175; Buckland *et al* 1980, 156-7; Elsdon 1982, 35, figure 9, nos. 26-32). The origin of this very individual type of pottery, which had no apparent relationship to the late Iron Age traditions of South Yorkshire or North Lincolnshire, and the reason underlying its sudden appearance in that region have hitherto proved difficult to explain (Rigby and Stead 1976, 181-5). Indeed, many of the forms cannot be precisely paralleled anywhere else in Britain at this period. However, in western Aquitaine (south-western France), there existed, in the late first and second centuries, a regional ceramic tradition, which embraced relatively fine reduced (grey) ware beakers, bowls, and flagons and flasks, with remarkably similar cordoned profiles, and decorative techniques that also included rouletting, roller-stamping and individually impressed stamps including frond-like and sausage-shaped motifs (nos. 167, 169-70, 172-4, 176-7; Vernou 1995; Buisson 1991; Santrot and Santrot 1979, forms 250, 259, 263, 269-71, 293, 302, 313). The known Gallic kiln sites occur mostly east and south of Saintes (capital of Aquitaine), and include Soubran, between Saintes and Bordeaux (Santrot and Santrot 1991; Michaud 1991). Could it be, that the ever enterprising Sarrius had also managed to attract to Rossington Bridge a potter (or potters) from this region?. There were certainly direct trading contacts between Bordeaux and ports in Britain from the second century onwards (Sireix and Convertini 1997, 28-29). Sarrius' action might have been with the express purpose of complementing the rather utilitarian repertoire of his potting-complex with the addition of fine tablewares, in the hope of reinforcing and expanding his tenuous toe-hold on long-distance trade with the army of the northern frontier zone.

The products of these South Yorkshire kilns did achieve some small measure of success in penetrating the military markets of northern Britain, but the absolute quantities appear to have been relatively small. Some uncertainty stems partly from the fact that the fabric of Rossington Bridge BB1 is often very difficult to distinguish macroscopically from its Dorset counterpart, though the former tends to be paler, harder and better sorted, with less sand and no shale inclusions (Williams 1977, 194). However, Bidwell has drawn attention to a distinctive type of chamfered bowl with a beaded rim and wide-spaced lattice (no. 153), a form absent from the Durotrigian BB1 repertoire, which is probably a Rossington Bridge product (Bidwell 1985, 194-5, figure 70, no. 135, from Vindolanda; Bidwell 1994, 226, 232-4, figure 8.8, nos. 6, and 15, from South Shields). The recording of this form in the future may help to illuminate the extent of the industry's long-distance trade and the chronology of its decline. Of the slightly more distinctive 'parisian ware', only very small amounts have been recognised so far, mainly in Antonine Scotland, and at South Shields (P Bidwell *pers comm*). The full publication of the Rossington Bridge kiln sites (Buckland *et al* 2001) may stimulate the recognition of plain as well as the more obvious decorated forms on other northern frontier sites. Not all the known 'parisian' stamped material came from the South Yorkshire workshops. Kilns at Market Rasen, Lincs were turning out slightly coarser versions of plain and stamped vessels (rouletted imitations of samian form Drag. 37, and a plain carinated bowl with a beaded rim and triple grooves just above the carination; material shown to the author by the excavator, K. Hunter). Another workshop somewhere in north-western Lincolnshire, possibly at Roxby or Dragonby, also included 'parisian ware' in its repertoire (Swan 1984, MF 3.457-8, 3.465; Elsdon 1982, 19-22, 35-40). Nevertheless, the long-distance exports seem more likely to have originated in the South Yorkshire potteries. Indeed, Rossington Bridge 'parisian' wares so rarely occur in the neighbourhood of their production, that they may well have been intended specifically for export (P Buckland *pers comm*).

It is unclear at present whether the ability of the South Yorkshire potters to break into northern military markets was due to Sarrius' first-hand contact with the military on the Antonine frontier, or whether the producers of commodities already being exported northwards from that same part of Yorkshire, had shipping facilities to share. Whatever the precise circumstances, the long-distance trade with the South Yorkshire potteries seems to have been relatively short-lived. Moreover, once the orders for the supply of northern military markets were lost, probably late in the second century, there was a marked contraction in the range of vessel-types made by these workshops (save for the mortaria); the production of BB1 ceased, and the repertoire of the South Yorkshire potteries in general came close to resembling that of an ordinary rural supplier (Buckland *et al* 1980: 163; P Buckland *pers comm*). This is, in fact, a telling illustration of the impact of military demand on the production of a pottery industry, and has implications for the interpretation of changes on other kiln sites which may have been involved in provisioning the military. Nevertheless, the manufacture of small quantities of mortaria alongside the ordinary cooking-wares (particularly bowls and jars of regional types, nos. 158-62, and grey imitations of BB1/BB2 forms, nos. 163-4) continued, at Cantley in particular, until well into the fourth century, mainly for relatively local consumers. Even York, less than 48 km away, never received more than very small amounts from the South Yorkshire potteries (Monaghan 1997, 893). The mortaria, however (no. 156), occasionally appear on the Northern frontier in the third and early fourth-centuries (K

Figure 15. (See page 60). The Holme-on-Spalding Moor pottery industry, East Yorkshire. All are in Romanised reduced (greyish) wares. Unless otherwise stated, the pots are from the Throlam or Hasholme kiln sites, probably date from about 220/230 onwards, and the line decoration is burnished. Scale 1:4.

195 Narrow-mouthed jar without handles; from the Well 1 at Dalton Parlours villa, West Yorkshire; probably mid/late third century (after Sumpter 1990).

196 Narrow-mouthed handled jar (after Corder 1930b); the cordoned mouth is almost bobbin-shaped (cf Crambeck product, no. 254); probably late third or early fourth.

197 Narrow-mouthed handled jar with 'coal-scuttle' mouth and coarse rouletting; probably third century (after Halkon and Millett 1999).

198 Narrow-mouthed handled jar with stabbed flange, from Well 1 at Dalton Parlours villa; probably mid/late third century (after Sumpter 1990).

199 Narrow-mouthed jar with flanged rim, an imperfect vessel from the well at Rudston villa, East Yorkshire; probably mid/late third century (after Rigby 1980).

200-201 Jars with looped handles (after Corder 1930b; after Hicks and Wilson 1975). These represent an intermediate stage in the series which eventually developed counter-sunk handles (cf. nos. 249-52). Typologically no. 201 is probably slightly earlier in date than no. 200, though both probably date to within the third century. The combed chevrons and wide-spaced lattice decoration seem more typical of this industry than of the other contemporary workshops such as Crambeck.

202 Beaker with pedestal foot and multiple shoulder-grooves; from Vindolanda fort (after Bidwell 1985). This form might conceivably have evolved from the earlier 'parisian ware' beakers, which were traded across the Humber to Brough-on-Humber and other sites in the neighbourhood.

203 Small S-shaped bowl with neck and girth-grooves (after Corder 1930b), possibly ancestral to the Crambeck form, no. 264; probably late third/early fourth century.

204-5 Carinated bowls with splayed foot-ring; no. 204 (after Halkon and Millett 1999), with the vertical wall and protruding carination (probably late third or early fourth century onwards), evidently developed from the earlier form, no. 205 (after Hicks and Wilson 1975) with the inward-leaning wall and simple carination (probably 220/230 onwards). Both versions were made at Norton (evidently not simultaneously), but they are virtually absent in the Crambeck repertoire.

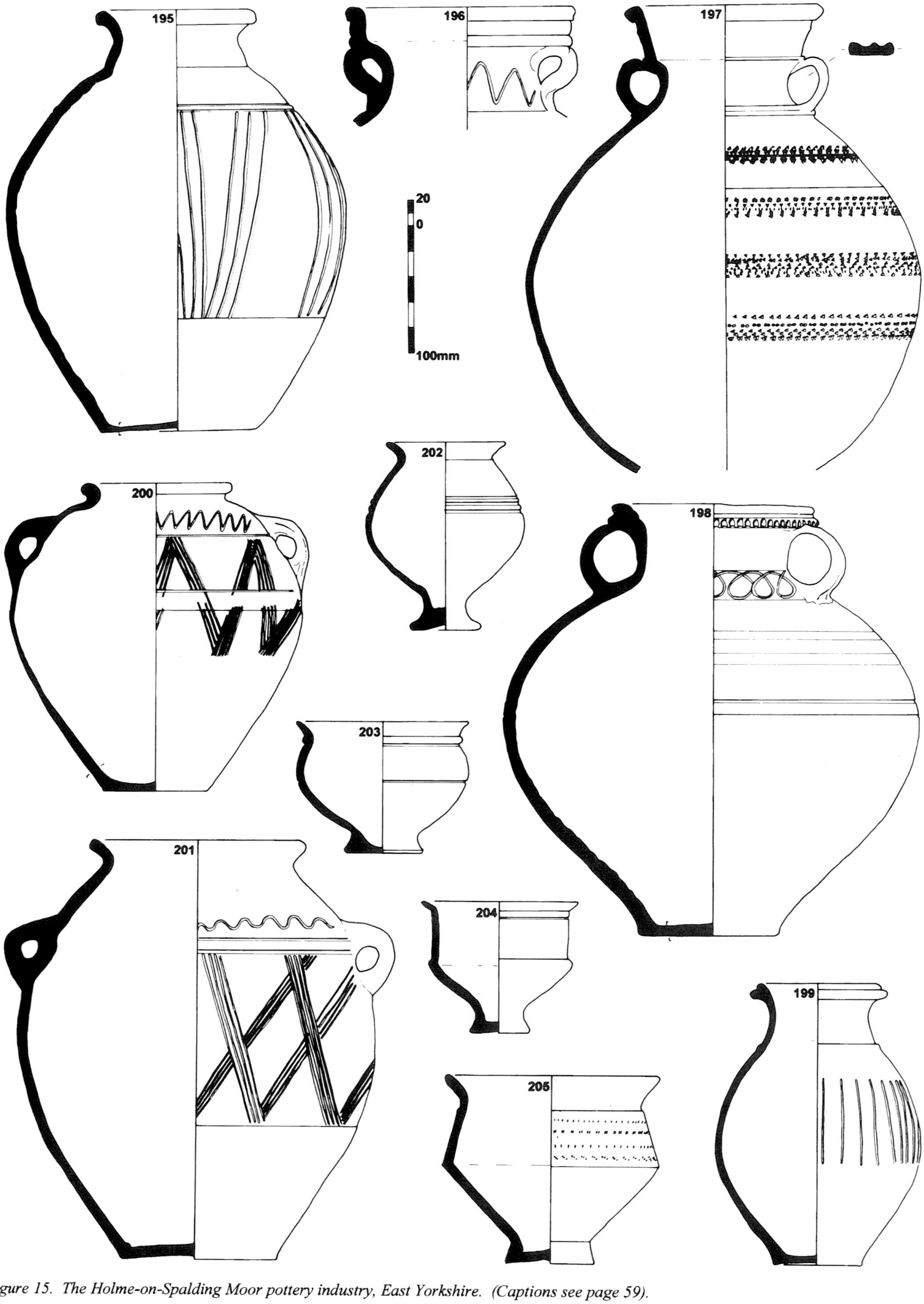

Figure 15. The Holme-on-Spalding Moor pottery industry, East Yorkshire. (Captions see page 59).

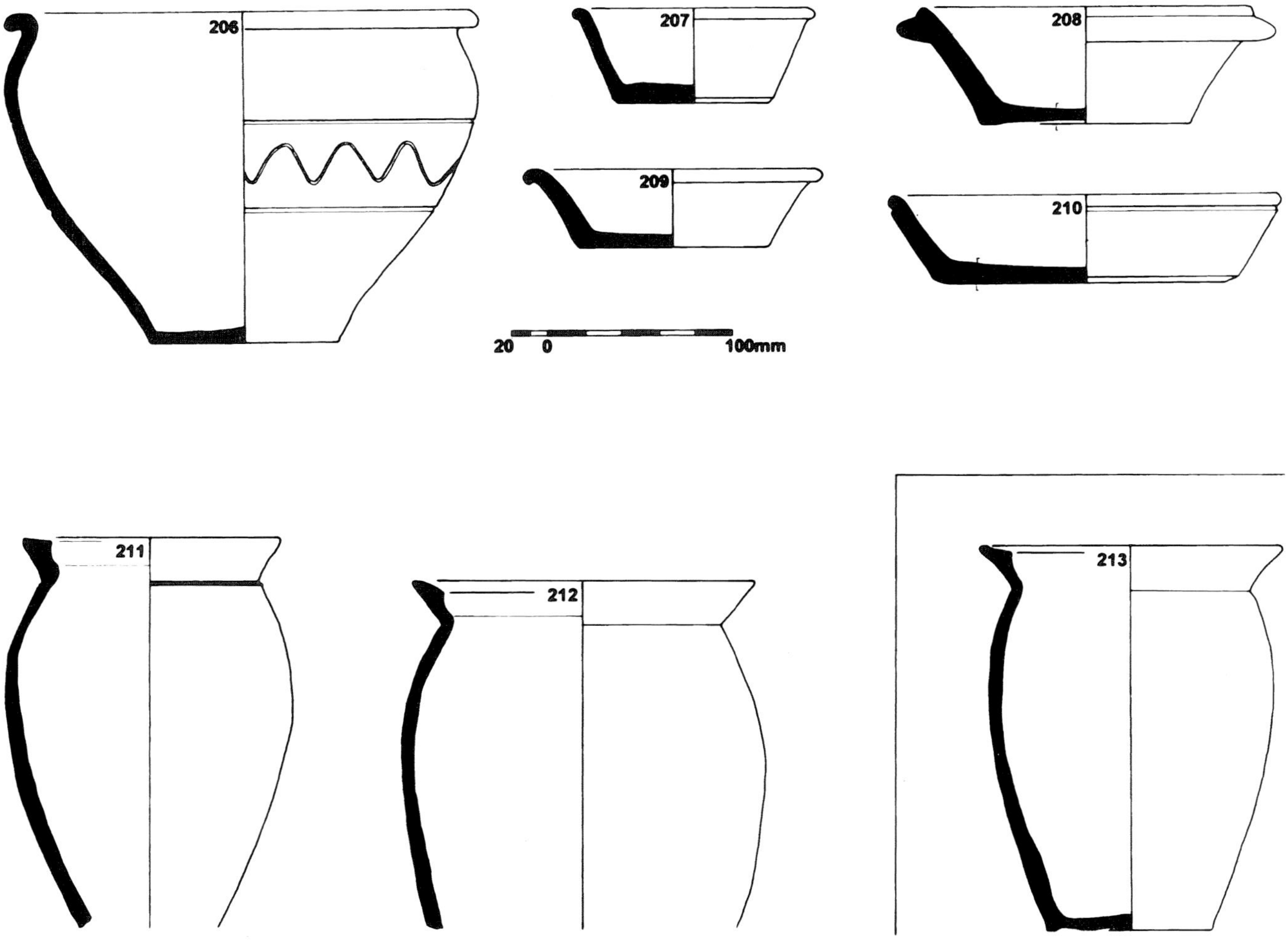

Figure 16. The Holme-on-Spalding Moor pottery industry, East Yorkshire (continued).
Nos. 206-212 are all in Romanised grey wares; unless otherwise stated, they come from the Throlam or Hasholme kiln sites, and the line decoration is burnished. Most vessels (except 208) fall within the main period of production (about 220/230-late third century). No. 213: Dales shelly ware, from North Lincolnshire. Scale 1:4.

206 S-shaped bowl (after Hicks and Wilson 1975); cf. similar products from the South Yorkshire, Norton and Crambeck workshops (cf. nos. 161-2, 227, 246, 258).
207 Bowl imitating approximately a BB2 vessel (after Corder 1930b).
208 Imitation of BB1 flanged bowl or dish, last third of third century or later (after Corder 1930b).
209 Imitation of BB2 dish (after Corder 1930b).
210 Imitation of BB2 dish, Gillam Type 317 (after Halkon and Millett 1999).
211-12 Cooking-pots with characteristic 'Dales-type' rim with double lid-seating (after Hicks and Wilson 1975).
213 Dales shelly ware, a cooking-pot with a distinctive rim profile, similar to previous vessels, but handmade in a fabric gritted with fossil shells, and produced at sites in North Lincolnshire such as Barnetby Top, Elsham (Samuels 1979); probably from the second/third decade of the third century onwards, from York (after Gillam 1968).

Hartley 1985, 183-4; K Hartley 1994, 210), though they are often difficult to distinguish from the products of late workshops at Catterick (no. 280), and Swanpool, just south of Lincoln (Hartley 1986, 149-50). Ultimately, in the course of the fourth century, competition from Crambeck wares (figure 19-20) confined the markets of the south Yorkshire products to a very restricted area immediately south of the Humber.

York: legionary production under the Severi

At York, it was another military event which resulted in a sudden revival of the participation of the army in local pottery production. Ebor oxidised, white-slipped and grey wares seem to have been made at Apple Tree Farm until shortly after the mid-second century. The absence there of imitations of BB2 could suggest that production did not last beyond 160s, when the troops on the northern frontier (who had been receiving such supplies) returned to the fortress after the evacuation of Scotland. It was, indeed, at about this period (the end of the second century) that the amount of BB2 in York began to increase significantly (Monaghan 1997, 864, 892). At this period, moreover, there were almost certainly other workshops (unlocated but probably in the general vicinity of the fortress), making grey imitations of black-burnished ware forms (Monaghan 1997, 900). Small quantities of East Yorkshire calcite-gritted wares, such as Knapton jars (no. 238), also now became apparent at York. Much of the legion must have been absent again in the 180s, dealing with troubles on the Northern Frontier (Dio lxxviii. 8. 1-2). In the 190s, most troops probably left for the Continent to fight in the civil war, which culminated in the battle of *Lugdunum* in 197, and the accession of the Emperor Severus. Though, no doubt, *Legio VI* returned much depleted the following year, it was not until a decade later, in response to continuing troubles in Britain, that the new Emperor decided to deal personally with the crisis (A Birley 1988, 170-87).

In 208, the Emperor Severus arrived in York, accompanied by his family, his wife, Julia Domna and their sons, Caracalla and Geta, and a huge personal retinue including a number of the *comites Augusti*. He also brought with him at least one, and probably several legionary vexillations, for what was clearly intended to be a massive imperial expedition to northern Britain. It can now be shown that some of these reinforcements were drawn from *Numidia*, (North Africa) and that others may have included men from *Gallia Narbonensis* (southern France). The detailed epigraphic and ceramic evidence has already been published elsewhere (Swan 1992). In brief, the old industrial area, just north-east of the fortress, became a works depot once more for a range of industrial processes, including the (official) counterfeiting of coins, and the manufacture of tiles, and pottery (Magilton 1986, 32-40; Swan and McBride 2002 and *forthcoming*). Glass-workers were active in an adjacent part of the *canabae*, probably at the same time, and some pots seem to have been made to order for use by them (Cool *et al* 1999, 150-54). There is ample evidence of the presence of soldier-potters from several different regions, some from outside the province, who were making utilitarian vessels in their own indigenous styles. Among such products were convex-walled platters with rilled, recessing or flanges on the mostly rounded bases, domed lids, and lid-seated casseroles with rounded, often rilled bases (nos. 182-191), all of North African type. The grooves, and other basal devices were probably intended to aid the stability of such vessels on a brazier, one of several modes of cuisine traditional in North Africa and the southern Mediterranean, but normally absent from the north-western provinces of the Empire. Crucially, such cooking-vessels, and other local products with North African affinities have been recorded in the York fortress, among rubbish dumped in the alley between the second and third barracks of the *Cohors Prima*, near the centurion's quarter of the third barrack (unpub assemblages from excavations under York Minster examined by the author *per* Dr B Heywood). These suggest that the occupants may have included some of the North African drafts brought in by Severus. Other local products of North African type at this period were head-vases, a Numidian tradition. Such vessels incorporated carefully modeled portraits (perhaps executed by sculptors), probably of the imperial family including Caracalla (no. 178) and Julia Domna (nos. 179-80), and through comparisons with their contemporary coin-portraits, may be dated to 209, or no earlier than 208 (Swan and Monaghan 1993). Among the other ceramic products probably belonging to this period, and apparently made in the same part of the *canabae*, were Ebor white-slipped, wall-sided mortaria of a type generally associated with workshops in the Lower/Middle Rhineland (no. 192; Swan and McBride 2002). A distinctive 'butt jar', locally made in Ebor oxidised and reduced (grey) wares, was also particularly common in the fortress for a similar short period (no. 193); it almost certainly represents an introduction from outside the region, perhaps from *Gallia Belgica* or the Rhineland.

Either now, or shortly after, a potter (or potters), seemingly from *Gallia Narbonensis*, introduced a type of cooking-jar with a distinctive and carefully formed triangular lid-seated rim in a sandy grey, or occasionally orange ware, also apparently made somewhere in the vicinity of *Eboracum* (no. 194). The fabric generally resembles that of reduced Ebor ware, but the evidence so far does not indicate that it was made in the same workshop as the African-type forms (Swan and McBride *forthcoming*). This particular product has often been viewed as an imitation of 'Dales Shelly ware' cooking-pots, distinctive, largely handmade, jars, in a fabric tempered with fossil shell (with late Iron Age ancestry), which were common in Humberside (no. 213; Gillam 1951; 1968, 66-70; Loughlin 1977). However, the author has argued that the sandy York version of the form could well owe its inspiration to the pottery of the Languedoc, *Gallia Narbonensis* (Swan 1992, 9; 1993). In fact, it seems to have appeared in York slightly earlier than the classic 'Dales Shelly ware' cooking-pot (Monaghan 1997, 899), and, more importantly, the form has no typologically close ancestor either among the earlier (pre-Dales) handmade shell-gritted wares of North Lincolnshire, or, indeed, among the small quantities of wheel-thrown pottery produced on the opposite side of the Humber in the second century. The very simple rim profiles of Rigby and Stead's so-called 'proto-dales ware' jars (1976, figure 81, nos. 46-9) are insufficiently close to those of the classic Dales ware cooking-pots for an independent and direct typological development to be comfortably accepted. The York ('Dales-type') jars thus seem more likely to have constituted the proto-type for the Dales Shelly ware cooking-pots. These new lid-seated jars evidently became fashionable beyond the fortress and colonia, and were, indeed, copied with varying degrees of precision, in grey wares, by many of the other workshops which apparently emerged in subsequent decades, in the vicinity of York, in north-west Lincolnshire, and in East Yorkshire (for example at the Holme-on-Spalding Moor potteries, nos. 211-2).

Third-century expansion of rural pottery manufacture in Yorkshire

The resurgence of military pottery production in York seems to have been short-lived, but the Severan episode heralded a time of increasing Romanisation and prosperity not only for the York *colonia*, but also its for its hinterland, particularly the hitherto more backward areas of the countryside, away from roads, forts and towns. One of the many factors underlying this may have been the fact that in 197, York had become the capital of the newly created province of *Britannia Inferior* (Mann 1997). Even so, a marked peak in the total number of known pottery kilns (with most situated in rural locations) and a significant decline in the number of kilns attached to urban settlements, is a phenomenon recorded for this period in many other regions of Roman Britain, including those areas far from the military zone (Swan 1984, 19, maps 5-6). In Yorkshire, it was at about this time or soon after that a number of Romanised potteries of some significance, not all substantially dependent on urban or military markets, began to expand or spring up *ab initio* just north of the Humber Estuary, where hitherto these had evidently been rare or very small.

The potteries in the Holme-on-Spalding Moor area

Of particular significance were the dispersed workshops, often attached to small settlements (eg. Hasholme and Bursea), which were situated within the landscape centred on Holme-on-Spalding Moor, often referred to as the Throlam industry (Corder 1930b, Hicks and Wilson 1975; Halkon 1983; Halkon and Millet 1999). These made ordinary reduced (grey-brown), wheel-thrown, mainly culinary vessels in Romanised fabrics, several imitating BB2 forms (nos. 207-10), and others of regional type (some originally influenced by the ceramic traditions of North Lincolnshire, nos. 201-6), with a particularly strong emphasis on large handled jars and *lagenae,* for storage and for use in wells (nos. 195-201). Such workshops seem to have burgeoned during the course of the third century, and were suppliers to Brough-on-Humber, the adjacent countryside of East Yorkshire and beyond, from around the 220/230s onwards. In fact, it now seems that production may have continued into the early decades of the fourth century, until it was largely stifled by competition from the Crambeck workshops (see below). This chronology may be suggested by the presence of probable Holme-on-Spalding jars alongside large quantities of Crambeck wares, in Well 1 of the Dalton Parlours villa, West Yorkshire (Sumpter 1990, 244, 236, Table 29), and by the absence of such wares in the East Yorkshire signal stations, which were established in the mid-380s (see below). Even so, the products of the Holme-on-Spalding industry never reached York in quantity, despite the relative ease of water transport (Monaghan 1997, 900-901). This failure may, indeed, account for the relatively restricted range of forms produced (or vice versa!); the classes of vessels made appear to have been largely geared to rural tastes. At this period, the needs of York were evidently met mainly by its own local civilian grey ware industries, though the location of these kilns is as yet unknown, and also by the Norton workshops (see below). Products of the latter centre can be seen as accessory vessels in the Trentholme Drive cemetery (nos. 221-2; Gillam 1968, figure 32, 1-8, vessels 205-12, figure 33, 6-7, vessel 223-4). An increasing reliance on local or regional production of grey wares may prove to be a feature of the third century, at *Eboracum* and in other parts of northern Britain. Croom and Bidwell (1998, 178-9) noted the same phenomenon in the vicus of the fort at Greta Bridge, Co Durham.

Norton (Malton)

Norton was another pottery industry which seems to have begun its rise to prominence at the end of the second or early in the third century. It was situated in the *vicus* at Norton, across the River Derwent from Malton fort, where third-century kilns were recorded during housing development (Hayes and Whitley 1950; Hayes 1988). It is not clear how the Malton garrison was supplied in the latter part of the second century, when the fort was re-occupied, following the abandonment of Antonine Scotland. Similarities in the fabrics of vessels of both periods, led Bidwell and Croom (1997, 101-02) to suggest that the workshops servicing the Flavio-Trajanic fort may have been located in the same general area as the later kilns, and that small-scale production may have trickled on throughout the second century at other Norton kiln sites, either those known to have been destroyed without record or unlocated examples (Swan 1984, MF 5.688-9 and 6.692). Several vessel forms from deposits in the Malton vicus may have pre-dated the main third-century products, for example, mortaria (no. 234), flagons, jugs, and imitations of BB2 cooking-pots and dishes (nos. 228, 230-1 and possibly 220). The third-century kiln structures recorded at Norton, with their chunky bars and long flues, as well as a proportion of their vessels, show some affinities with those found at the regional workshops of North Lincolnshire, perhaps indicative of the arrival of non-local potters, civilian craftsmen or even artisans recruited to the garrison (Swan 1984, 109-10). However, some of these later products also exhibit retrospective traits, suggestive of a fusion of new and pre-existing traditions. For example, the bowls and dishes imitating BB2 have triangular rims (nos. 228-9, 232; after 180 such rims had begun to be superseded by rounded rims in the BB2 industry, though there seems to have been a decade or two of overlap). The same Norton vessels also sometimes have wavy-line decoration (a feature also in decline on the late second-century BB2 prototypes, and which may have largely disappeared by 210). Bidwell and Croom (1997, 101-3) have highlighted several other forms which may have had second-century antecedents, and Corder (1950, 36-7) remarked on the typologically early appearance of some of the beakers from the Norton kiln site. In fact, indented beakers likely to be of Norton origin were noted at York in deposits which might even pre-date AD 200 (Monaghan 1997, 985), and possible Norton products have recently been recorded in a Severan context at South Shields (P Bidwell *pers comm*). In sum, the main period of late production at Norton may have begun at, or very soon after, the beginning of the third century, up to two decades earlier than Corder (1950) had originally suggested, and, probably continued until at least after the end of the third century (Bidwell and Croom 1997, 101-03), clearly overlapping with the earliest phase of production of the nearby Crambeck potteries, and contrary to previous assumptions (see below). These wares not only served the Malton fort, but appear to have travelled to York (nos. 221-2), and beyond, with small quantities recorded on the northern frontier. However, the Norton output is difficult to distinguish from other ordinary grey wares, except when the vessels have distinctive or unusual profiles or decoration, such as the 'smith pots' with applied motifs (no. 217), the indented beakers (no. 222), and the cauldron-like bowls with ring-handles (no. 221).

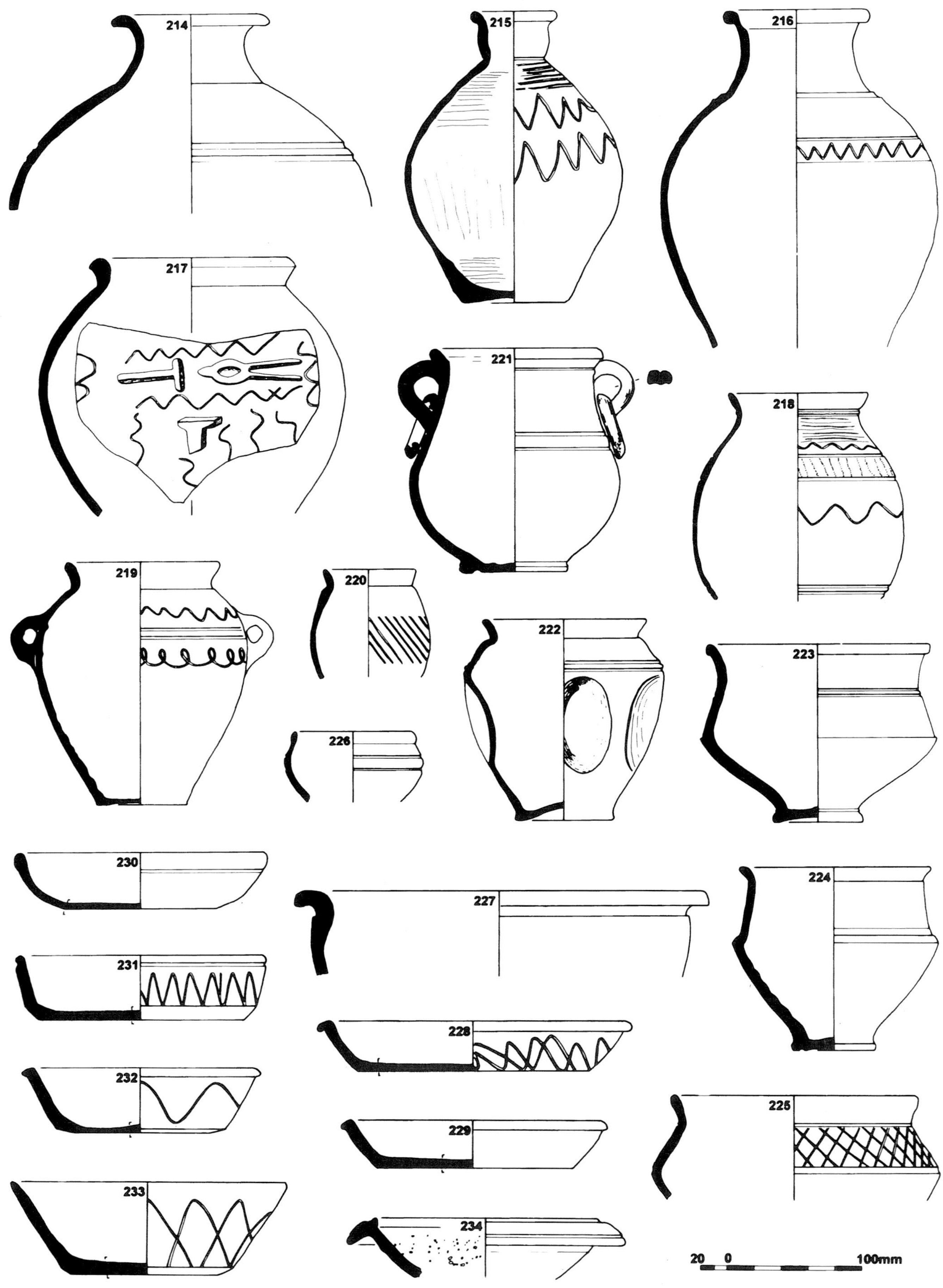

Figure 17. Captions opposite

Figure 17. The Norton pottery industry, East Yorkshire.All vessels are in Romanised grey fabric and decoration is burnished, unless otherwise stated. Most come from, or can be paralleled at, the two Norton kiln sites known from excavations, Model Farm (after Hayes and Whitley 1950) and Grove Bungalow (after Hayes 1988); these should therefore date to within the main period of production, c 200/210-270. Nos. 230-31 (not recorded at known kiln sites), and possibly also 221-3, 228, 232 and 234 probably immediately pre-date this period. No. 224 probably belongs to the end part of the main production period, or a decade or two later. Scale 1:4

214-5 Narrow-mouthed jars from Model Farm and Grove Bungalow kiln sites.
216 Narrow-mouthed lid-seated jar from Malton fort vicus (after Bidwell and Croom 1997).
217 'Smith pot', with applied tools, hammer, tongs, and portable anvil, from Malton fort vicus (after Mitchelson 1964).
218 Ovoid jar with comb-stabbed decoration, from Grove Bungalow.
219 Cooking-pot with looped handles from Model Farm, a form representing an early/intermediate stage in the regional series which eventually developed counter-sunk handles (eg. nos. 249-52 in Crambeck reduced ware).
220 Jar, perhaps imitating a BB1 beaker of cooking-pot form (Gillam 1976, no. 20); from Malton fort vicus (after Bidwell and Croom 1997).
221 Cauldron-like jar with loop-handles and suspension rings imitating metal prototypes (Gillam Type 175); from York (after Gillam 1968).
222 Indented beaker from York (after Gillam 1968).
223 Biconical bowl from Model Farm; the concave-walled form is typologically early in the series (cf. later versions, nos. 224-5).
224 Carinated bowl with near-vertical wall and short everted rim, from Model Farm. Typologically this is the latest version in the Norton series; cf. the later vertical-walled version from Holme-on-Spalding Moor (no. 204).
225 Biconical bowl with inward-leaning wall and long everted rim, from Model Farm. Typologically earlier than 223 and later than 223, cf. similar form at Holme-on-Spalding Moor (no. 205).
226 Cordoned cup from Grove Bungalow
227 S-shaped bowl (from Model Farm), a widespread north-eastern regional type (cf. nos. 161-2, 206, 227, 245, 258).
228-9 Dishes imitating BB2 vessels (Gillam Types 311/312); from Malton fort vicus (after Bidwell and Croom 1997).
230 Dish perhaps loosely imitating BB form; the curved wall suggests a second-century date; from Malton fort vicus (after Bidwell and Croom 1997).
231 Dish, probably an imprecise imitation of the BB2 form, Gillam Type 317; (after Bidwell and Croom 1997).
232 Bowl imitating BB2 form, Gillam Type 223; from Malton fort vicus (after Bidwell and Croom 1997).
233 Bowl probably imitating BB1 form (Gillam 1976, no. 81). The markedly leaning wall suggests this may post-date most of the other Norton imitations of BB forms; from Grove Bungalow.
234 Mortarium, from Malton fort vicus, c 180-250 (after Bidwell and Croom 1997).

'Native' pottery production: late second to third century

In the course of the third-century, the indigenous producers of handmade pots also seem to have increased their output. This particularly applied to the East Yorkshire workshops near Malton, and probably elsewhere, which were turning out cooking-vessels gritted with mineral calcite. Such pots included the so-called Knapton ware jars (no. 238), named after a probable production site about 11 km east-north-east of Malton (Corder and Kirk 1932, 96-99). These vessels began to appear in York from the end of the second century, though initially quantities were quite small.

The third century: the York area and the Northern Frontier

During the course of the third century, the workshops of the Humber basin and the hinterland of York not merely benefited from a general increase in the use of pottery by the indigenous population, but ultimately were also able to take advantage of changing supply procedures in the northern frontier zone. Alterations in the imposition of the *annona* may have constituted one factor (A Birley 1981). Moreover, because the East Yorkshire workshops were the nearest significant Romano-British potteries to the Northern Frontier zone, the costs of their wares may have been able to undercut those of the products of other potteries further south. It has been appreciated for some time now that ceramic evidence often constitutes a 'marker' for trade in other products which survive only rarely. The steady expansion of agriculture (perhaps partly in response to the demands of the York garrison), which seems to have occurred in tandem with the Romanisation of the region, almost certainly resulted in the shipment of grain from Yorkshire to the North-East, from at least the late-third or early fourth century, as the evidence at the South Shields supply-base suggests (van der Veen 1994, 251-58; M. van der Veen *pers comm*), and probably earlier. Local pots might well have formed components of mixed cargoes which must have been so frequent in the Roman period.

Another factor which probably contributed to the burgeoning of pottery workshops in the hinterland of the York fortress after Severus may have been changes in recruitment patterns. Caracalla's extension of the citizenship to all free inhabitants of the Empire in *c*.213 (A Birley 1988, 190) would probably have significantly widened the scope of recruitment to *Legio VI*. Hitherto, the legion must inevitably have drawn many (or latterly most) of its recruits from among the sons of serving

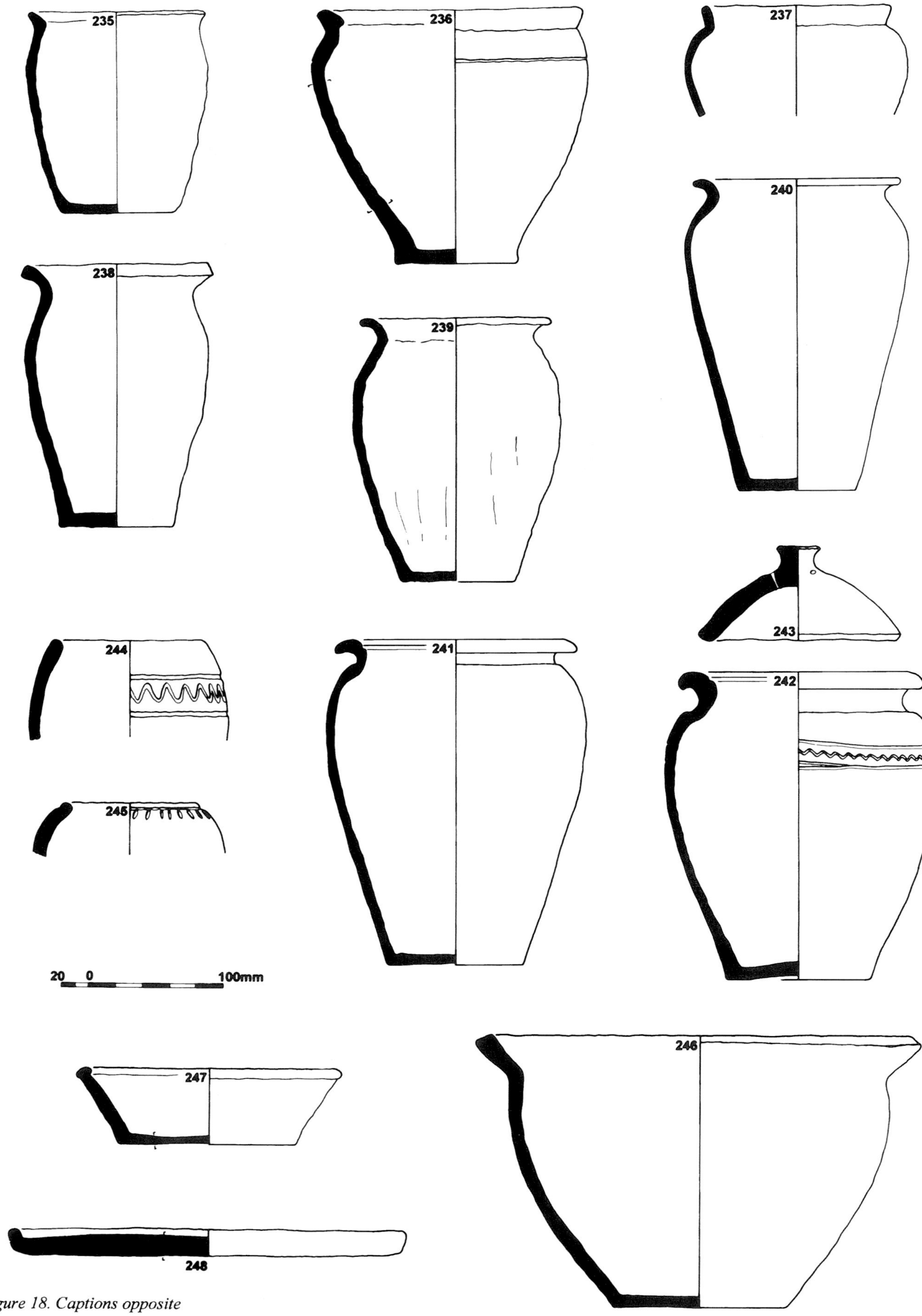

Figure 18. Captions opposite

Figure 18. The development of native indigenous pottery in East Yorkshire.Cooking-vessels tempered with calcitic mineral particles. Many forms can only be dated very approximately, particularly those pre-dating the late third century. Scale 1:4.

235 Handmade jar with low shoulder and simple out-turned rim, from Brough-on-Humber; first or second century (after Corder and Romans 1938).

236 Handmade bowl or jar with thickened everted rim, flattened and undercut on the interior, from Brough-on-Humber; probably first to second century (after Corder and Romans 1937). Occurring on both sides of the Humber, and it is just possible that the profile may have been ultimately influenced by that of the butt-beakers imported into the region before the Roman conquest.

237 Handmade jar, from Hayton fort (after Johnson 1978); the high shoulder and neat, proportionately slim, gently everted rim probably represent a local native imitation of the contemporary Roman vessels used by the garrison; Flavio-Trajanic on this site.

238 Knapton ware cooking-pot: handmade, with characteristic low shoulder and short, blunt-ended gently flaring rim; late first to mid/late third century, from probable production centre at Knapton, near Malton; (after Corder and Kirk 1932).

239 Cooking-pot transitional between the previous earlier Knapton type and the later proto-Huntcliff type (no. 240); the rim, now slim and taping, has begun to hook over; the shoulder of the vessel is more rounded, but the whole jar is still handmade; probably mid/late third century, from York (after Monaghan 1997).

240 Cooking-pot of Proto-Huntcliff type; the more high shouldered body, with a long even taper to the base, is handmade, but unlike no. 239, the hooked rim has been added to the neck of the vessel on a wheel; mid/late third century onwards; from Malton fort vicus (after Bidwell and Croom 1997).

241 Plain Huntcliff-type cooking-pot; the characteristic rim with a lid-seating groove has been added on a wheel to the otherwise handmade vessel; fourth decade of the fourth-century onwards, from Dalton Parlours villa (after Sumpter 1990).

242 Huntcliff-type cooking-pot with incised decoration and a sharply marked shoulder where the grooved rim has been added to the handmade vessel on the wheel; fourth decade of the fourth-century onwards, from Elmswell, East Yorkshire (after Corder 1940).

243 Knapton ware lid with steam-hole, a form could have been used with the lid-seated Huntcliff-type jars; second to fourth century, from Knapton (after Corder and Kirk 1932).

244-5 Late handmade jars with incised or stabbed decoration, in a sand-tempered calcite-gritted fabric; from the signal stations of Scarborough and Filey respectively; fourth century, possibly second half (after Hull 1932).

246 Handmade native copy of S-shaped bowl; (cf. Romanised examples from other centres, nos. 161-2, 206, 227, 258); the rim resembles that of the Knapton cooking-pot; probably third century, from the Langton villa (after Corder and Kirk 1932).

247 Dish, probably ultimately imitating a BB1 type, but the slightly concave wall is characteristic of such native copying; second to fourth century, from Langton villa (after Corder and Kirk 1932).

248 Crude Knapton ware platter of type found in the Yorkshire signal stations; fourth century, possibly second half; from Langton (after Corder and Kirk 1932).

soldiers or veterans, because their citizenship made them eligible to join, and a career in the legion with regular pay and other bonuses was still an attractive proposition. Such men, born and brought up *in castris*, would undoubtedly have had certain pre-conditioned expectations of the quality and range of their ceramic supplies. That the origins of the soldier, whether consumer or potter, had an impact on the character of the pottery requisitioned by, and supplied to a unit, will have already been evident in this paper. With the citizenship, and therefore recruitment to the legion, open to all free men after *c* AD 213, the number of legionaries drawn from the countryside around York must have grown substantially. This increasing new element in the garrison would have been accustomed to a more frequent use of reduced (grey) vessels in coarser fabrics, and fewer classes of vessels in daily use, and this may indeed be exemplified by a general decrease in the demand for tablewares (except for drinking-vessels), a greater emphasis on jars, deep bowls and storage-containers, and a lessening of the requirement for specialist utilitarian forms, mortaria excepted. Ultimately this may well have contributed both to the steady decline in the total repertoire of utilitarian vessel-types made for and supplied to Roman army of the North, and to the burgeoning of potteries making vessels in native fabric traditions. Indeed, from the evidence of the pottery from the York fortress (and from other military sites), the very gradual progress of both phenomena can be charted continuously from the early third century until the army ceased to be paid and most Roman pottery production collapsed in the early fifth century.

The third century: changes in military supply

A hint of changes in the long-distance supply of pottery, that were to become increasingly apparent as the third century progressed, can be glimpsed at Carpow, the Severan (or Commodan to Severan) fortress on the Tay. There, among the small amount of very diverse pottery published, are several forms which probably emanated from East or South Yorkshire or Humberside. For example, the S-shaped bowls (R Birley 1963, figure 8, nos. 1 and 16) are not dissimilar to vessels made in the Holme-on-Spalding Moor and Cantley potteries (Corder 1930b, figure 11: Annable 1960, figure 10, no. 27); the lid-seated jar (R Birley 1963, figure 7, no. 13) has

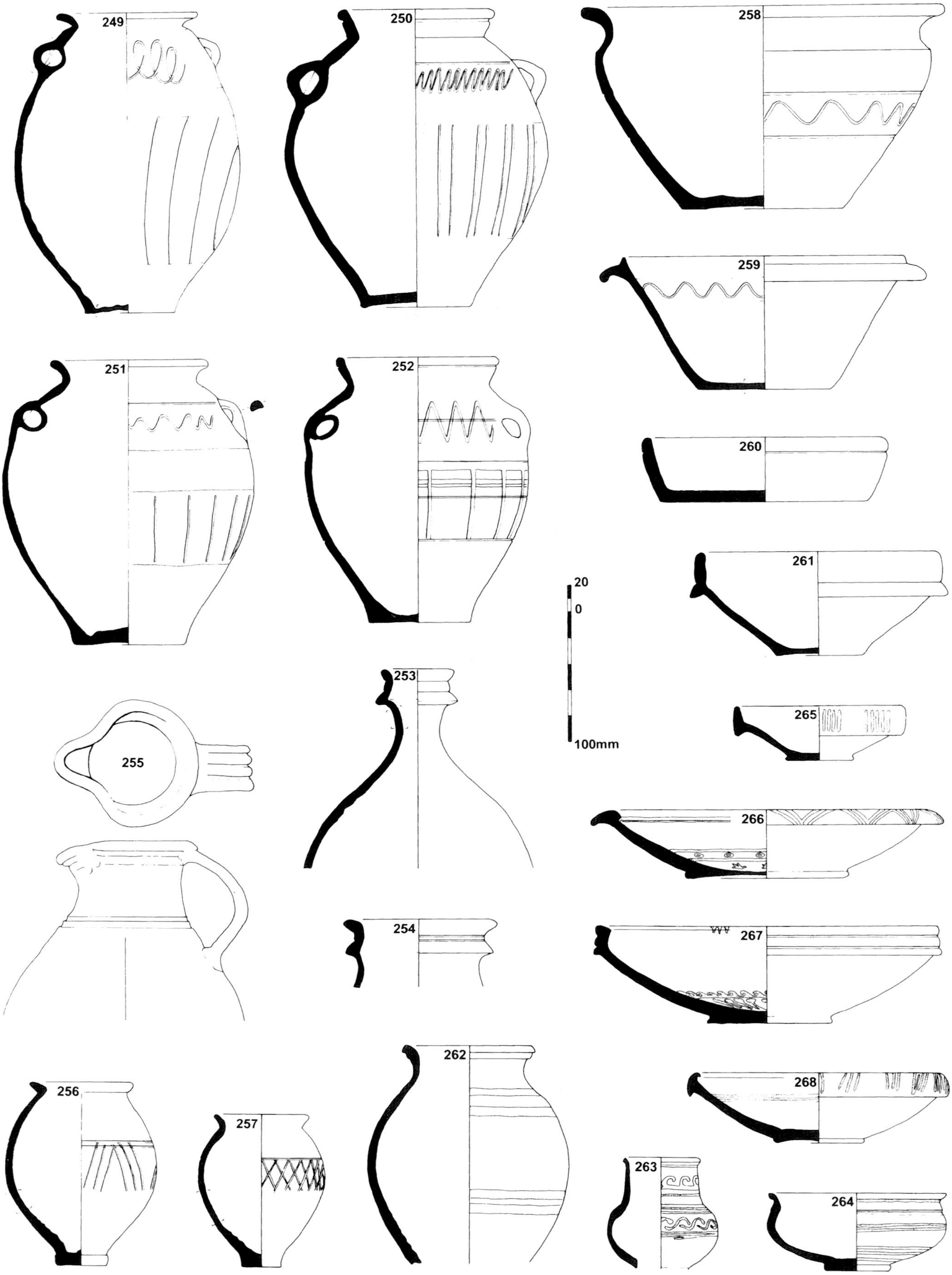

Figure 19. Captions opposite and continued on page 70

Figure 19. The Crambeck pottery industry: culinary vessels and tablewares. Utilitarian vessels in reduced (greyish) ware, and tablewares in oxidised orange ware, white ware or parchment ware (white fabric with painted decoration). Unless otherwise stated, decoration is burnished on the reduced wares, and red/brown painted slip on the parchment wares. Scale 1:4. The reduced, oxidised, and plain white wares mainly date from about 270/280 and the parchment wares from about 340/50, except where indicated. *(Captions continued on page 70)*

249 Reduced ware jar (after Hodgson et al 1938); the form is typologically early in the Crambeck series, characterised by a very sharply everted rim, proportionately tall body, and handles beginning to be recessed into the shoulder of the vessel (incipient counter-sunk handles); from Bewcastle and therefore probably dating to between about 270 and 312/4. Cf. a probable pre-250/270 Norton predecessor with looped handles (no. 219).

250 Reduced ware jar, typologically later than previous being proportionately less tall and characterised by a less sharply everted rim; from Langton (after Corder and Birley 1937, Crambeck Type 3a).

251 Reduced ware jar; as 250, but perhaps slightly later as the handles are almost fully countersunk; from Rudston villa well (after Rigby 1980).

252 Reduced ware jar, one of the latest types in the series, more squat and with a near-upright rim, and handles wholly counter-sunk into the shoulder of the vessel, with no interruption in the external profile; from Scarborough signal station (after Hull 1932).

253 Reduced ware flask with 'bobbin-shaped' flanged mouth, Crambeck Type 14 (after Corder 1928).

254 Reduced ware lagena with bobbin-shaped rim, Crambeck Type 14a (after Corder 1928); cf. the typologically slightly earlier Holme-on-Spalding Moor form (no. 196).

255 Reduced ware jug with simple pinched spout and typologically late slack profile; from Crambeck (after Corder 1928).

256-7 Reduced ware beakers with everted rims; Crambeck Type 11 series (after Corder 1928).

258 Reduced ware S-shaped bowl, Crambeck Type 4 series (after Corder and Birley 1937); cf. examples of this form from other production centres (nos. 161-2, 206, 227, 246).

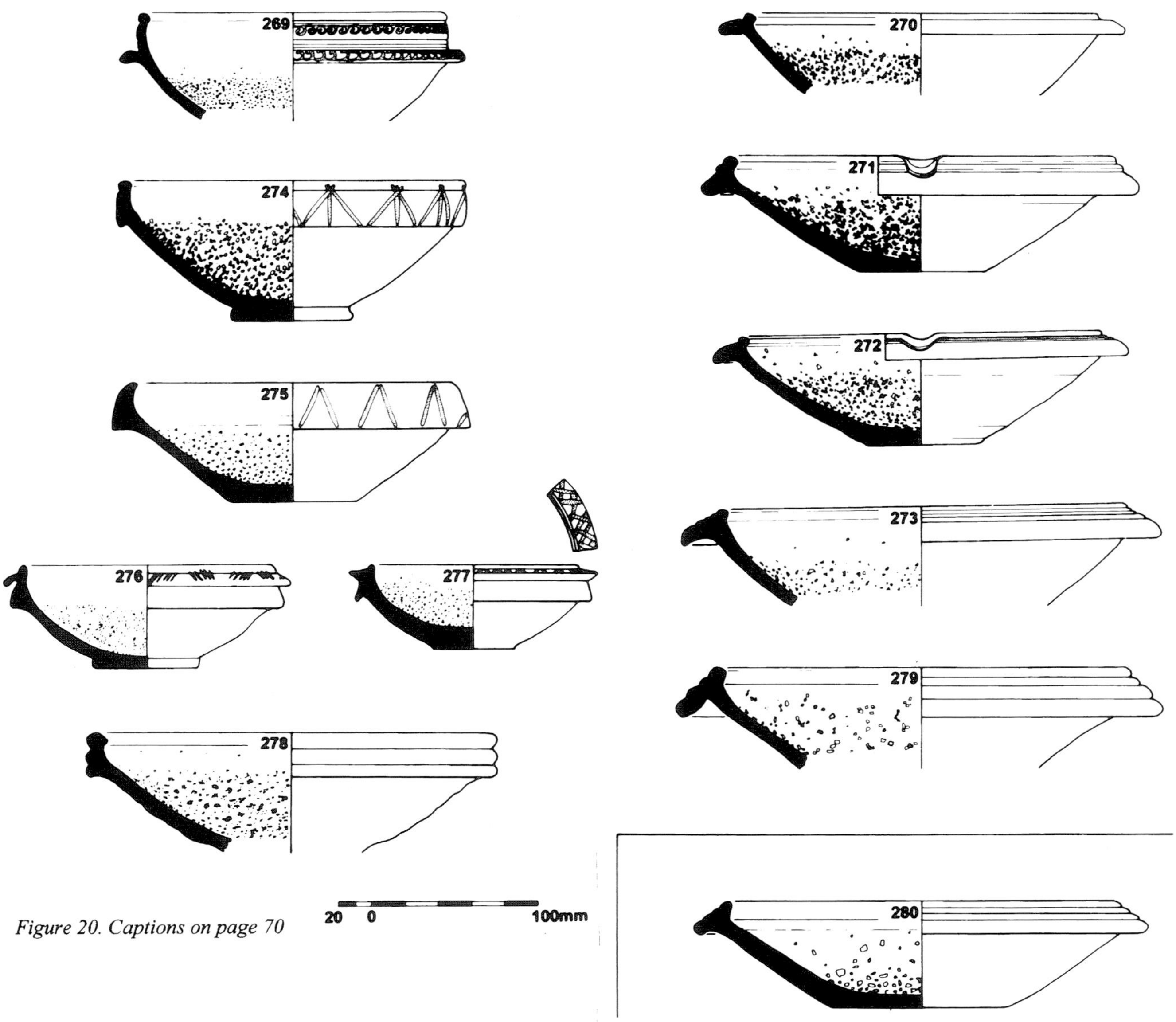

Figure 20. *Captions on page 70*

Figure 19. The Crambeck pottery industry. Captions continued from page 69.

259 *Reduced ware flanged bowl imitating a BB1 form; the internal burnished wavy is a late indicator, suggestive of a date possibly post-350 at least; from Malton fort vicus (after Bidwell and Croom 1997).*
260 *Reduced ware dish, an imprecise imitation of a BB type; Crambeck Type 2a (after Corder and Birley 1937).*
261 *Oxidised orange ware bowl, ultimately an imitation of samian form Drag. 38; Crambeck Type 5a (after Corder 1928).*
262 *Narrow-mouthed jar or flask in parchment ware; from Crambeck (after Corder 1928).*
263 *Funnel-mouthed beaker in parchment ware, with characteristic 'hooked' motif; from York (after Monaghan 1997).*
264 *Shallow bowl in parchment ware (Crambeck Type 13b); from York (after Monaghan 1997).*
265 *Wall-sided bowl in parchment ware; from Malton fort vicus (after Mitchelson 1964).*
266 *Dish in parchment ware (Crambeck Type 10); from Malton fort (after Corder 1930a).*
267 *Bowl in parchment ware (Crambeck Type 9), from Scarborough signal station (after Hull 1932); the form was also used for mortaria.*
268 *Wall-sided bowl in parchment ware (Crambeck Type 7); from York (after Monaghan 1997).*

Figure 20. The Crambeck pottery industry (continued): Mortaria (269-79). All are in white ware or parchment ware (white with red/brown slip-paint); nos. 270-73 probably date from c 270/280, and the remainder from c 340/50. Catterick Vicinity mortarium (280). Scale 1:4.

269 *Mortarium in parchment ware with characteristic hooked motif, from York (Crambeck Type 5b; after Hartley 1995). The form, ultimately copies a Drag. 38 samian bowl.*
270 *Mortarium in Crambeck white ware, one of a group of related forms (nos. 271-3), produced early in the life of the industry, and long referred to as Crambeck Type 6. More recently, Hartley has published a more precise classification with letter subdivisions (1995, Crambeck Type 6a), from York.*
271 *Mortarium in Crambeck white ware (Hartley 1995, Crambeck Type 6b); from York.*
272 *Mortarium in Crambeck white ware (Hartley 1995, Crambeck Type 6c); from York.*
273 *Mortarium in Crambeck white ware (Hartley 1995, Type 6d); from York.*
274 *Wall-sided mortarium in parchment ware (Crambeck Type 7 series); the profile, loosely based on the samian mortarium form Drag.45, was also used for Crambeck bowls; from Crambeck (after Hartley 1995).*
275 *Wall-sided mortarium in parchment ware (Crambeck Type 7 series); cf. an ungritted bowl version (no. 268) from Malton fort vicus (after Mitchelson 1964).*
276-7 *Mortaria with hooked flanges in parchment ware, Gillam's 'General de Gaule' profile, (Crambeck Type 8); from York (after Monaghan 1997); the form was also used for Crambeck bowls.*
278 *Mortarium in Crambeck white ware (Crambeck Type 9); cf. the ungritted bowl version (no. 267); from York (after Hartley 1995).*
279 *Mortarium with slim reeded hammerhead rim in Crambeck white ware; the profile is unusual for a Crambeck product and more reminiscent of mortaria made in the Lower Nene Valley; from York (after Hartley 1995).*
280 *Mortarium in Catterick Vicinity White-slipped ware (orange-buff fabric with thin slip and iron slag trituration grits); the form, fabric and gritting are reminiscent of Cantley products; from South Shields (after Bidwell 1994).*

parallels at Cantley and Norton (Cregeen 1957, figure 3, no. 112; Hayes 1988, figure 54, nos. 16-17). Mention has already been made of the presence of possible Norton wares at Severus' supply-base at South Shields. To be sure, these were not the first Yorkshire or Humberside products to find their way to the Northern frontier zone. A calcite-gritted vessel (presumably of second-century date) is recorded in Turret 34A (Charlesworth 1973, figure 5, no. 5), but must have arrived in the baggage of a soldier who had been transferred or recruited from that region; it need not be viewed as an object of trade. East Yorkshire calcite-gritted wares made their first significant recorded appearance in the Hadrian's Wall zone, at *Vindolanda* fort, from about the mid 230s, and Yorkshire grey wares are present there from the 250s, both in very small amounts, but thereafter steadily increasing in quantity, until, after the late third century, they had become major components in that fort's supply (Bidwell 1985, 73, 178-79). At South Shields, the supply-base for the eastern and central sectors of Hadrian's Wall (and beyond), the picture is similar; a minute trickle of Dales, calcite-tempered and other gritty wares arrived before c AD 235, but the East Yorkshire grey wares did not figure in appreciable quantities until the last quarter of the third century (Bidwell 1994, 224, 232, table 8.10).

AD c 270: a political and economic water-shed?

Changes in the provisioning of the northern frontier, which may also have had a significant effect on the Yorkshire pottery industries, seem to have occurred at in the early 270s (as yet the dating is not precise). At South Shields supply-base, kitchen wares from the Thameside kilns (pottery allied to BB2, from South Essex), which had dominated the supply of the eastern and central sectors of Hadrian's Wall from early in that century, ceased to be imported towards 300 (P Bidwell *pers comm*). In addition, from about the 280s the production and export of mortaria from the Midland workshops at Hartshill/Mancetter went into a rapid decline (K Hartley *pers comm*). Thereafter the bulk of the pottery on the northern frontier was drawn increasingly from producers in East and North Yorkshire. The reasons for this drastic change are unclear. Was the safety of shipping from the Thames Estuary being endangered by pirates and raiders?

Alternatively, had there been significant rises in transport costs, due to changes in the imposition of *portaria* (A Birley 1981)? It is noteworthy that no threat seems to have prevented Nene Valley wares continuing to be shipped (in modest quantities) from the Wash, up the east coast to the Humber and Tyne estuaries, until the late fourth century. Indeed, from the 270s onwards, for several decades, exports of mortaria from that industry were to be more frequent (K Hartley 1994, 210, 213, Table 8.4). Moreover, at about this time, there was an observable increase in the supply of BB1 to the Northern Frontier (probably shipped up both the west and the east coasts, the latter utilising London as an entrepôt), though it had been near-absent from the central and eastern sectors of Hadrian's Wall since early in the third century (Bidwell 1985, 173, table vii; Bidwell 1994, 223-24). This in turn had a direct effect on the forms of the Black-burnished ware imitations produced by many workshops in Yorkshire (eg. nos. 208, 259). It can hardly be pure chance, that at this time of radical re-orientation in the Northern Frontier's sources of pottery-supply (and by implication probably the supply of other commodities), there were serious troubles in adjacent provinces of the western Empire (the collapse of the Gallic Empire and barbarian invasions in Gaul). All these events were also approximately coincident with an important change in the nature of settlement along Hadrian's Wall and most of the North. In the early 270s, the *vici* attached to forts on Hadrian's Wall, for which the numismatic evidence is known (excluding Corbridge and Carlisle which had become towns) were evidently abandoned. Elsewhere in the North-West and North-East (north of the Tees), few *vici* seem to have survived much into the fourth century (Bidwell 1991, 14). It has always been assumed that *vici* played a significant role in the supply of forts. Thus their demise, almost *en bloc* must surely have stemmed from some unified military plan (Bidwell 1999, 29-30). A concern to bring the *annona* into the protection of fort defences seems to be implicit in at least one documentary source of this general period from Egypt (Bell 1962, 73), but it is not known whether the policy applied equally to Britain. Bidwell has suggested (1991, 12) that the towns proper, which had developed in the military zone (eg. Corbridge, Carlisle, and Catterick), now began to play a significant part in its provisioning. Noteworthy too at this period, is the marked contrast between the area receiving ceramic supplies (i.e. the northern parts of the military zone) and the increasing affluence of the general region further south in which these, and perhaps other basic commodities, were produced. The *vicus* of Malton fort, far from being abandoned, saw a period of lavish rebuilding, beginning in the early fourth century (Wenham and Heywood 1997); many of the villas between York and the Tees too, seem to have prospered in the later third and fourth centuries. Possibly some sort of centralised (state) policy underlay the change in the sources of pottery supplies to the frontier, and the increasing dependence on a mere handful of manufacturers and distributors in the nearby towns, which was based purely on economic and logistical factors, cheaper transport costs (perhaps shared with commodities grown in the agricultural hinterlands of the York fortress and with supplies destined for civilian markets in the towns), and, in particular, the ability of the Yorkshire potters for the first time to satisfy the ceramic demands of the military.

Late third to fourth century 'native' production

Whatever the underlying causes, there are clear indicators, towards the late third century, of increased commercialisation and streamlining in the indigenous potteries of East Yorkshire, which were making mineral calcite-gritted wares. The simple everted rims of the Knapton-type cooking-pots (no. 238), which had hitherto been handmade like the rest of the vessel, were gradually superseded by hooked rims, added to the shoulder of the vessel at first by hand (no. 239), but soon on a potter's wheel (no. 240), though the body of the jar remained hand-formed. These new forms now seem to have emanated not from the Knapton area, but from a production centre further to the east or north-east, possibly at the eastern end of the Vale of Pickering (Evans 2000, 40, fig 26). Early stages in this change of profile are exemplified by several jars from the lowest level in the well of the villa at Rudston (East Yorkshire), a deposit with a *terminus post quem* of *c* 293. Here the rims of such vessels were already wheel-made, but as yet, were all only tending towards a hooked profile (Rigby 1980, figures 49-50, nos. 254, 257-60). However, the forms with fully hooked rims, usually termed proto-Huntcliff type jars (no. 240), must have been fully developed and common well before 312/4, the date that the outpost forts of Hadrian's Wall seem to have been abandoned (Casey and Savage 1980), since a number of examples are recorded at Bewcastle outpost fort (Hodgson and Richmond 1938, figure 27, nos. 70, and 72; Austen 1991, figure 12, no. 103; Gillam *et al* 1993, figure 21, nos. 43-5). These were the precursors of the ubiquitous Huntcliff-type cooking-pots and bowls with the characteristic lid-seating groove (nos. 241-2). Incidentally, the use of the term Huntcliff ware (as opposed to Huntcliff type) for these forms is erroneous, since the Huntcliff signal-station was merely the place where this type of vessel was first distinguished, not where it was made. The evidence for the date of the emergence of the fully-fledged (ie. lid-seated) Huntcliff-type cooking-pots is still subject to debate. A single example of the form occurred at Bewcastle, but might conceivably have reached the site after the departure of the garrison in 312/4. Recent excavations elsewhere have now shown that these jars were possibly current by 340/350, a date 20 to 30 years earlier than previously assumed. The evidence comes from the Hadrian's Wall fort of Birdoswald, from a sealed context under a re-laid granary floor which was securely dated by 27 coins which ended in 348 (Wilmott 1997, 203-08). In this deposit, jars of proto-Huntcliff type (as no. 240; Gillam Type 161) were exceeded by the Huntcliff-types (as nos. 241-2; Gillam Type 163), by 93% to 118% (on the basis of rim percentages), suggesting that the latter were already more popular before the middle of the fourth century (Hird and Perrin 1997, 247-49, figure 167). The evidence is corroborated by the presence of large numbers of Huntcliff-type jars at the Dalton Parlours villa, where the coins suggest a cessation of occupation by 370, if not slightly earlier (Sumpter 1990, 135-46). Nevertheless, there must have been a very substantial overlap in the production of these two profiles, since cooking-pots of proto-Huntcliff type were still present, though 'uncommon' (Hull 1932, 242-43) in the Yorkshire signal stations. Recent work (Ottaway 2000) has now confirmed Casey's (1979) suggestion that this defensive system was not established until 383/7 or later, not, as long assumed, after the Picts' War of 367 (once a chronological benchmark in the dating of pottery). The Huntcliff-type jars, together with other calcite-gritted vessels such as dishes, platters, bowls and lids (nos. 243, 246-8; Hull 1932, nos. 242-9), were to steadily increase their share of the market, until, by the end of the fourth century, they were the most common type of pottery in Northern Britain (Monaghan 1993, 744). A few other minor potteries existed, probably elsewhere in East Yorkshire, and working in generally similar traditions, but with slight variations in their fabrics. These included a late sandy, burnished, calcite-gritted ware, exemplified by the lattice-decorated jar (Gillam Type 164), but it never seems to have achieved more than a relatively small share of the market beyond the East Yorkshire signal-stations and neighbouring territory (Evans 1996a, 71-77; 2000, 41, fig. 29).

Late third to fourth century: Catterick

Sometime in the mid to late third century there was a brief revival of mortarium production at Catterick or its vicinity (no. 280; Tomber and Dore 1998, 195). The workshop is so far unlocated and how widely it traded also remains to be established. There are difficulties in distinguishing some of its products from mortaria made at Cantley (South Yorkshire), but probable examples occur at South Shields (K Hartley 1994, 210, 213). The recent discovery of a kiln at Catterick, probably dating from the late third or early fourth century, provides evidence for the production of close imitations of Black-burnished ware (BB1), presumably mainly for this increasingly prosperous town and late military establishment, and generally not traded very far (Evans 1996b; 2000, 40).

Late third to fourth century: Crambeck

Another major development in the Malton region, which also occurred in about 270 or soon after (the dating is not precise as yet), was the emergence of a new pottery industry in the area around Crambeck, about 7 km south-west of Malton and Norton (Corder 1928; Corder and Birley 1937). Its main products comprised fine quality kitchen wares, in a 'gun-metal' grey with a paler, often almost whitish, core (nos. 249-60), and small quantities of orange-red oxidised vessels, mostly imitations of late samian forms, such as Drag. 38 (no. 231). In addition, however, because of the proximity of a seam of the rarer, whitish Oxford clay with very low iron content, as well as the local boulder clay with relatively high iron content, Crambeck was able to produce white mortaria (nos. 269-79) and (particularly later) fine white tablewares (nos. 262-8). In fact the ubiquitous pale grey fabrics were also the result of mixing both clays (Evans 1989, 54-56).

It has been assumed by some that the Crambeck workshops had been set up by craftsmen from the Norton potteries, replacing the latter with little chronological overlap, and that disparities in some of the products were the result of chronological differences in the industries. However, the vessels held in common between the two centres, such as the plain-rimmed dishes and flanged bowls imitating Black-burnished ware types (nos. 259-60), S-shaped bowls (no. 258) and bobbin-shaped flasks (no. 254), mostly comprise forms which were widely produced, either province-wide or in the Humber basin and East Yorkshire, and are not necessarily indicative of a link. It now seems possible that the earlier phase of production at Crambeck may have closer links than have hitherto been appreciated with the traditions of the Holme-on-Spalding Moor group of potteries, particularly those exemplified by the third-century kilns at Hasholme (Hicks and Wilson 1975), Bursea House (Halkon and Millett 1999) and Skiff Lane (excavations by R Mackey in 1996; R Mackey *pers comm*). Indeed, there seem to have been as many (possibly more) types held in common between this industry and Crambeck, as there are between Norton and Crambeck (eg. nos. 206/258, 2-8/259, 200-201/249-51, 196/254, 203/264). Moreover, a recent examination by the author, of the clay debris and kiln furniture from Hutton Hill, a newly discovered, probably relatively early, Crambeck grey ware production site, has shown for the first time, that the internal structural details of Crambeck kilns (as well as their elongated shape), were more like those at Hasholme, than those recorded at Norton (Swan and Evans forthcoming). Potters in antiquity in general were always more conservative in their kiln-technology than in the choice of their ceramic repertoire. Kiln structures may, therefore, be more indicative of the intrinsic traditions of a workshop and its connections with other potteries, than are its actual products.

At first, Crambeck's output appears to have been limited, both in the range of vessel types and in quantities exported. The early mortarium profiles were relatively simple 'bead and flange' affairs (nos. 270-73; Crambeck Type 6 series), imitating the products of other industries, such as Hartshill/Mancetter, the Nene Valley, and perhaps Cantley. Vessels in grey Crambeck ware appear at South Shields soon after about 286, but at first the amounts were far from large (P Bidwell *pers comm*). There seems to have been a gradual increase in production and distribution, until by the 340s, these potteries became a dominant force in ceramic supplies to Yorkshire and the northern frontier, eventually wiping out all competition from the East Yorkshire grey ware industries (such as the Holme-on-Spalding Moor grouping), the Dales ware cooking-jars, the Cantley mortaria, as well as the kitchen wares from the Dorset BB1 workshops. These last had already shown a significant decline in the early fourth century, initially in the central and eastern sectors of Hadrian's Wall, and in Yorkshire (eg. Vindolanda and Castleford), but by the mid fourth century had even lost their major markets in the North-West. After about 370, BB1 was shipped no further than Carnarfon (*Segontium*), North Wales (Webster 1993, 252-53), resulting in the complete dominance of East Yorkshire calcite-gritted products and Crambeck grey wares in the military assemblages of the North-West, as well as the North-East. Indeed, the distinctive, typologically latest products of the BB1 workshops, which post-date the middle of the fourth century in south-western Britain, are totally absent in the whole of the North (Holbrook and Bidwell 1991,

94).

Changes in the Crambeck repertoire, c 340/350

As Crambeck wares became dominant and widespread, so the industry seems to have diversified and upgraded its products. This now seems to have happened by at least the 340s, though hitherto it was thought to have occurred in about 367 (the date of the 'Picts' War'). The range of Crambeck white ware forms was now enormously increased to include tableware forms, their surface finish was improved by slipping and careful polishing, and red-painted (parchment ware) decoration became common (nos. 262-8); new mortarium forms too were produced in quantity and existing forms began to be decorated (nos. 269-78). Again the Birdoswald granary deposit provides the earliest dating so far for painted parchment wares (Hird and Perrin 1997, 247-9), and this earlier chronology may also be supported by the evidence from the Dalton Parlours well (Sumpter 1990, 244). Can it be pure chance that the emergence of the lid-seating groove on Huntcliff-type cooking-pots (nos. 241-2) coincided very approximately with the upgrading of the Crambeck repertoire? Whether or not these marked changes in the range of products made by the Crambeck potteries indicate the arrival of new potters from outside the region, or even outside Britain, cannot yet be determined for certain. The newly introduced 'parchment ware' and other vessels shared relatively few forms with the other major contemporary Romano-British potteries, such as the New Forest and Oxfordshire centres. Most of the forms are universally popular types, such as copies of late samian, including Drag. 38s (no. 261), and also shallow S-shaped bowls (no. 264), and funnel-mouthed beakers (no. 263). Indeed at least one not very common Oxfordshire parchment ware dish (Young P16) was probably itself a copy of a Crambeck dish (Crambeck Type 9; no. 267). Evans (1989, 52) concluded that there was insufficient evidence to postulate a link with the Nene Valley potteries, though K Hartley (1995, 313) has drawn attention to mortaria typologically very similar to Lower Nene Valley products. More recently , Dövener (2000, 136, Abb. 228) has drawn attention to a jug in Crambeck parchment ware, with an applied female face on the neck. Details of the hair in particular suggest the mould used for the face was apparently the same as that employed in the Lower Nene Valley for the manufacture of similar face-jugs. Such Crambeck forms are, however, extremely uncommon and a trade in moulds seems more probable than the transfer of a potter. In fact, more parallels to common Crambeck forms can be found in the repertoire of the Argonne industry, North France, than in workshops elsewhere in Roman Britain. There are squat S-shaped bowls with foot-rings (Argonne pl. xiii, nos. 316a/b = Crambeck Type 13), flagons with a bobbin-mouth, some with closely similar body proportions (Argonne figure 40, no. 5 = Crambeck Type 14a); and a variety of very shallow almost wall-sided conical dishes with foot-rings and straight gently-sloping bottoms (Argonne pl. xii, nos. 305-9, pl xxv, no. 1, xxvii, nos. 236, 285 = Crambeck I, Forms 54-8 & 62; Crambeck Type 9). The last basic vessel types and their variants are particularly prominent in both the Argonne and the Crambeck industries, but rare elsewhere in Britain. However, there are some significant absences of obvious parallels to Crambeck forms and detailing. Some of the most popular painted motifs on Crambeck vessels, such as the repetitive hanging hooks (nos. 263, 267, 269), are not known elsewhere. Equally idiosyncratic is the common Crambeck mortarium form (Gillam Type 290, his so-called 'General de Gaule' profile!; nos. 276-7). There is clearly scope for more research on the affinities of the products. A full and detailed corpus of all known forms and their variants, with their corresponding decorative motifs, and an indication of their dated contexts should be an aim for the future. Now that the life-span of the production of Crambeck tablewares has been extended so radically, there ought to be evidence of changes in the morphology or decoration of the products during that period. Already it seems (from the Birdoswald evidence) that the burnished wavy-line, present on the interior of some Crambeck grey ware flanged bowls (no. 259), may not have become current until some time after the marked diversification of Crambeck forms, perhaps not until the 360s or 370s.

Between the 340s and the final collapse of the major Romano-British pottery industries engaged in long-distance trade, presumably early in the fifth century when coinage ceased to circulate, the Crambeck potteries and the workshops making Huntcliff-type jars and other calcite-gritted vessels not only increasingly excluded virtually all other competing products from Northern Britain (north of the Humber), but their wares even reached North Wales (Casey *et al* 1993, 316, figure 17.29. no. 859). It is difficult to see this as the sudden result of a military contract, as has previously been suggested (Evans 1989, 78-80). Even so, it is not impossible that changes in the sources of supplies to the military, albeit less sudden in their execution than was once assumed, could have been made within the context of some sort of official policy-guidelines. On the other hand, the overwhelming supremacy of these two industries in the north, and the apparent contraction or collapse of so many other production centres should also be viewed against the ceramic background of Roman Britain as a whole. It has been clearly shown (Swan 1984, 13, 19), that in the wholly civilian areas of Roman Britain (the Midlands, and the South and the East of England), there was a marked decline in the total number of kiln sites in the course of the fourth century, and that a very small number of major centres, making both fine and coarse wares, came to dominate the pottery supply of huge regions.

There is undoubtedly a great need for the publication of many more dated excavated ceramic sequences (both from deeply stratified urban sites, such as York, and from military sites excavated under research conditions, such as South Shields), and for further work on Yorkshire kiln sites supplying the North, but it is instructive to reflect on this paradox. At the start of the Roman occupation, the native pottery producers of Yorkshire could barely make any contribution to military supply; by the twilight of Roman Britain, the whole of the North was dependent on them!

ACKNOWLEDGEMENTS

The information in this paper summarises the results of the work of a number of researchers in Roman pottery, too many to acknowledge individually, but their contributions will be evident from the bibliography; here, and in the more recent excavation reports, can be found much quantitative data underpinning the author's discussion. However, I would like to thank those who provided information on their unpublished work: Ian Betts (York tiles), Paul Bidwell (South Shields and Wallsend excavations), Maggi Darling (Brough-on-Humber pottery production), Jeremy Evans (pottery from the Crambeck kiln site at Hutton Hill), Kay Hartley (Bearsden and other mortaria), Brenda Heywood (York Minster Excavation Archive), Louise Hird and Andrew Fitzpatrick (Piercebridge pottery), Edna King (Ebor ware from the Borthwick Institute kiln dumps, York), Ian Lawton (pottery from Scoreby, Stamford Bridge and Apple Tree Farm, York), Rod Mackey (Skiff Lane kiln, Holme-on-Spalding Moor), and Paula Ware (1996 excavations at King's Pool, Peaseholme Green, York). I am also grateful to those who provided the author with access to the museum collections: Lindsay Allason-Jones (Mus of Antiq, Newcastle Univ: Turret 26a pottery), John Pickin (formerly Bowes Museum: Piercebridge pottery), Alison Sheridan and Fraser Hunter (Royal Nat Mus Scotland, Edinburgh: Carpow pottery from R Birley's excavations) and to colleagues at the Northern Frontier Seminars for their stimulating discussions. Above all, Paul Bidwell has been a constant source of information, advice and encouragement, and kindly commented on a late version of the paper to the author's great benefit. The author, however, must bear the responsibility for errors and misconceptions that remain.

The first version of this paper was originally submitted to the editors in October 1995 (without illustrations, since the author was awaiting cataract surgery). It was revised in 2000 to take account of newly published chronological information, particularly from the Northern frontier, and to update the bibliography. However, the bulk of the paper is as originally submitted, and it was not until the final stages of revision that the author discovered that her tentative conclusions on the origins of 'parisian wares' (developed over a period of years during her annual attendances at SFECAG pottery congresses in France), had also been reached quite independently by Valery Rigby (formerly British Museum) in the course of her soon to be published study of the Rossington Bridge 'parisian' kiln products (Buckland *et al* 2001). No alterations to the original 1995 content of author's paper have been made as a result of Rigby's work.

The illustrations follow the conventions established by John Gillam (Gillam Types), whereby a line resembling a square bracket is inserted in the drawn section of each vessel where the profile has been restored, its open end turned towards the restored portion. All such restorations are based on more complete examples of analogous types.

BIBLIOGRAPHY AND ABBREVIATIONS

Anderson, A C, and Anderson, A S, (eds) 1981 *Roman Pottery Research in Britain and North-West Europe: Papers Presented to Graham Webster.* Oxford: Brit Archaeol Rep, Int Ser **123**

Annable, F K, 1960 *The Romano-British Pottery at Cantley Housing Estate Doncaster. Kilns 1-8.* Doncaster: Mus Pub **24**

Argonne 00: illustrated pottery in Chenet, G, 1941 *La Céramique Gallo-Romaine d'Argonne de IV Siècle, et la Terre Sigillée décorée à la Molette,* Fouilles et Documents d'Archéologie antique en France, Maçon.

Austen, P S, 1991 *Bewcastle and Old Penrith: a Roman Outpost Fort and a frontier vicus.* Kendal: Cumberland Westmorland Antiq Archaeol Soc Res Ser **6**

Bell, H I, (ed) 1962 *The Abbinaeus Archive*, Oxford.

Betts, I M, 1985 *A scientific investigation of the brick and tile industry of York to the mid-eighteenth century.* Unpub PhD Thesis, Univ Bradford

---- , 1998 The Roman tile. In H E M Cool and C Philo (1998), *Roman Castleford, Excavations 1974-85, I, the Small Finds.* Wakefield: Yorks Archaeol **4**, 225-32

Bidwell, P T, 1985 *The Roman Fort of Vindolanda, at Chesterholm, Northumberland.* London: Hist Building Monu Comm Archaeol Rep 1

----, 1991 Later Roman barracks in Britain. In V A Maxfield, and M J Dobson (1991), *Roman Frontier Studies 1989, Proceedings of the XVth International Congress of Roman Frontier Studies (Canterbury 1989).* Exeter: Univ Press

----, 1994 The Pottery. In Bidwell and Speak (1994), 206-242

---- , (ed) 1999 *Hadrian's Wall 1989-1999: a summary of recent Excavations and Research prepared for the Twelfth Pilgrimage of Hadrian's Wall 14-21 Aug 1999.* Kendal

---- , and Croom, A T, 1997 Coarse Wares. In Wenham and Heywood (1997), 61-103

---- , Snape, M E, and Croom, A T, 1999 *Hardknott Roman Fort, Cumbria, including an account of the excavations by the late Dorothy Charlesworth.* Kendal: Cumberland Westmorland Antiq Archaeol Soc Res Ser **9**

---- , and Speak, S, 1994 *Excavations at South Shields Roman Fort Vol 1.* Newcastle: Soc Antiq Newcastle upon Tyne Mono Ser **4**

Bird, J, (ed) 1998 *Form and Fabric, Studies in Rome's Material Past in Honour of B R Hartley.* Oxford: Oxbow Mono **80**

Birley, A R, 1971 *VI Victrix* in Britain. In Butler (1971), 81-96.

---- , 1981 The economic effects of Roman Frontier Policy. In A King and M Henig (1981), *The Roman West in the Third century: contributions from Archaeology and History.* Oxford: British Archaeol Rep, Int Ser **109**, 39-53

---- , 1988 *The African Emperor Septimius Severus.* London: Routledge

---- , 1997a Supplying the Batavians at Vindolanda. In Groenman-van Waateringe *et al* (1997), 273-80

---- ,1997b *Hadrian: the restless emperor.* London: Routledge

Birley, R E, 1963 Excavations of the Roman Fortress at Carpow, Perthshire, 1961-2. *Proc Soc Antiq Scotland,* **96**, 184-207

Bishop, M C, 1996 *Finds from Roman Aldborough: a Catalogue of*

Small Finds from the Romano-British town of Isurium Brigantium, Oxbow Mono **65**, Oxford

Bogaers, J E, 1967 Die Besatzungstruppen des Legionslagers von Nijmegen im 2. Jahrhundert nach Christus. In *Studien zu den Militärgrenzen Roms, Vortrage des 6. Internationalen Limeskongresses in Süddeutschland.* Köln/Graz: Beihefte der Bonner Jahrbucher **19**, 54-76

---- , and Haalebos, J-K, 1977 Die Nijmegener Legionslager seit 70 nach Christus. In D Haupt and H G Horn (1977), *Studien zu den Militärgrenzen Roms II*, Köln, 93-108

Breeze, D J, 1977 The fort at Bearsden and the supply of pottery to the Roman army. In Dore and Greene (1977), 133-45

Brunsting, H and Steures, D C, 1997 The brick stamps and the occupation history of the legionary fortress at Nijmegen. In Groenman-van Waateringe *et al* (1997), 324-29

Buckland, P C, 1976 A Romano-British pottery kiln site at Branton, near Doncaster, *Yorks Archaeol J* **48**, 69-82

---- , 1986 *Roman South Yorkshire: A Source Book.* Sheffield: Dept Archaeol Prehist, Univ Sheffield

---- , and Dolby, M J 1980 *A Roman pottery kiln site at Blaxton Quarry, Auckley.* Archaeol Doncaster **4.1**. Doncaster: Museum

---- , and Magilton, J R, 1986 *The Archaeology of Doncaster 1. The Roman Civil Settlement.* Oxford: Brit Archaeol Rep, Brit Ser **148**

---- , Hartley, K F and Rigby, V, 2001 The Roman Pottery Kilns at Rossington Bridge: excavations 1956-1961. *J Roman Pottery Stud* **9**, 1-96

---- , Magilton, J R, and Dolby, M J, 1980 The Roman Pottery Industries of South Yorkshire: a Review. *Britannia,* **11**, 145-64

Buisson, J-F, 1991 Les décors à la molette des céramiques communes de Poitou-Charentes. *Société Française d'Étude de la Céramique Antique en Gaule. Actes du Congrès de Cognac, 8-11 Mai 1991, 33-9.*

Butler, R M, (ed) 1971 *Soldier and Civilian in Roman Yorkshire: essays to commemorate the nineteenth centenary of the foundation of York.* Leicester: Univ Press

Cam. 00: vessel types, in C F C Hawkes, and M R Hull, 1947 *Camulodunum*. London: Rep Res Comm Soc Antiq **14**

Caruana, I D, 1992 Carlisle: Excavation of a section of the Annexe Ditch of the First Flavian Fort, 1990. *Britannia*, **23**, 45-109

Casey, P J, 1979 Magnus Maximus in Britain: a reappraisal. In P J Casey (ed) (1979), *The End of Roman Britain: Papers arising from a conference, Durham 1978.* Oxford: Brit Archaeol Rep, Brit Ser **71**, 66-79

---- , 1987 The coinage of Alexandria and the chronology of Hadrian. In P Huvelin, M Christol, and G Gautier (eds) (1987), *Mélanges de Numismatique Offerts à Pierre Bastien à l'Occasion de son 75ème Anniversaire.* Wetteren, Belgium 65-72.

---- , Davies, J L, with Evans, J, 1993 *Excavations at Segontium (Caernarfon) Roman Fort, 1975-1979.* London: CBA Res Rep **90**

---- , and Hoffmann, B, 1995 Excavations at Alstone Cottage, Caerleon, 1970. *Britannia*, **26**, 63-106

---- , and Savage, M, 1980 The coins from the excavations at High Rochester in 1852 and 1855. *Archaeol Aeliana*, 5th ser **8**, 75-87

Charlesworth, D, 1973 A re-examination of two turrets on Hadrian's Wall. *Archaeol Aeliana*, 5th ser **1**, 97-99

CIL = Corpus Inscriptionum Latinarum. Berlin

Congreve, A L, 1938 *A Roman and Saxon site at Elmswell, East Yorkshire. Second Interim Report, 1937.* Hull: Mus Pub **198**

Conspectus 00 = forms of Italian-type sigillata, in E Ettlinger, B Hedinger, *et al,* 1990 *Conspectus Formarum Terrae Sigillatae Italico Modo Confectae.* Frankfurt am Main: Römisch-Germanische Kommission des Deutschen Archäologischen Instituts, Materialen zur Römisch-Germanischen Keramik **10**

Cool, H E M, Jackson, C M, and Monaghan, J, 1999 Glass-making and the Sixth Legion at York. *Britannia*, **30**, 147-61

Corder, P, 1928 *The Roman Pottery at Crambeck, Castle Howard.* York: Roman Malton District Rep **1**

---- , 1930a *The defences of the Roman Fort at Malton.* Leeds: Roman Malton District Rep **2**

---- , 1930b *The Roman Pottery at Throlam, Holme-on-Spalding Moor, East Yorkshire.* Leeds: Roman Malton District Rep **3**

---- , 1935 Excavations at Brough, East Yorkshire. Second Interim Report, 1934. *Trans East Riding Antiq Soc*, **28**, 1-35

---- , 1940 *Excavations at Elmswell, East Yorkshire*, 1938. Hull: Univ College

---- , 1950 The date of the Norton kilns. In Hayes and Whitley (1950), 35-37

---- , 1958 Parisian Ware. *Yorkshire Archaeol J*, **39**, 48-52

---- , and Birley, M, 1937 A Pair of Fourth-Century Romano-British Pottery Kilns near Crambeck. *Antiq J*, **17**, 392-413

---- , and Kirk, J L, 1932 *A Roman Villa at Langton near Malton, East Yorkshire.* Leeds: Roman Malton and District Rep **4**

---- , and Romans, T, 1937 Excavations at Brough, E. Yorkshire. Fourth Interim Report, 1936. *Trans East Riding Antiq Soc,* **28**, 84-153

---- , and ---- , 1938 Excavations at Brough-Petuaria. Fifth Report, 1937, *Trans East Riding Antiq Soc*, **28**, 173-239

Crambeck I, Forms 00 = Crambeck kiln products in Corder 1928

Crambeck Type 00 = Crambeck ware type-series in Corder and Birley 1937

Cregeen, S M, 1957 The Romano-British Excavations at Cantley

Estate, Doncaster: the Pottery from Kilns 9-25. *Yorks Archaeol J*, **39**, 364-88

Croom, A T, and Bidwell, P T, 1998 The pottery. In P J Casey and B Hoffmann (1998), Rescue Excavations in the Vicus of the Fort at Greta Bridge, Co. Durham, 1972-4. *Britannia*, **29**, 161-80

---- , McBride, R M, and Swan, V G, *forthcoming* The Roman pottery from excavations at Stamford Bridge in 1998

Darling, M J, 2000 The Roman pottery. In K Hunter-Mann with M J Darling and H E M Cool. Excavations on a Roman extra-mural site at Brough-on-Humber, East Riding of Yorkshire, U.K. *Internet Archaeol* **9** (Dec 2000), sect 4 (= http//intarch.ac.uk/journal/issue9/hunter-mann/section 4.html).

Dickinson, B M, and Hartley, K F, 1971 The evidence of Potters' Stamps on samian ware and on mortaria for the trading connections of Roman York. In Butler (1971), 127-42

Digest = Justinian I (Emperor of the East), *Corpus Iuris Civilis.*

Dobson, B, and Mann, J C, 1973 The Roman army in Britain and Britons in the Roman army. *Britannia*, **4**, 191-05

Dore, J N, 1983 The Coarse ware. In R Miket, *The Roman Fort at South Shields: Excavation of the Defences 1977-1981.* Tyne and Wear Museums, (undated but probably 1983), 61-107

---- , and Greene, K, (eds) 1977 *Roman Pottery Studies in Britain and Beyond: Papers Presented to John Gillam, July 1977.* Oxford: Brit Archaeol Rep, Supp Ser **30**

Dövener, F, 2000 *Die Gesichtskr üge der römischen Norfwestprovinzen.* Oxford: Brit Archaeol Rep, Int Ser **870**

Elsdon, S M, 1982 *Parisian ware: a Study of Stamped Wares of the Roman Period in Lincolnshire, Humberside and South Yorkshire.* Highworth: Vorda Res Ser **4**

Evans, J, 1989 The Development of a Major Northern Pottery Industry. In P R Wilson 1989, *The Crambeck Roman Pottery Industry.* Leeds: Yorkshire Archaeol Soc Roman Antiq Sect, 43-90

---- , 1995 Later Iron Age and 'native' pottery in the north-east. In B Vyner (ed) (1995), *Moorland Monuments: studies in the Archaeology of north-east Yorkshire in honour of Raymond Hayes and Don Spratt.* York: CBA Res Rep **101**, 46-68

---- , 1996a The Roman Pottery. In D S Neal 1996, *Excavations on the late Roman Villa at Beadlam, Yorkshire*, Leeds: Yorks Archaeol Rep **2**, 69-92

---- , 1996b The pottery. In P A Busby, J Evans, J P Huntley, and P R Wilson (1996), A Pottery Kiln at Catterick. *Britannia*, **27**, 288-93

---- , 2000 The end of Roman pottery in the North. In T Wilmott and P R Wilson (eds) (2000), *The Late Roman Transition in the North: Papers from the Roman Archaeology Conference, Durham 1999.* Oxford: Brit Archaeol Rep, Brit Ser **299**, 39-46

Filtzinger, P, 1972 *Novaesium V: Die Römischen Keramik aus dem Militärbereich von Novaesium.* Berlin: Limesforschungen Band II

Fitzpatrick, A P, and Scott, P R, *forthcoming Report on the Excavations by the late Peter Scott on the Roman sites at Piercebridge, County Durham*

Frere, S S, 2000 M Maenius Agrippa, the *Expeditio Britannica* and Maryport. *Britannia* **31**, 23-28.

Gillam, J P, 1951 Dales Ware: a distinctive Romano-British Cooking-pot. *Antiq J*, **31**, 154-64

---- , 1968 Coarse Ware. In L P Wenham (1968), *The Romano-British Cemetery at Trentholme Drive, York.* London; Min Pub Buildings Works Archaeol Rep **5**, 56-86.

Gillam Type 00 = vessel types in J P Gillam 1970, *Types of Roman Coarse Pottery Vessels in Northern Britain* (3rd edn). Newcastle upon Tyne: Oriel

---- , 1973 Sources of Pottery found in Northern Military Sites. In A P Detsicas (1973), *Current Research in Romano-British Coarse Pottery.* London: CBA Res Rep **10**, 53-62.

---- , 1976 Coarse Fumed Ware in North Britain and Beyond. *Glasgow Archaeol J*, **4**, 57-80

---- , Jobey, I M, and Welsby, D A, 1993 *The Roman Bath-house at Bewcastle,* Cumbria. Kendal: Cumberland Westmorland Antiq Archaeol Soc Res Ser **7**

Greene, K, 1977 Legionary pottery, and the significance of Holt. In Dore and Greene (1977), 113-32

Groenman-van Waateringe, W, van Beek, B L, Willems, W J H, and Wynia, S L, (eds) 1997 *Roman Frontier Studies 1995, Proceedings of the XVIth International Congress of Roman Frontier Studies.* Oxford: Oxbow Mono **91**

Haalebos, J K, 1992 La céramique Belge à Nimègue, *Société Français d'Étude de la Céramique Antique en Gaule, Actes du Congrès de Tournai, 28-31 Mai 1992,* 17-27

---- , and Koster, A, 1981 Marbled Wares from the Netherlands. In Anderson and Anderson (1981), 69-92

Halkon, A P M, 1983 Investigations into the Romano-British Industries of Holme-on-Spalding Moor. *East Riding Archaeol,* **7**, 15-24

---- , and Millett, M, (eds) 1999 *Rural Settlement and Industry: studies in the Iron Age and Roman archaeology of lowland East Yorkshire.* Leeds: Yorks Archaeol Rep **4**

Hargrove, W, 1818 *History and Description of the Ancient City of York II, ii.* York

---- , W, 1838 *The New Guide for Strangers and Residents in the City of York.* York

Hartley, B R, 1971 Roman York and the northern military command to the third century AD. In Butler (1971), 55-69

Hartley, K F, 1975 The Mortarium Stamps. In A Robertson, M Scott, and L Keppie 1975, *Bar Hill: a Roman Fort and its Finds.*

Oxford: Brit Archaeol Rep, Brit Ser, **16**, 142-47

----, 1985 Mortaria. In Bidwell (1985), 182-84

----, 1986 Mortarium stamps [and fabrics] from the civil settlement, Doncaster. In Buckland and Magilton (1986), 143-53

----, 1994 The mortaria (identifications). In Bidwell (1994), 208-14

----, 1995 Mortaria. In Phillips and Heywood (1995), 304-23.

----, 1998 The incidence of stamped mortaria in the Roman Empire, with special reference to imports to Britain. In Bird (1998), 199-217

----, 2000 The 'Pottery Shop' Mortaria [and] The Mortarium Stamps. In Rush *et al* (2000), 183-90

----, *forthcoming* The mortaria from Bearsden. In D J Breeze *forthcoming*, *Excavations at the Antonine Wall Fort of Bearsden*. Edinburgh: Soc Antiq Scotland Mono Ser

Haupt, D, 1984 Römische Töpfereibezirk bei Soller, Kreis Düren. *Beiträge zur Archäologie des Römischen Rheinlands 4*, Rheinische Ausgrabungen **23**, 391-497

Hayes, R H, 1988 *North-East Yorkshire Studies: Archaeological Papers by Raymond H. Hayes*. Leeds: Yorkshire Archaeol Soc Roman Antiq Sect

----, and Whitley, E, 1950 *The Roman Pottery at Norton, East Yorkshire*. Leeds: Roman Malton District Rep **7**

Hicks, J D, and Wilson, J A, 1975 Romano-British Kilns at Hasholme. *East Riding Archaeol*, **2**, 43-70

Hird, L, *forthcoming* The Coarse Pottery from Piercebridge. In Fitzpatrick and Scott *forthcoming*

----, and Perrin, J R, 1997 The Roman Ceramics: the Coarse wares. In Wilmott (1997), 233-56

Hodgson, K S, and Richmond, I A, 1938 Coarse pottery from Bewcastle. In I A Richmond, K S Hodgson, and K St Joseph, 1938 The Roman fort at Bewcastle. *Trans Cumberland Westmorland Antiq Archaeol Soc*. ns **38**, 219-30

Holbrook, N, and Bidwell, P T, 1991 *Roman Finds from Exeter*. Exeter: Exeter Archaeol Rep **4**

Holwerda, J H, 1941 *De Belgische Waar in Nijmegen*. Beschrijving van de Verzameling van het Museum G.M. Kam te Nijmegen

----, 1944 *Het in de Pottenbakkerij van de Holdeurn Gefabriceerde Aardewerk uit de Nijmegeegsche Grafvelden*. Leiden: Beschrijving van de Verzameling van het Museum G.M. Kam te Nijmegen

Hull, M R, 1932 The Pottery from the Roman Signal Stations on the Yorkshire Coast. *Archaeol J*, **89**, 220-53

ILS = Dessau, H, (ed) 1892-1916 *Inscriptiones Latinae Selectae*. Berlin

Jarrett, M G, 1976 An Unnecessary War. *Britannia*, **7**, 145-151.

Johnson, S, 1978 Excavations at Hayton Roman Fort, 1975, *Britannia*, **9**, 145-51

Jones, M U, 1971 Aldborough, West Riding, 1964: Excavations at the South Gate and Bastion and Extra-Mural Sites. *Yorkshire Archaeol J*, **43**, 39-78

Keene, L, 1970 Roman red-painted ware from Yorkshire. *Antiq J*, **50**, 88-93

Keppie, L J F, 1989 The Fate of the Ninth Legion - a Problem for the Eastern Provinces. In D H French and C S Lightfoot (eds), *The Eastern Frontier of the Roman Empire, Proceedings of a colloquium held at Ankara in September 1988*. Oxford: Brit Archaeol Rep, Int Ser **553**, 247-55

----, 2000 *Legio VIIII* in Britain: the Beginning and the End. In R J Brewer (ed), *Roman Fortresses and their Legions*. Cardiff: Nat Mus Galleries Wales/London: Soc Antiq, 83-100

King, E, 1974 Roman Kiln Material from the Borthwick Institute, Peaseholme Green. In P V Addyman, Excavations in York, 1972-1973, *Antiq J*, **54**, 213-17

Lawton, I G, 1989 *Roman Kiln Study*. BA Diss Univ York, Jan 1989

----, 1992-3 Apple Tree Farm 1987-1992: an Ebor ware kiln site. *Yorks Archaeol Soc Roman Antiq Sect Bull*, **10**, 4-8

Loughlin, N, 1977 Dales Ware: a contribution to the Study of Roman Coarse Pottery. In Peacock (1977), 85-146

Magilton, J R, 1986 A Roman Building and Roman Roads in Aldwark. In D Brinklow, R A Hall, J R Magilton, and S Donahey 1986, *Coney Street, Aldwark, and Clementhorpe, Minor Sites and Roman Roads*. Archaeol York **6/1**. York: CBA, 32-48

Mann, J C, (ed M Roxan) 1983 *Legionary Recruitment and Veteran Settlement during the Principate*. Univ London Inst Archaeol Occ Pub **7**

----, 1997 The Division of Britain in AD 197. *Zeitschrift für Papyrologie und Epigraphik*, **119**, 251-4

Marsh, G, 1981 London's samian supply and its relationship to the development of the Gallic samian industry. In Anderson and Anderson (1981), 173-238

May, J, 1996 *Dragonby: report on excavations at an Iron Age and Romano-British Settlement in North Lincolnshire*. Oxford: Oxbow Mono **61**

----, Gregory, A K, and Swan V G, 1996 The pottery kilns at Dragonby. In May (1996), 103-6, 134-5, 574-83

May, T, 1911 The Roman pottery in the York Museum, II. *Yorks Phil Soc Ann Rep for MCMX*, 14-44

----, 1912 The Roman pottery in the York Museum, III. *Yorks Phil Soc Ann Rep for MCMXI*, 2-48

----, 1922 *The Roman Forts of Templebrough near Rotherham*. Rotherham

McCarthy, M R, 1991 *The Roman Waterlogged Remains and later features at Castle Street, Carlisle: Excavations 1981-2*. Kendal: Cumberland Westmorland Antiq Archaeol Soc Res Ser **5**

Michaud, A, 1991 Les ateliers de potiers saintais de la rive droite de la Charente, *Société Française d'Étude de la Céramique Antique en Gaule. Actes du Congrès de Cognac, 8-11 Mai 1991,* 41-51

Mitchelson, N, 1964 Roman Malton: Excavations in Orchard Field, 1949-1952. *Yorkshire Archaeol J,* **41**, 209-61

Monaghan, J, 1993 *Roman Pottery from the Fortress: 9 Blake Street.* Archaeol York **16/7**. York: CBA

----, 1997 *Roman Pottery from York.* Archaeol York **16/8**. York: CBA

Myres, J N L, Steer, K A, and Chitty, A M H, 1959 The Defences of *Isurium Brigantium* (Aldborough). *Yorkshire Archaeol J,* **40**, 1-77

Ottaway, P, 1993 *English Heritage Book of Roman York.* London: English Heritage/Batsford

---- , 2000 Excavations on the site of the Roman signal station at Carr Naze, Filey, 1993-94. *Archaeol J,* **157**, 79-199

Peacock, D P S, (ed) 1977 *Pottery and Early Commerce: Characterisation and Trade in Roman and Later Ceramics.* London: Longman

Perrin, J R, 1977 'Legionary' ware at York. In Dore and Greene (1977), 101-12

---- , 1981 *Roman Pottery from Skeldergate and Bishophill.* Archaeol York **16/2**. London: CBA

----, 1990 *Roman Pottery from the Colonia 2: General Accident and Rougier Street.* Archaeol York **16/4**. London: CBA

----, 1995 Roman Coarse Pottery. In Phillips and Heywood (1995), 324-45

Phillips, D, and Heywood, B, 1995 *Excavations at York Minster 1, From Roman Fortress to Norman Cathedral.* London: RCHME

Purdy, J G, and Manby, T G, 1973 Excavations at the Roman tilery at Grimescar, Huddersfield 1964. *Yorkshire Archaeol J,* **45**, 96-107

RCHME 1962 = Royal Commission on Historical Monuments England *An Inventory of the Historical Monuments in the City of York 1. Eburacum: Roman York.* London: HMSO

RIB I = Collingwood, R G, and Wright, R P, 1965 *The Roman Inscriptions of Britain.* Oxford: Clarendon Press

RIB II = Frere, S S, Tomlin, R S O, *et al,* 1990-1995 *The Roman Inscriptions of Britain,* II, *Instrumentum Domesticum,* fasc 1-8. Sutton: Stroud

Richardson, B, 1986 The waterfront group: coarse wares and non-samian fine wares. In L Miller, J Schofield, and M Rhodes, 1986, *The Roman Quay at St Magnus House, London. Excavations at New Fresh Wharf, Lower Thames Street, London 1974-78,* London Middlesex Archaeol Soc Spec Pap **8**, 106-38

Rigby, V, 1980 The Coarse Pottery. In I M Stead 1980, *Rudston Roman Villa.* Leeds: Yorkshire Archaeol Soc, 45-94

---- , 1998 Where did Cen, Reditas and Sacer produce pots? A Summary of the Range and Distribution of Romano-British Stamped Wares. In Bird (1998), 191-97

----, and Stead, I M, 1976 The Coarse Pottery. In Stead 1976, 136-90

Rupp, V, 1988 *Wetterauer Ware Eine römische Keramik im Rhein-Gebiet.* Bonn. Schriften des Frankfurter Museums Für Vor-und Frühgeschichte Archäologisches Museum.

Rush, P, 2000 The Coarse wares. In Rush *et al* (2000), 89-166

---- , Dickinson, B, Hartley, B, and Hartley, K F 2000 *Roman Castleford, Excavations 1974-85, III, the Pottery.* Wakefield; Yorks Archaeol **6**

Samuels, J, 1979 The excavation of two Romano-British pottery kilns at Barnetby Top, South Humberside. *Lincolnshire Hist Archaeol,* **14**, 11-18

Santrot, M-H, and Santrot J, 1979 *Céramiques Communes Gallo-Romaines d'Aquitaine.* Paris: CRNS

---- , and ---- , 1991 Soubran et Petit-Niort (Charente-Maritime), concurrence 'organisée' entre potiers d'ateliers ruraux spécialisés. *Société Française d'Étude de la Céramique Antique en Gaule. Actes du Congrès de Cognac, 8-11 Mai 1991,* 83-98

Saxer, R, 1967 *Untersuchungen zu den Vexillationen des Römischen Kaiserheeres von Augustus bis Diokletian.* Köln/Graz: Epigraphische Studien 1, Beihefte der Bonner Jahrbücher **18**

SHA Hadr = Scriptores Historiae Augustae, Hadrian

Sireix, C and Convertini, F, 1997 La céramique à l'éponge de la région bordelaise: la céramique marbrée d'Aquitaine, *Société Française d'Étude de la Céramique Antique en Gaule. Actes du Congrès du Mans, 8-11 Mai 1997,* 321-33

Stead, I M, 1976 *Excavations at Winterton Roman Villa and Other Roman Sites in North Lincolnshire,* 1958-1967. London: Dept Environment Archaeol Rep **9**

Sumpter, A B, 1990 Roman pottery. In S Wrathmell and A Nicholson (eds) (1990), *Dalton Parlours, Iron Age Settlement and Roman Villa.* Wakefield: Yorkshire Archaeol **3**, 135-46, 235-45

Swan, V G, 1984 *The Pottery Kilns of Roman Britain.* London: RCHME Supp Ser **5**

---- , 1992 *Legio VI* and its Men: African Legionaries in Britain. *J Roman Pottery Stud,* **5**, 1-33

----, 1993 Une Tradition de Gallia-Narbonensis à York au Début du IIIème Siècle, *Société Français d'Étude de la Céramique Antique en Gaule, Actes du Congrès de Versailles, 20-23 Mai 1993,* 371-80

----, 1996 The dating and wider context of the pottery from Kilns 3, 4, and Pit F 2567. In May (1996), 574-7, 578-9, 582

----, 1997 Vexillations and the garrisons of Britannia in the second and early third centuries: a ceramic view-point. In Groenman-van Waateringe *et al* (1997), 289-94

---- , 1999 The Twentieth Legion and the history of the Antonine Wall reconsidered. *Proc Soc Antiq Scotland,* **129**, 399-480

---- , *forthcoming* The structure of the Piercebridge pottery kiln. In Fitzpatrick and Scott *forthcoming*

---- , *in prep The pottery from the early timber forts at Vindolanda.* Vindolanda Res Rep

---- , and Bidwell, P T, 1998 Camelon and Flavian troop-movements in southern Britain: some ceramic evidence. In Bird (1998), 21-30

---- , and Evans, J, *forthcoming* Hutton Hill: the kiln evidence and its wider implications

---- , and McBride, R M, 2002. A Rhineland potter at the Legionary Fortress of York. In M Aldhouse-Green and P Webster (eds.). *Footprints in the Sands of Time: Papers on Aspects of the Celtic and Roman Worlds.* Cardiff: University of Wales Press.

---- , and McBride, R M, forthcoming The pottery from the Roman Legionary kiln dumps excavated in 1995 at Peaseholme Green, York, In P Ware *et al forthcoming, Report on the Excavations at Layerthorpe Bridge, York, in 1995*

---- , and Monaghan, J, 1993 Head-pots: a North African Tradition in Roman York. *Yorkshire Archaeol J,* **65**, 21-38

---- , and Philpott, R A, 2000 *Legio XX VV* and Tile production at Tarbock, Merseyside. *Britannia,* **31**, 55-67

Tab Vindol II = A K Bowman and J D Thomas, 1994 *The Vindolanda Writing Tablets (Tabulae Vindolandenses II).* London: British Museum Press

Tomber, R, and Dore, J, 1998 *The National Roman Fabric Reference Collection: a Handbook.* London: Mus London Archaeol Serv

Tuffreau-Libre, M, 1995 Céramique Gallo-Romaine du Haut-Empire en Atrébatie. In Tuffreau-Libre and Jacques (1995), 11-28.

---- , and Jacques, A, (eds) 1995 *La Céramique du Haut-Empire en Gaule Belgique et dans les Régions Voisines: faciés régionaux et Courants Commerciaux*, Nord-Ouest Archéologie **6** (for) 1994

Van der Veen, M, 1994 Report on the Biological Remains. In Bidwell and Speak (1994), 243-69

Vermeersch, D, 1995 La céramique Gallo-Romaine du nord de l'Ile-de-France: l'exemple de Beaumont-sur-Oise, production et consommation. In Tuffreau-Libre and Jacques (1995), 131-62

---- , Jobic, F and Jobelot, N, 1993 Les différentes productions céramiques du *vicus* de Beaumont-sur-Oise. In *Trésors de Terre: céramiques et potiers dans l'Ile-de-France gallo-romaine.* Versailles, 115-23.

Vernou, C, 1995 Les Productions de Céramique Gallo-Romaine du Haut-Empire en Saintonge. In Tuffreau-Libre and Jacques (1995), 321-5

Wacher, J S, 1969 *Excavations at Brough-on-Humber 1958-1961.* London; Rep Res Comm Soc Antiq **25**

---- , 1974 *The Towns of Roman Britain.* London: Batsford

Ward, M, 1998 Some finds from the Roman Works-depot at Holt. *Studia Celtica,* **32**, 43-84

Webster, P V, 1993 Coarse pottery. In Casey *et al* (1993), 250-316

Wenham, L P, 1968 Two Excavations: 2(ii) Appletree Farm, Bad Bargain Lane. *Yorkshire Phil Soc Ann Rept Trans for 1967,* 50-60

---- , and Heywood, B 1997 *The 1968 to 1970 excavations in the vicus at Malton, North Yorkshire.* Leeds: Yorkshire Archaeol Rep **3**

Williams, D F, 1977 The Romano-British Black-Burnished Industry: a essay on Characterisation by Heavy Mineral Analysis. In Peacock (1977), 163-220

---- , 1985 Fabric Examination. In L F Pitts, and J K S St Joseph (1985), *Inchtuthil: the Roman Legionary Fortress Excavations 1952-65.* London: Britannia Mono Ser **6**, 339-40

---- , and Carreras, C, 1995 North African amphorae in Roman Britain: a re-appraisal. *Britannia* **26**, 231-52

Wilmott, T, 1997 *Birdoswald: Excavations of a Roman fort on Hadrian's Wall and its successor settlements, 1987-92.* London; English Heritage Archaeol Rep **14**

Woodfield, C, 1965 Six Turrets on Hadrian's Wall. *Archaeol Aeliana,* 4th ser **43**, 87-200

Young 00 = pottery type-series in C J Young 1977 *The Roman Pottery Industry of the Oxford Region.* Oxford: Brit Archaeol Rep, Brit Ser **43**

BROKEN BOTTLES AND QUARTZ-SAND: GLASS PRODUCTION IN YORKSHIRE AND THE NORTH IN THE ROMAN PERIOD

JENNIFER PRICE

Although glass working took place in virtually every area of the Roman world, the evidence for glass production is generally rather difficult to identify and interpret. There is a marked contrast between the wide range of vessels and objects, which survive complete in burials or, more frequently, are found in fragments on settlements, and the ephemeral and episodic nature of the evidence for the production processes by which they were formed. In part, this can be explained by the processes themselves, as most Roman glass vessel and object production probably took place in association with very small furnaces, and the various categories of waste glass created by the production processes were normally recycled, leaving little diagnostic material to be identified.

Several categories of evidence for glass production in the Roman period have been recorded in military, rural and urban settlements in Yorkshire and northern Britain, and this evidence forms part of the wider picture of production in the province of Britain and other north-western provinces. This paper will present a brief précis of the evidence for the nature of glass production in the Roman world as the background for this regional study; it will then examine and evaluate the strands of evidence identified in Yorkshire and northern Britain against this background; and will conclude with a consideration of the reasons why so much of the evidence is invisible or difficult to interpret.

Evidence for Roman glass production in the Roman world

Archaeology is always the principal source of information for investigations into Roman glass production, as it is for other craft processes at this period. The archaeological evidence is sometimes augmented by information from literary works,[1] or epigraphic records,[2] or iconography,[3] but these provide only very limited detail. In particular, the literary surviving accounts have a strong bias towards the Mediterranean provinces.

It is now accepted that much of the production was divided into two distinct groups of activities which could, but did not have to, take place in the same localities. These were ***primary production***, in which glass was made from the basic raw materials, using sand as the former, lime as the stabiliser and soda as the flux; and ***secondary production***, in which glass already made was reheated and worked to produce glass vessels and objects.

Primary production

Although the direct evidence for primary production is very limited, it seems that this was generally a specialised pyrotechnical process, requiring access to supplies of suitable raw materials, good stocks of fuel, and a furnace capable of sustaining high temperatures for the duration of the melting process. Of the raw materials, the sand and lime were very widely available.[4] Soda, however, was more difficult to obtain. Deposits of mineral soda (natron) were rare in the Roman world, the greatest concentration being in Egypt, in the Wadi Natrun and elsewhere. Literary references indicate that natron was traded during the early empire around the coasts of the east Mediterranean, perhaps to India, and to southern Italy,[5] but it is uncertain whether this material, which was both soluble in water and corrosive, was regularly an empire-wide item of trade.

Few primary production sites have been identified, and most have been found in the same regions of the East Mediterranean, Syria-Palestine and Egypt (Figure 1), as the raw materials mentioned above. It is, however, noteworthy that where these sites can be dated they belong to the sixth-seventh and later centuries, and that no examples dating from the first-fifth centuries have yet been noted (*cf.* the discussion of lumps of glass, below).

Several primary production sites have been recorded in Israel. These sites, with rectangular furnaces estimated to be producing up to eight or nine tons of glass at each firing, were engaged in glass manufacture on a large scale. At Bet Eliezer, close to the Mediterranean coast near *Caesarea Maritima*, a complex of seventeen rectangular furnaces built in mud-brick, each measuring approximately 4 x 2 metres internally was excavated in 1992 (Gorin-Rosen 1995; Freestone and Gorin-Rosen 1999, 105-06, fig 3). The furnaces were aligned south-west to north-east, presumably to use the prevailing wind, with a double flue at the south-west end. The structure of the furnaces indicates that the glass in them was made from the raw materials as a one-stage process, without a preliminary fritting phase.[6] The slabs of glass were then broken up and moved away, leaving only small pieces. Similar furnaces have been recorded at Dor and *Apollonia*, which are also close to the Mediterranean coast (Freestone and Gorin-Rosen 1999, 108), and a large rectangular glass slab which failed to melt completely, measuring 3.4 x 1.95 x 0.45 metres and estimated to weigh about nine tons, is known from Bet She'arim in lower Galilee. However, it has been argued that this episode of primary production may be Islamic rather than Roman or Byzantine, perhaps dating from the early ninth century (Freestone and Gorin-Rosen 1999). Evidence of large-scale primary glassmaking has also been recorded in Egypt, at *Taposiris Magna*, south west of Alexandria, and at several

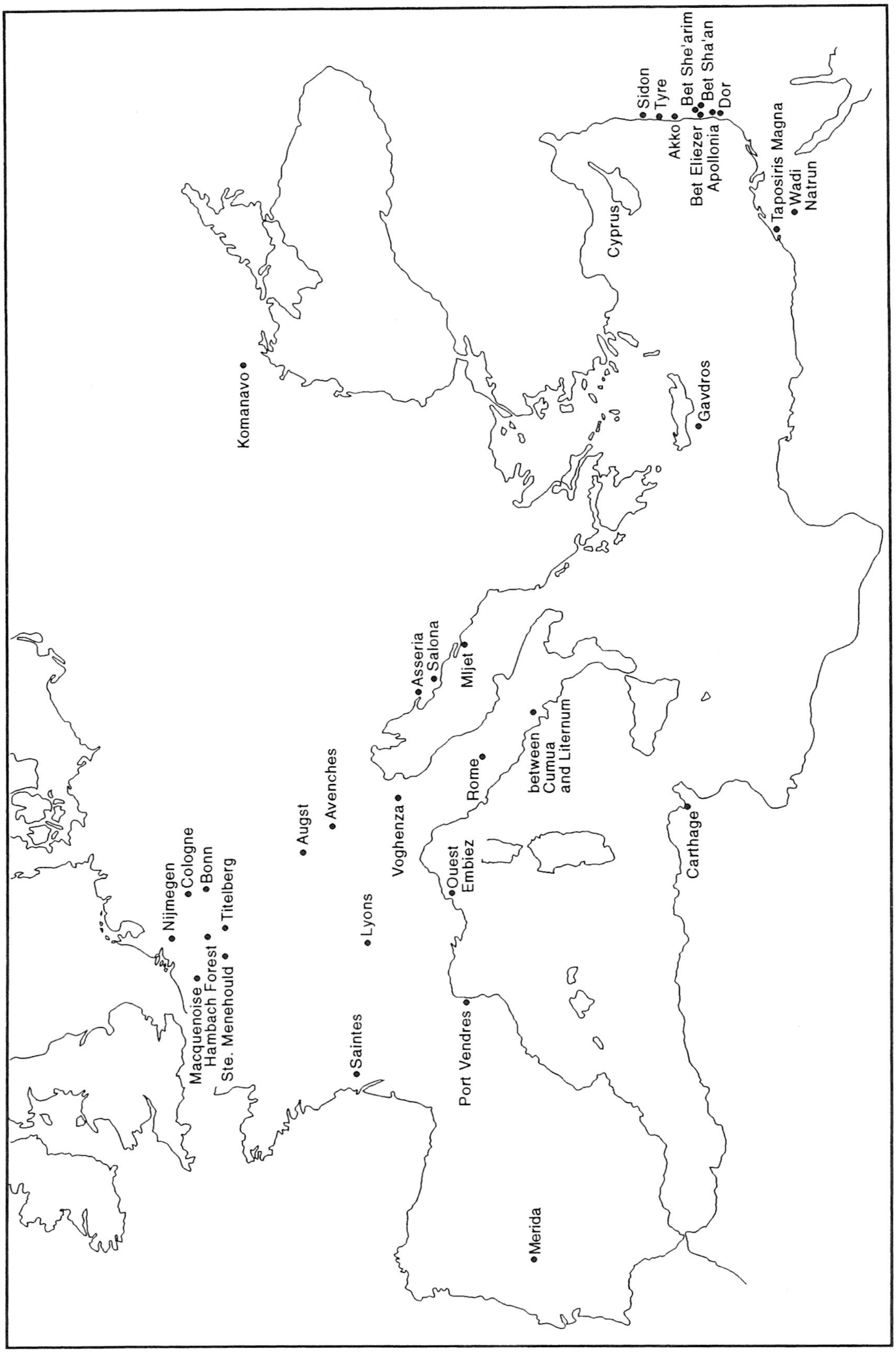

Figure 1. Map: sites outside Britain mentioned in the text

sites in Wadi Natrun, north west of Cairo (Saleh et al 1972; Nenna *et al* 1997). The sites have been identified by field survey which have found glass blocks, fragments of tank furnaces and other debris, but furnace installations have not yet been noted and the sites are not closely dated within the Roman period.

After the slabs of glass had been broken up they were presumably moved away, either by land or by sea. The positions of most of the primary production installations described above would have been very convenient for loading the lumps of glass onto ships, as they are situated close to the Mediterranean coast. The sea-borne transport of lumps of glass within the Mediterranean region is attested from chance finds and shipwrecks. A lump of blue-green glass from a wreck was found in the sea to the north of Gavdos, an island off the south west coast of Crete (Weinberg 1963, 107), and similar finds have been noted off the coast of Israel (Gusta Jacobson *pers comm*). These finds are undated, but shipwrecks with lumps of glass as part of the cargo indicate that this practice occurred in the Roman world. A wreck dating from the later first century AD found close to the island of Mljet, Croatia, contained about 100kg of blue-green lumps (Radic and Jurisic 1993, 122, fig 7.2), and wrecks dated to the early third and early fifth century are known off the coast of southern France, such as Ouest Embiez 1 which contained colourless lumps, and Port Vendres 1, which contained yellowish green lumps (Foy 1997). The glass lumps may have been carried in the ships as a form of ballast, like lead ingots or building stone, as well as being cargo to be traded at various ports.

Secondary production

Much more is known about sites concerned with secondary production. Glassworking sites have been found throughout the Roman world, and they provide most of the archaeological evidence relating to the production of vessels and objects (Figure 1). These sites may produce the remains of furnaces or annealing ovens, or crucibles, or tools and other equipment. Glass production sites are also identified by the presence of certain characteristic categories of waste glass, such as the surplus pieces of glass and drips and trails remaining from the production processes; broken vessels or objects produced at the site; deposits of broken vessel glass collected from the production site or brought from elsewhere for re-cycling; and blocks of glass from the crucibles and lumps of raw glass.

Furnaces The furnaces and installations such as annealing ovens are the features most likely to be recognised, though little more than the ground plans usually survive. These structures are generally very small. For example, the furnaces in the first-century glassworking site at Avenches (*Aventicum*), Switzerland had internal diameters of 0.50-0.65 metre (Morel *et al* 1992, 5-6, figs 3-7), and circular and rectangular glassworking structures at Eigelstein, in Köln, Germany, were nearly all less than 1.0 metre in internal dimensions (Follman-Schulz 1991).

Finds of the clay superstructures are very fragmentary and much of the current information about the appearance of the furnaces derives from contemporary representations on two pottery lamps and an Egyptian terracotta group.[7] The pottery lamps, which date from the later first century AD were found at Asseria in Dalmatia (Abramic 1959) and at Voghenza, Ferrara, in north-east Italy (Baldoni 1987); they show the same glassworking scene which includes a low, domed furnace structure with two apertures, indicating two levels within the furnace (Figure 2), while the terracotta group (Price 1988a), dated to the first or second century AD, shows a tall conical structure, also with two apertures (Figure 3). The tall cone presumably served as a chimney to assist in raising and maintaining the temperature to heat the glass.

Neither of these scenes illustrates the arrangements for annealing[8] which was an essential process in the production of blown glass. Depictions of medieval glass furnaces show that an annealing chamber was sometimes sited in the third level within the furnace (for examples, see Charleston 1978, 11, 13, figs 1-3; Foy and Sennequier 1989, 111 no 48, 113 no 49a-b pls 1, 4 and fig), or attached to the side of the furnace furthest away from the flue (Charleston 1978, 22, fig 16), but evidence for similar arrangements in Roman furnaces survives very rarely. It is possible that separate structures for annealing may have existed at glass production sites, but little is known about these. Alternatively, a quite different arrangement which did not involve any kind of structure, such a pit filled with ash and/or sand may have been used for the controlled cooling of the vessels.[9]

Crucibles The archaeological and iconographic evidence for the size of the furnaces described above indicates that they would generally have held only one container for the glass at a time. Whenever containers have been found, they are made of fired clay, and two kinds have been recognised, pots with circular mouths and convex or straight sides, and rectangular, trough-shaped containers, often known as tanks.

The first appear to have been open bowls. An object of this

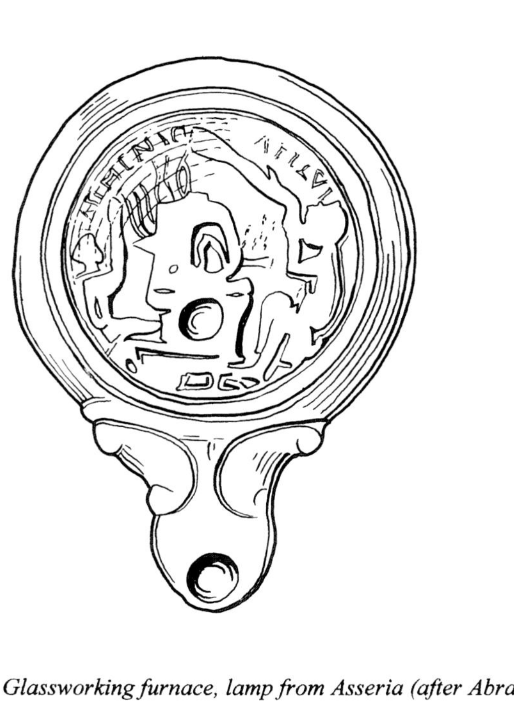

Figure 2. Glassworking furnace, lamp from Asseria (after Abramic 1959)

kind is held in the right hand of the winged figure standing before the furnace structure on the terracotta group (Figure 3), and archaeological finds indicate that these containers were usually of ordinary earthenware pottery of kinds also used in contemporary domestic and funerary contexts. The examples in the workshop in Avenches were forms well-known in other contexts in the city (Heidi Amrein *pers comm*), as were those from the late second - early third- century workshops at Augst (Rütti 1991, 151-53, figs 96-98, pl 218.07), while the examples from the late Roman workshop at Titelberg in Luxembourg (Sternini 1995, 77, fig 90; van Geesbergen 1999, 112-13, fig. 5) and from the Hambach Forest in Germany (Gaitzsch 1999, 135-39, figs 14-17), were Argonne ware and Mayen ware bowls.

The second form of containers was constructed from large tiles which formed the base and slabs of clay which formed the sides, and they were built into the structure of the furnace. These would have held a much larger quantity of glass than the coarseware pots. Examples found in the legionary fort at Bonn in lower Germany, were 0.9 metre long and 0.55-0.60 metre wide (Follman-Schultz 1991, 36, fig 6; 1992, 100, fig).

Tools Roman glassblowers are assumed to have used a range of metal and wooden tools somewhat similar to those used today, such as blowing irons, pontil irons, shears, pincers, moulds and other tools for manipulating the hot glass, but very few have been recognised. The wooden tools have generally disappeared without trace, and most of the metal tools are not sufficiently diagnostic to be identified unless they occur in association with other evidence for glassworking. The long iron blowpipe with an expanded terminal for gathering the glass from the crucible is the most recognisable tool associated with the glassblowing process. The scenes on the lamps and terracotta group described above show tools likely to be blowpipes, and a few examples of these and other tools have been recognised from glassworking sites. Several fragmentary iron tubes with expanded gathering terminals, a pair of iron pincers or shears and a solid iron rod, perhaps a pontil iron (Figure 4), were recovered in association with glass waste characteristic of blowing and fragments of fourth- to fifth-century vessels at Mérida (*Augusta Emerita*), western Spain (Price 1974, 80-84, figs 4-5; Lang and Price 1975), an iron tube found at Salona, Croatia was identified as a blowpipe (Auth 1975, 167), a piece of iron tube 0.832 m long from Sainte-Menehould, northern France is very probably a blow pipe (Foy and Sennequier 1989, 104 no 36, fig), and fragments of solid iron rods, perhaps pontil irons, and iron pincers or shears were found with a knife, a hammer, another iron tool and a terracotta mould in a third- or fourth-century context at Komarawo, Ukraine (Stern 1995, fig 8).

Single- and multi-piece moulds were used both to shape vessels during the blowing process and to decorate their surfaces. Some examples made in terracotta, stone or metal have survived from the north west provinces and elsewhere in Roman world. Stone body and base moulds for square bottles are known at Augst (Rütti 1991, 162-64, fig 103, pl 218.05-06), Lyons (Foy and Sennequier 1989, 100-01), Saintes (Hochuli-Gysel 1993, 87, figs 5-7), and a half-mould of fired clay for forming a grape-flask is known from Macquenoise, Belgium (Chambon 1955, 3 and fig), and one-piece stone and metal moulds are also found occasionally (see Foy and Sennequier 1989, 102-03 nos 51 and 53 for examples from France).

Waste glass Additional evidence relating to glassworking tools such as the blowpipe, shears and pincers comes from various categories of glass production waste. The waste most diagnostic of glass blowing is the moile.[10] Two main forms are found, short cylindrical moiles from vessels with hot-finished rims which were separated at the time of blowing, and lid-shaped moiles from vessels with cold-finished rims which were separated after annealing. Moiles frequently have traces of iron scale or staining on the inside surface of the end attached to the blowpipe, and are

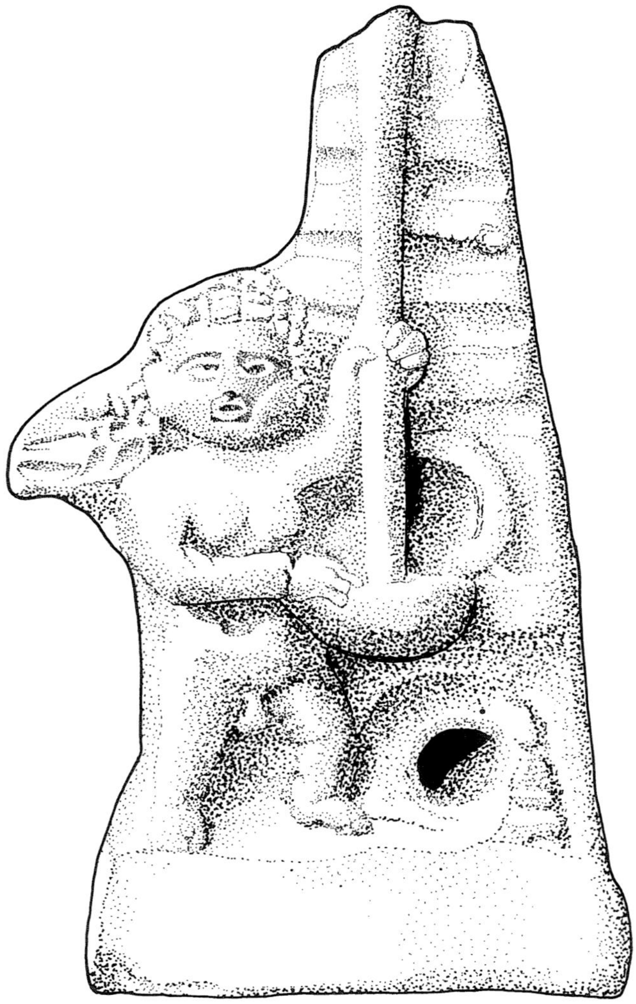

Figure 3. Winged figure holding blowing iron and crucible in front of glassworking furnace, Egyptian terracotta group (after Bird in Price 1988)

Figure 4. Iron glassworking tools from Mérida in western Spain (photo: J Price)

sometimes complete enough to provide information about the diameter of the gathering end of the pipe. Other categories of waste glass generated at production sites include trails, some twisted and showing the imprint of pincers, and drawn threads, and a range of drips and blobs, and pieces of glass detached from crucibles, as well as deposits of cullet.[11] Such material is found at glassworking sites in many parts of the Roman world, as in the *canabae* at Nijmegen (Isings 1980), Augst (Rütti 1991, 156-62, fig101), the Hambach Forest (Gaitzsch 1999, figs 6-7, 12, 15) and elsewhere.

Evidence in Britain

No written evidence has direct relevance for glass production in Britain. The sole literary reference to glass in Britain occurs in Strabo (*Geography* iv, 6,3) in a passage which lists glass vessels among the items imported from Gaul in the reign of Augustus. Similarly, no iconographic or unequivocal epigraphic evidence has been recorded. It is conceivable that future excavations might produce a funerary monument of a glassworker, or a vessel with the name, initials or insignia of a town[12] or military unit, or a note about glass production written on a wooden tablet, but this has not happened yet!

The archaeological evidence in Britain indicates that glass production was usually sited within or close to military bases and towns. Glass workshops are often close to other industrial activities such as metalworking, boneworking or leatherworking, and in several cases there is a close association between glass and pottery production. With the exception of Coppergate in York, which is discussed below, all the glassworking finds in Britain (Figure 5) relate to secondary production. Recent surveys have outlined the evidence for production in Britain (Price and Cool 1991; Price 1998) and this is summarised below.

Apart from a rectangular structure at Caistor by Norwich, the known furnaces are circular in plan with a single flue, and they are very small. The furnace structure dating from the middle of the second century or later which was found among pottery kilns at Mancetter in Warwickshire, was initially about 0.8m in internal diameter and was reduced in size thereafter by successive relinings (Figure 6). The furnace dating from the later third century in the corner of the Market Hall at Leicester was similar in size (Wacher 1978, pl. 30), and the one found among pottery kilns at Castor, Water Newton in Cambridgeshire, known only from a drawing (Figure 7), was probably comparable in dimensions (Artis 1828, pl XXV, 4-5). Very little is known about the furnaces in the industrial settlement at Wilderspool in Lancashire, but the published plans suggest they may have been similar in outline to the circular structures known elsewhere in Britain.

As in other provinces, the pots are open coarseware bowls. A third- or fourth-century Black-burnished flanged bowl was used as a crucible for glass in a workshop at Deansway, Worcester, and fragments of at least four coarseware bowls containing glass were found in late fourth-century levelling deposits in building 2, *insula* XXVII at Verulamium. Other fragmentary examples are known from Silchester, Castor, Water Newton, and Norton Folgate in London. The rim diameters of the largest of these vessels are around 0.27-0.30 metre and the base diameters about 0.18 metre but as their heights are not known their capacity is uncertain, although

Figure 5. Map: glassworking evidence in Britain

Boon (1974, 280-01) has calculated that the Silchester example may have been about 0.30 metre high, in which case it could have held 15 kilograms of glass. The only examples of tanks recorded in Britain have come from sites in London, where they are commoner than pots (Shepherd and Heyworth 1991, 14).

Glass waste is more commonly found. In southern Britain, cylindrical and lid-shaped moiles have been noted in London, Colchester, Leicester, Mancetter, Wall in Staffordshire, Caerleon and Wroxeter, and pincered and twisted trails, drawn threads, rounded and misshapen lumps, glass from crucibles, wasters and cullet (Figure 8) at these sites and Castor, Water Newton, Wilderspool and Worcester. The quantities of this material are generally very small. However,

Figure 6. Glassworking furnace at Mancetter, Warwickshire (photo: K F Hartley)

an early second-century deposit of blowing and working waste and cullet weighing 50 kilos has been found in London, close to the fort at Cripplegate (John Shepherd *pers comm*), which indicates that glassworking may have been taking place on a large scale. By contrast with the evidence for reuse of glass at glassworking sites, there is little sign of the presence of lumps of raw glass, apart from one at Colchester (Cool and Price 1995, 209-10 no 2269, fig 12.1).

The surviving evidence for the size of the furnaces and the pots, and the quantity of waste glass found, gives an overall impression of small-scale glass production in Britain, though this may be rather misleading. The process of forming and finishing glass vessels and objects, unlike the firing of pottery vessels, takes place outside the furnace, so the size of the furnace and the pot only limits the amount of glass to be heated and worked at any one time, and the pots may well have been refilled many times before they were broken or replaced. In addition, it was common practice either to recycle all waste glass in subsequent melts in the same workshop or to collect this for remelting elsewhere, so the recovery of large quantities of cullet, as at Cripplegate, is exceptional and may well indicate the deposition of material at the ending of an episode of production.

Yorkshire and the North

In this region, as elsewhere in Britain, virtually all the archaeological evidence for glass production in the Roman period has been found in close association with military or urban settlements. The largest centre of population in the area was York. The legionary fortress housing in succession the

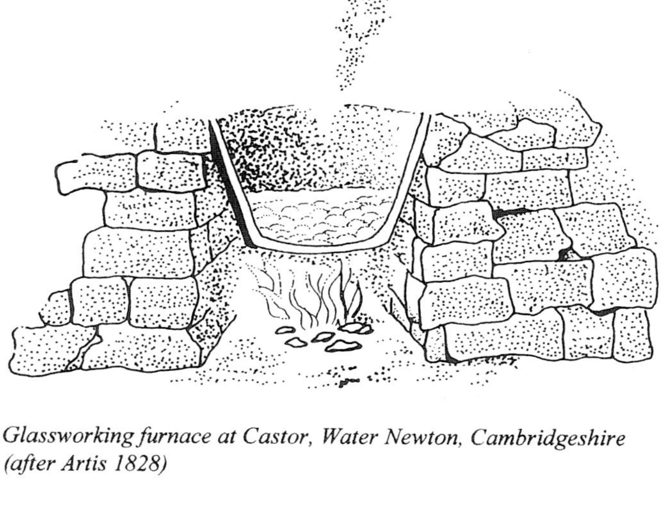

Figure 7. Glassworking furnace at Castor, Water Newton, Cambridgeshire (after Artis 1828)

Ninth *Hispana* and Sixth *Victrix* legions was sited there, and a city with chartered status was established on the opposite bank of the River Ouse.[13] It is therefore unsurprising that York has produced the widest range of evidence for glass production. Nonetheless, some relevant information has also come from other settlements, such as Castleford and Roxby, in Yorkshire, Binchester in County Durham, Carlisle in Cumbria, and from Newstead, Traprain Law, Camelon and Cramond in lowland Scotland (Figure 5).

The glass production evidence in York (see also Cool, this volume) comes from several find-spots. The Coppergate site, in the *canabae* outside the south-east corner of the legionary fortress, has produced the widest range of production evidence; this includes crucibles containing part-melted raw materials, crucibles with melted glass, a little evidence for glass blowing, and vessel fragments which may have been collected as cullet. Recent excavations in the grounds of the Royal York Hotel, outside the north corner of the *colonia*, have produced further evidence for glass melting in crucibles and a little evidence for blowing. There is also a very small amount of evidence for glassworking within the legionary fortress at Blake Street and the Minster site, where a trail with pincer marks and coloured glass cubes were found, and perhaps also in the *colonia* at Rougier Street, where a group of melted vessel fragments and small lumps may represent the collection of cullet.

Of these finds, the most extraordinary is the evidence for primary production at Coppergate. The Coppergate material was examined soon after excavation (Bayley 1987). When discovered, it was thought to belong to the Anglo-Scandinavian period, but was subsequently recognised as Roman material redeposited in later contexts (Bayley 1987, 254). This assemblage of glass production evidence has been re-examined in detail recently (Jackson *et al* 1998; Cool *et al* 1999).[14] A total of 187 potsherds, weighing approximately 3 kilograms, came from buff-coloured and reddish pottery vessels dated to the late second or early third century. These had been used as crucibles and retained colourless or greenish glass on the inside or outside surface, ranging from a thin coating of glass on the upper parts to a thickness of 10 millimetres or more in the base. They represented a minimum of 36 vessels, mostly in Ebor Ware which was made in York (Monaghan 1997, 869-80, Swan, this volume). Twenty-four of the vessels were identified as bowls, and nine as either bowls or jars, with rim diameters between *c* 0.22 and *c* 0.38 metre, and the forms included one interpreted as a specialist vessel made for industrial use (see Cool, this volume, figure 4 for the range of vessel forms). It is noteworthy that the clay of the Ebor Ware vessels has good refractory properties and can be heated to at least 1150° centigrade without deforming.

By itself, this evidence does not indicate production from raw materials in York, although it points to glass melting taking place on a large scale. Cool *et al* (1999, 153) have calculated that the crucibles represented by the Coppergate fragments may have contained 0.32 cubic metre of glass. However, at least one of the crucibles had pieces of glass partially formed from raw materials (semi-reacted batch material) attached to the rim and outside surface, and similar pieces, which had presumably become detached, were found separately. The batch material consisted of partially fused quartz interspaced with glass or blocks of glass, with white frothy layers on top

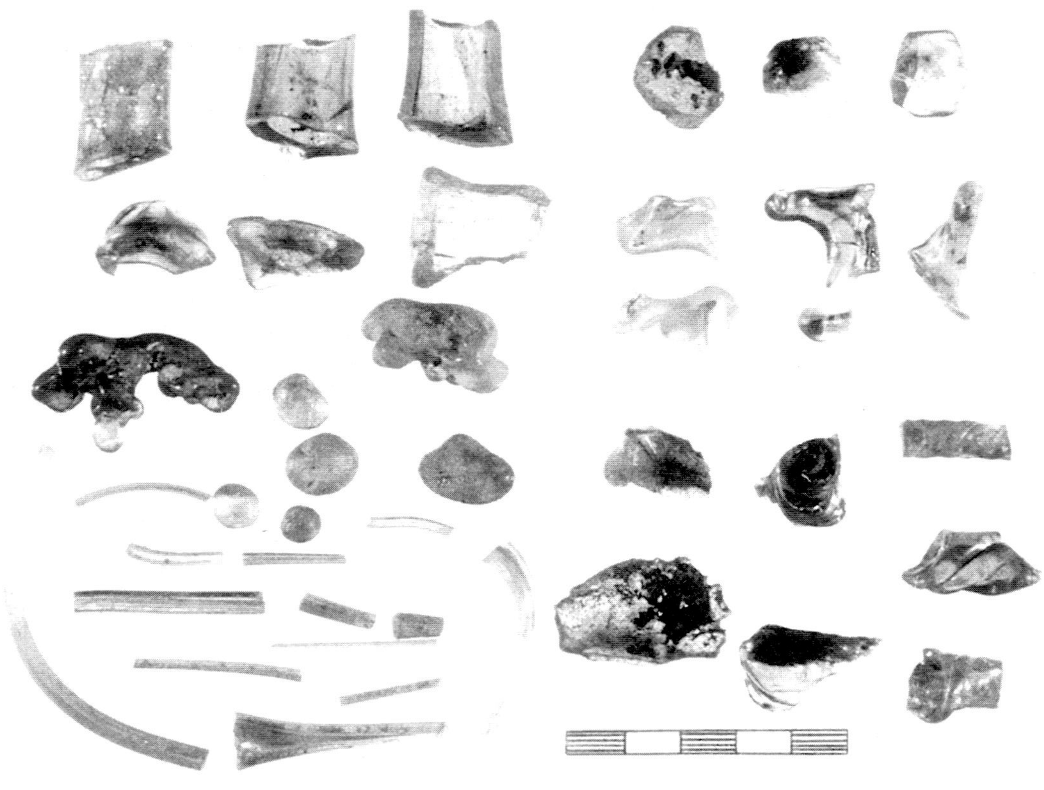

Figure 8. Moiles and other glassworking waste from Wroxeter (Photo: Univ Leeds Photographic Service).

or inside the blocks. The batch material indicates that raw materials were used to make the glass, either in an unsuccessful attempt to make glass as a one-stage process or with the intention of fritting the raw materials as part of a two-stage process. The similarity of composition between the glass in the batch material and the glass in the crucibles, and the significant variations between both of these (slightly increased levels of iron, alumina and potassium) and the typical composition of Romano-British natron-based glass, has led Jackson *et al* (1998) and Cool *et al* (1999) to argue that the glass in the crucibles is also the result of primary manufacture of glass.

This material, which appears to be unparalleled in the Roman world, represents a very different kind of primary production from that recorded at Bet Eliezer and the other sites in the east Mediterranean region discussed earlier. It was small in scale, and involved the use of pottery vessels as crucibles to heat the raw materials. The semi-reacted batch had been heated to temperatures close to 1250°C, which may indicate that it represents a failed attempt to make glass by a one-stage rather than a two-stage process. Fritting would have required a much lower temperature (in the region of 600-750°C), and a pottery bowl would not perhaps have been a very practical form of container for burning off the impurities from the raw materials. In this respect, it is interesting that pottery vessels were used as crucibles for making the glass, rather than tanks. Tanks were shallow open containers, would have held larger quantities of glass and could presumably have been constructed from tiles made in the legionary kilns at York, so the preference for pots may be connected with the superior refractory properties of the Ebor Ware.

Bluish green glass, the commonest colour in the first to third centuries, was not present in the crucibles, and the distinctive composition of the glass contained in the crucibles was not otherwise recorded among the vessels or waste at Coppergate, which raises questions about the uses of this glass. It has been suggested that it may have been produced to make window panes, though no compositional links with window glass found in York have been noted. A small quantity of glassblowing debris was also found at Coppergate, consisting of 2 moiles and 3 other fragments, but this material was not related to the episode of glass making described above. In addition, a large quantity of vessel fragments which may have been collected for recycling was present at the site.

In 1999, excavations directed by Nick Pearson in the grounds of the Royal York Hotel produced pottery vessels with melted glass which were very similar to the Coppergate finds, as well as a little glass blowing waste (Hilary Cool *pers comm*; see also Cool, this volume)

Apart from these two sites, only a very small amount of evidence of glass working has been recognised. A single pincered trail was found at Blake Street (Cool *et al* 1995, 1592, 1661 no 6253, fig 752; see also Cool, this volume), but it is uncertain whether this represents the production of vessels or objects. Two small opaque blue cubes were also found at Blake Street (Cool *et al* 1995, 1592-93, 1661, nos 6251-52, fig 752), and two others, in opaque blue and opaque turquoise, at the Minster site (Price 1995, 353, 368 nos 93-95, fig 146). Similar cube-shaped lumps, which may be evidence for some kind of glassworking or another industrial process, are discussed below in connection with material found at Castleford.

The practice of collecting broken glass for recycling is almost certain to have been in operation in York, and glass for recycling may also have been collected in settlements in the vicinity and then brought to the city. As already mentioned, many of the vessel glass fragments found at Coppergate may have been cullet, but deposits of this kind are generally difficult to identify in the absence of more diagnostic glassworking waste. A collection of at least 220 bluish green fragments, droplets and thin slivers of glass weighing 0.31 kilogram was found in one context at Rougier Street. Many of these were noticeably uniform in size - at least 90 measured approximately 40 x 25 millimetres - and in the degree of melting, and in the absence of black specks and charcoal embedded in the melted surfaces. They may perhaps have been cullet prepared for the production of objects, though this remains uncertain (see Cool, this volume).

Elsewhere in Yorkshire, two other settlements are known to have produced some evidence for glassworking. About fifty small cubes and melted lumps, one opaque yellow and the rest opaque blue, were found in the *vicus* at Castleford (Cool and Price 1998, 193-94). Small cube-shaped glass lumps, often known as *tesserae*, have been found quite frequently on Romano-British sites, and they have frequently been assumed to come from late Roman mosaic pavements, such as one at Aldborough, North Yorkshire (Neal 1981, 38-39, fig 12). However, many of the sites producing glass cubes are early in date, and others have no evidence for the presence of mosaic pavements, so these objects may have quite different purposes. For example, small blue glass cubes were sometimes attached as decoration to pottery vessels. These have been found in Italy (Paul Roberts *pers comm*), but have not been noted in Britain. Alternatively, the cubes may be small quantities of raw glass for use in the production of glass objects. Guido (1978, 47, 100) suggested they were employed in the production of blue glass melon beads, though this may not be the case as most glass cubes are more or less opaque whereas the glass of melon beads is translucent.

Nearly all of the glass cubes at Castleford were found at one site (Site 1(74)) in the *vicus,* in mid second-century and later contexts. It is noteworthy that the same site also produced a very large proportion of the glass beads found at Castleford, and that most of the beads came from mid second-century contexts. Many of the bead forms are quite unusual, and the possibility of bead production at or near to the site in the middle of the second century has been discussed (Cool and Price 1998, 181-82). It is however, difficult to associate the glass cubes directly with any such bead production.

Another possibility is that the glass of the cubes might have been mixed with bluish green glass to produce streaky blue glass. Several categories of objects at Castleford were made in glass of this kind, including melon beads (nos 102-03), an annular bead (no 106) and a bangle (no 221), but evidence for association with the cubes is equivocal, as most of the objects either came from earlier phases of occupation than the glass cubes, or from the military settlement rather than the *vicus.* Finally, the glass lumps may have been intended for use as enamel on copper alloy objects. There is good evidence for the presence of a workshop producing champlevé enamelled vessels at Castleford (Bayley and Budd 1998, 203-22; and Bayley, this volume) though it was a considerable distance from the *vicus*. It is arguable that some production of glass or enamelled objects took place in or close to the *vicus* in the mid second century but the nature of the activity is as yet unidentifiable.

Excavations at the late Iron Age-early Roman settlement at Roxby, on the North York Moors produced a small fragment of an oval-sectioned blue and green glass rod in House 2 (Inman *et al* 1985, 199-200, fig 5). This may be debris from some kind of glassworking, though what was being produced is not known. It is the only piece of evidence of the kind to be recorded from a round house in the region, and it is noteworthy that evidence for other craft activities, principally iron-working and jet-working, was also recorded in House 2 (see Wilson, this volume).

Lumps of dark green, dark blue and opaque white raw glass and glass waste were found at Bursea House during fieldwork and excavation in the parish of Holme on Spalding Moor in East Yorkshire in the past two decades. These were formerly thought to be Roman, but have now been shown to be not of Roman date; because of the presence of arsenic in the mix they are likely to come from post-medieval production (Henderson 1999; Halkon, this volume)

North of Yorkshire, the fort at Binchester in County Durham has produced twelve fragments of glassworking waste (Price and Worrell *forthcoming*). One greenish moile with iron-staining was noted, indicating that vessels were blown using an iron blowpipe, and the remainder were bluish green, greenish and colourless drips and trails with pincer marks. These finds came from late Roman and post Roman contexts, but the colour and quality of the glass of the moile suggests that the glass blowing occurred in the late third or fourth century. At Carlisle, a small fragment from the lower body and base of a pottery crucible containing bluish green glass was found at the Lanes, redeposited in a post-Roman context. Trails with pincer marks and rounded lumps were also recorded, providing further indications of some glass working activity (unpublished).

In lowland Scotland, two fragments of moiles have been noted from a pit in the annexe of the fort at Camelon, at least one of which may be lid-shaped. In addition, some vessel fragments were too distorted to have been used and some which did not appear to have been annealed. This material indicates an episode of glass blowing at the fort, probably during the late first-century occupation of the fort. It is noteworthy that the base of a furnace similar in plan to the Mancetter and Leicester glass furnaces, though larger in size, was also found in the annexe, but there is no certain association between the glassblowing waste and the furnace (Valerie Maxfield *pers comm*). There is a little evidence for glass melting at Cramond in Antonine and Severan contexts, where four groups of burnt clay fragments with deposits of blue/green vitreous material were found, but it is unclear whether this melting was intentional (Price *forthcoming*) Evidence for glass working has also been claimed at two other sites, the fort at Newstead and the hill top settlement at Traprain Law, but the material is unpublished and no detailed information is currently available.

Interpretation of the evidence

Examined critically, the glass working evidence from Yorkshire and the North described above represents one episode of primary production in the late second or early third century at Coppergate, York, but no identifiable products from this activity; and at least three episodes of production of blown vessels, at Coppergate, York where the date is not known, at Camelon probably in the late first century, and at Binchester, probably in the late third or fourth century. All the remaining evidence points only to the melting of glass and perhaps to the forming of objects.

This is, however, most unlikely to be an accurate reflection of the extent of glass production in the region. Both the military and the civil populations used glass extensively. Glass vessels are present in virtually all settlements both as tablewares for drinking and serving liquids and as containers for liquid and semi-liquid foodstuffs, cosmetics, medicines and other substances; a wide variety of glass objects and window panes are also found. Some of these were undoubtedly brought into the region from elsewhere, but some vessels, many of the objects and all of the window panes are likely to have been produced close to where they were used. It is thus likely that some form of glassworking took place at many of the settlements in the Roman North.

The army must have been involved in some of this production. Every auxiliary fort would have required window panes, at least for the high-status residential buildings and the bathhouse, and the glazing requirements of the legionary fortress would have been considerably greater. However, the extent to which the military personnel themselves were the glassworkers remains unclear. Little is recorded about soldiers engaged in glassworking, apart from a third-century tombstone at *Carnuntum* that commemorates a glazier in the Fourteenth *Gemina* legion (Bormann 1914, 336-40). The siting of glassworking activities in the *canabae* in York, in the fort annexes at Camelon and Newstead and within the fort at Binchester indicates close military control of the processes. The army in northern Britain is also likely to have been involved in the collection of broken bottles and other vessels as cullet for recycling in local workshops. The distribution patterns of the glass fragments in the Flavian legionary fortress at Inchtuthil suggest that the broken glass had been collected up before the abandonment of the site (Price 1985, 304), and the small size and fragmentary nature of the glass finds at the Minster site (Price 1995, 346) and at Coppergate points to the systematic collection of cullet.

Production in the towns presumably catered for the civil population and this may well have been organised independently of the military production. The civil settlements would have been a ready source of broken glass for recycling and there is some evidence that organised collection of this took place in towns in the region. For example, the differences in the size and number of fragments of broken vessels deposited at different periods at Blackfriars Street, Carlisle suggest that the organised collection of broken glass commenced after the end of the first century or beginning of the second (Price 1990, 164).

Nothing is known in detail about the operation of the glass production sites of northern Britain, but it is possible that they operated intermittently rather than continuously, perhaps using the blowing and forming expertise of itinerant glassworkers to supply the local markets. In addition to the workshops known within or close to forts and towns, some glass production may perhaps have taken place in association with centres of pottery production, as has been noted elsewhere in Britain, although no evidence for this has yet been recognised.

There is very little information about the products of the workshops in the region. Window panes and many of the everyday tablewares and household vessels were probably produced locally; these probably did not travel long distances, either because they were unlikely to survive intact or because their value was less than the cost of transport, but it is generally not possible to recognise where they were produced. In the Roman world, the archaeological record of successful glassworking generally contains very few clues as to what was being produced, even if vestiges of the furnace and crucibles, or evidence for blowing and other production waste have survived. Many of the activities are likely to be completely invisible, because they did not involve the use of furnaces or crucibles.

The production of many categories of objects, such as beads, bangles, counters, finger rings and hairpins, belong to this 'invisible' category. Modern glassworkers have demonstrated that a wide range of beads can be produced by heating a lump of glass in a domestic fire or open hearth and manipulating it on a metal rod (Gam 1993, 261-267), and that seamless glass bangles and rings can be formed in a similar manner, using a second rod to widen the aperture symmetrically (Haevernick 1960, 23-28; Korfmann 1966). Similarly, plano-convex counters can be formed by dropping a small quantity of melted glass onto a flat surface, and hairpins by heating and manipulating glass rods. Therefore any discussion of individual centres of production for these and other objects is dependent on concentrations of finds and on the recognition of diagnostic manufacturing details. For example, the possibility of bead production in the *vicus* at Castleford discussed above was based on a significant concentration of unusual bead types. A similar case has been made for production of bangles with twisted cord decoration 'somewhere in the vicinity of York', based on the local concentration of finds (Price 1988b, 342-47), though this is unlikely to have been the only place of production, as similar bangles are found in southern Britain in the third quarter of the first century, and throughout the Roman North in late first- and early second-century contexts.

It is also possible that small and simple vessels could have been blown without using either a furnace or a crucible, either by heating a lump of glass in an open hearth, attaching it to the end of a blowing iron and blowing it to shape, or by fusing broken vessel fragments together to form a larger sheet of glass, attaching this to the end of a blowing iron and then re-heating and blowing it to form the vessel (George Scott *pers comm*). Each of these methods is technically possible using the sources of heat available in the Roman world, although there is no certainty that either was in fact employed.

To sum up, the evidence for glass production in Yorkshire and northern Britain, most of which has come to light in the past two decades, is broadly comparable with that found elsewhere in the Roman world. The record is scattered and ephemeral, and many of the activities certain to have taken place in the region are simply not visible to the modern researcher. Nonetheless, some episodes of glass blowing and forming taking place at settlements in the region at various times from the later first to the fourth century have been identified, and more substantial information may well be discovered in the future.

Acknowledgements

I have been assisted in preparing this paper by many colleagues and friends. Heidi Amrein, Donald Bailey, Justine Bayley, Hilary Cool, Ian Freestone, Yael Gorin-Rosen, Kay Hartley, Caroline Jackson, Gusta Jacobson, Valerie Maxfield, Marie-Dominique Nenna, Paul Roberts, George Scott, John Shepherd and Sally Worrell have discussed glass production with me and/or provided information about finds, Catherine Johns has read and commented on a draft of the text; and Yvonne Beadnell and Trevor Woods have produced the illustrations. I am very grateful to all of them.

Endnotes

1. Literary references to Roman glass production from the late first century BC onwards are numerous, and of varying value. The works of Strabo (*Geography*) and Pliny (*Natural History*) are well known, (see notes 4-6 and 11 below), and many other writers also make reference to aspects of glass production. A study of these has been made by Trowbridge (1930).

2. Epigraphic records provide a limited range of information about glassworkers. Funerary records indicate the place of death, and give the names and sometimes the place of origin, as on the early third-century tombstone of Julius Alexsander at Lyons, who was of African origin and a citizen of Carthage (Foy and Sennequier 1989, 61-62 no 8). A few inscriptions on the body or base of mould-blown vessels or stamped into the handles of blown vessels also give both the name and place of origin, such as Aristeas from Cyprus (Constable-Maxwell 1979, no 280), Artas and Philippos from Sidon (Fremersdorf 1938), and Sentia Secunda of Aquileia (Harden 1969, 49, 73, pl 4B) but these do not always indicate that the vessels were produced at the places named. It has been argued (Fremersdorf 1965/66, 29, pl 10.2) that the letters CCAA on the bases of bottles and flasks originate in the colonia at Köln (*Colonia Claudia Ara Agrippinensium*). Most frequently, only a personal name or an abbreviation is given, in which case any identification of a possible place or region of production is dependent on concentrations of finds, as is the case with bottles with abbreviations of Frontinus which are widely distributed in the middle and lower Rhineland, northern Gaul and Britain (Price 1978, 76, fig 61).

3. Very few representations of Roman glassworking have survived. Scenes on pottery lamps and a terracotta group (Abramic 1959; Baldoni 1987; Price 1988a) are discussed in this paper.

4. Although literary evidence indicates that some Mediterranean glass workers in the early Roman empire favoured particular sources of sand. Strabo (*Geography* 16, 758) recorded a source of sand between Ptolemais (Acre) and Tyre used by glassmakers at Sidon, and Pliny noted the glassmaking qualities of the sand from the mouth of the River Belus near Ptolemais (*Natural History*, 36, 190-92) and from the River Volturno found on the shore between Cuma and Literno in Campania (*Natural History*, 36, 194).

5. See Pliny, *Natural History*, 36, 192-94

6. This contrasts with Pliny's description of glassmaking. Writing in the third quarter of the first century AD, he indicated that glass in Campania was melted in a series of furnaces, being formed into lumps and then re-melted, which implies a fritting process (*Natural History* 36, 193-94).
Fritting is the preliminary heating of the raw materials to fuse them partially and to burn off impurities. The frit produced is then broken up and the best pieces are remelted to a much higher temperature to produce glass.

7. Other depictions have been claimed as representations of glass furnaces but the identifications are less secure. For example, a scene on a fragment from a fifth-century lamp found in Carthage was recently interpreted as a glass furnace and a glass blower (Caron and Lavoie 1997; Stern 1999, 445, fig 22). However, this interpretation does not take account of all the elements of the scene found on the complete lamps, which show that the structure and the human figure are sited in the top of a tree! (Donald Bailey *pers comm*; see also Bailey 1999)

8. Annealing is the process of reducing the temperature of glass vessels gradually to ensure uniform cooling of the surfaces to minimise the risk of them cracking or breaking; this may have taken place in a heated chamber or in a pit.

9. A pit of this kind was found in the late Antique glass workshop at Bet Sha'an (*Scythopolis*) in Israel (Yael Gorin-Rosen *pers comm*).

10. A moile is the glass between the end of the blowpipe and the top of the vessel which is left after the vessel has been blown and the rim finished off.

11. Cullet is broken vessel glass collected for remelting. The collection of broken glass is noted in Roman literature; Martial (*Epigrams* 1.41.1-5) and Statius (*Silvae* 1.6.70-74) both record this practice in Rome in the late first century AD, presumably for recycling (Leon 1941; Whitehouse 1999, 78).

12. It has been suggested (Price 1978, 70; Price and Cool 1991, 23) that the letters CCV found on the bases of some rectangular and square bottles in Britain (*RIB* II.2, no 2419.96-99) may indicate that they were produced in Colchester (*Colonia Claudia Victricensis*). However, this interpretation has been questioned, as the title *Claudia* is not directly attested in connection with Colchester (*RIB* II.2, 110).

13. Ottaway (1993, 64-66, fig 31) notes that Aurelius Victor, writing in the fourth century, described York as a *municipium* at the time of visit of Septimius Severus in AD 209, and it apparently became *colonia ebur[acensium]* before AD 237 as it is thus described on the altar of M. Aurelius Lunaris.

14. I have also looked through the Coppergate material, but much of the information in this account is derived from the published accounts (Jackson *et al* 1998; Cool *et al* 1999).

Bibliography

Abramic, M, 1959 Eine römische Lampe mit Darstellung des Glasblasens. *Bonner Jahrbücher* **159**, 149-51

Artis, E T, 1828 *The Durobrivae of Antoninius identified and illustrated in a series of plates, exhibiting the excavated remains of that Roman station, in the vicinity of Castor, Northamptonshire*. London: privately printed

Auth, S H, 1975 Roman glass. In C W Clairmont (ed.), *Excavations at Salona, Yugoslavia, 1969-1972*. New Jersey: Noyes Press, 145-67

Bailey, D, 1999 Glasshouse or tree house? *J Glass Stud* **41**, 167-68

Baldoni, D ,1987 Una lucerna romana con raffigurazione di officina vetraria: alcune considerazioni sulla lavorazione del vetro soffiato nell'antichità. *J Glass Stud* **29**, 22-29

Bayley, J, 1987 Viking glassworking - the evidence from York. *Annales du 10e Congres de l'Association Internationale pour l'Histoire du Verre, Madrid-Segovia 1985*, 245-54

----, and Budd, P, 1998 In Cool and Philo (1998), 195-222

Boon, G C, 1974 Silchester, the Roman town of Calleva. Newton Abbot and London: David and Charles

Bormann, E, 1914 Die Grabungen in Standlager zu Carnuntum: Epigraphischer Anhang. *Die römische Limes in Österreich*, **XII**, 315-42

Caron, B, and Lavoie, C, 1997 Un fragment du lampe représentant un four verrier. *J Glass Stud* **39**, 197-98

Chambon, R, 1955 *L'histoire de la verrerie en Belgique du 2e siècle à nos jours*. Brussels

Charleston, R J, 1978 Glass furnaces through the ages. *J Glass Stud* **20**, 1-33

Constable-Maxwell, 1979 *Sale Catalogue of the Constable-Maxwell Collection of Ancient Glass*. London: Sothebys

Cool, H E M, Jackson, C M, and Monaghan, J, 1999 Glass-making and the Sixth legion at York. *Britannia* **30**, 147-61

----, Lloyd-Morgan, G, Hooley, A D, 1995 *Finds from the Fortress*. Arch York. The Small Finds **17/10**. York, CBA

---- , and Philo, C, 1998 *Roman Castleford Volume 1. The Small Finds*. Wakefield: Yorks Archaeol **4**

----, and Price J, 1995 *Roman glass from excavations in Colchester, 1971-1985*. Colchester: Colchester Archaeol Rep **8**

----, & ----, 1998 The vessels and objects of glass. In Cool and Philo (1998), 141-94

Follmann-Schulz, A-B, 1991 Fours de verriers romains dans la province de Germanie Inférieure. In Foy and Sennequier (1991), 35-40

---- , 1992 *Die römischen Gläser in Rheinischen Landesmuseum Bonn*. Rheinland-Verlag GmbH Köln

Foy, D, 1997 Commerce et technologie du verre antique: le temoignage de l'èpave 'Ouest Embiez 1'. *Congrés de la Nationale Societe pour Histoire et Sciences*, 120, Aix-en-Provence 1995, 121, Nice 1996, Archéologie - Mediterranée, 121-34

---- , and Sennequier, G, (eds) 1989 *À travers le verre*. Rouen: Museé departmental des Antiquités

---- , and ---- , (eds) 1991 *Ateliers de Verriers de l'Antiquité à la période Pré-industielle*. Rouen: Association Française pour l'Archéologie du Verre

Freestone, I C, and Gorin-Rosen, Y, 1999 The great glass slab at Bet She'arim, Israel: an early Islamic glassmaking experiment? *J Glass Stud* **41**, 105-16

Fremersdorf, F, 1938 Römische Gläser mit buntgefleckter Oberfläche. *Festschrift für August Oxé*. Darmstadt, 116-121

----, 1965/66 Die Anfänge der römischen Glashütten Kölns. *Kölner Jahrbuch* **8**, 24-43

Gaitzsch, W, 1999 Spatromische Glashutten im Hambacher Forst. Die Werkstatt des ECVA-Produzenten. In Polfer (1999), 125-49

Gam, T. 1993, Experiments in glass: present and future. Annales du 12e Congrès de l'Association Internationale pour l'Histoire du Verre (Vienna 1991), 261-270

Gorin-Rosen, Y, 1995 Hadera, Bet Eli'ezer. *Excavations and Surveys in Israel* **13**, 42-43

Guido, M, 1978 *The glass beads of the prehistoric and Roman periods in Britain and Ireland.* London: Rep Res Comm Soc Antiq London, **35**

Haevernick, T E, 1960 *Die Glasarmringe und Ringperlen der Mittel- und Spätlatènezeit auf dem europäischen Festland.* Bonn: Habelt

Harden, D B, 1969 Ancient Glass, II: Roman. *Archaeol J* **126**; 44-77

Henderson, J, 1999 The debris from a glass-making industry. In P Halkon and M Millett, *Rural Settlement and Industry: Studies in the Iron Age and Roman Archaeology of Lowland East Yorkshire.* Leeds: Yorks Archaeol Rep **4**, 165-67

Hochuli-Gysel, A, 1993 Römisches Glas aus dem Südwesten von Frankreich. *Annales du 12e Congrès de l'Association Internationale pour l'Histoire du Verre (Vienna 1991)*, 79-88

Inman, R, Brown, D R, Goddard, R E, Spratt, D A, 1985 Roxby Iron Age settlement and the Iron Age in North-East Yorkshire. *Proc Prehist Soc* **51**, 181-213

Isings, C, 1980 Glass from the canabae legionis at Nijmegen. *Berichten van Rijksdienst voor het Oudheidkundig Bodenmonderzoek* **30**, 281-346

Jackson, C M, Cool, H E M, and Wager, E C W, 1998 The Manufacture of Glass in Roman York. *J Glass Stud* **40**, 55-61

Korfmann, M, 1966 Zur Herstellung nahtloser Glasringe. *Bonner Jahrbuch* **166**, 48-61

Lang, J, and Price, J, 1975 Iron tubes from a late Roman glassmaking site at Mérida (Badajoz), in Spain. *J Archaeol Sci* **2**, 289-96

Leon, H J, 1941 Sulphur for broken glass. *Trans American Philological Assoc* **72**, 233-36

Monaghan, J, 1997 *Roman Pottery from York.* Archaeol York, The Pottery **16/8**. York: CBA

Morel, J, Amrein, H, Meylan, M-F, and Chevalley, C, 1992 Un atelier de verrier du milieu du 1er siècle apr. J-C à Avenches. *Archäologie der Schweiz*, **15.1**, 2-17

Neal, D S, 1981 *Roman Mosaics in Britain.* London: Britannia Mono **1**

Nenna, M-D, Vichy, M, and Picon, M, 1997 L'atelier de verrier de Lyon, du Ier siècle apres J.-C., et l'origine des verres 'romains'. *Revue d'Archéométrie* **21**, 81-87

Ottaway, P, 1993 *Roman York.* London: Batsford/English Heritage

Pliny, *Natural History* Eichholz, D E, (trans) 1962 *Pliny: Natural History, X.* London: Heinemann

Polfer, M, (ed) 1999 *Artisanat et productions artisanales en lieu rural dans les provinces du nord-ouest de l'Empire romain.* Montagnac: editions Monique Mergoil; Monographies Instrumentum **9**

Price, J, 1974 Some Roman glass from Spain. *Annales du 6e Congrès de l'Association Internationale pour l'Histoire du Verre (Cologne 1973)*, 65-84

---- , 1978 Trade in glass. In J du Plat Taylor and H Cleere (eds), *Roman Shipping and Trade: Britain and the Rhine Provinces.* London: CBA Res Rep **24**, 70-78

---- , 1985 The Roman glass. In L F Pitts & J K St Joseph, *Inchtuthil. The Roman legionary fortress.* London: Britannia Mono **6**

---- , 1988a An Egyptian terracotta group showing Eros beside a glass furnace. *Antiq J* **68**, 317-19

---- , 1988b Romano-British Glass Bangles from East Yorkshire. In J Price and P R Wilson (eds), *Recent Research in Roman Yorkshire - Studies in Honour of Mary Kitson Clark (Mrs Derwas Chitty).* Oxford: Brit Archaeol Rep, Brit Ser 193, 339-66

---- , 1990 Roman vessel glass and window glass. In M R McCarthy, *A Roman, Anglian and Medieval site at Blackfriars Street, Carlisle.* Kendal: Cumberland Westmorland Antiq Archaeol Soc Res Ser **4**, 163-79

---- , 1995 Roman glass. In D Phillips and B Heywood, *Excavations at York Minster, Volume I; part 2, the finds.* London: HMSO, 346-71

---- , 1998 The Social Context of Glass Production in Roman Britain. In P McCray (ed), *The Prehistory and History of Glassmaking Technology.* Cincinnati: Ceramics and Civilisation vol **VIII**, 331-48

---- , *forthcoming* The Roman glass. In N Holmes (ed), *Excavations at Cramond*

---- , and Cool, H E M, 1991 The evidence for the production of glass in Roman Britain. In Foy and Sennequier (1991), 23-30.

---- , and Worrell, S, forthcoming The Roman Glass. In I M Ferris and R F J Jones, *Excavations at Binchester.*

Radic, I, and Jurišic, M, 1993 Das antike Schliffswrack von Mljet, Kroatien. *Germania* **71**, 113-38

RIB II.2 = Collingwood, R G, and Wright, R P, (S S Frere and R S O Tomlin (eds)) *The Roman Inscriptions of Britain. Volume II Instrumentum Domesticum, fascicule 2: weights, gold vessel, silver vessels, bronze vessels, lead vessels, pewter vessels, shale vessels, glass vessels, spoons.* Stroud: Sutton

Rütti, B, 1991 *Die römischen Gläser aus Augst und Kaiseraugst.* Forschungen in Augst Band **13**. Augst: Römersmuseum

Saleh, S A, George, A W and Helmi, F M, 1972 Study of glass and glass-making processes at Wadi el-Natrun, Egypt in the Roman period 30 B.C. to 359 A.D. *Stud in Conservation* **17**, 143-72

Shepherd, J, and Heyworth, M, 1991 Le travail du verre dans Londres romain (Londinium): un état de la question. In Foy and Sennequier (1991), 13-22

Stern, E M, 1995 *Toledo Museum of Art: Roman Mold-Blown Glass; the 1st through 6th centuries.* Rome: "L'Erma" di Bretschneider

---- , 1999 Roman glassblowing in a cultural context. *American J Archaeol* **103**, 441-84

Sternini, M, 1995 *La Fenice di Sabbia: storia e tecnologia del vetro antico.* Bari: Edipuglia

Strabo, *Geography* Jones, H L, (trans) 1930 *The Geography of Strabo, VII.* London: Heinemann

Trowbridge, M L, 1930 *Philological Studies in Ancient Glass.* Urbana: Univ Illinois Press

van Geesbergen, D, 1999 Les ateliers de verriers dans le nord de la Gaule et en Rhenanie (1er-IVe s. ap. J.-C.) In Polfer (1999), 105-24

Wacher, J, 1978 *Roman Britain.* London: Dent

Weinberg, G D, 1963 Découvertes archéologiques récentes de verres. Grèce. *Bulletin des Journées Internationales du Verre* **2**, 104-07

Whitehouse, D, 1999 Glass in the Epigrams of Martial. *J Glass Stud* **41**, 73-81

COPPER ALLOYS IN ROMAN YORKSHIRE

DAVID DUNGWORTH

Introduction

This paper will examine the production and use of copper alloys in Yorkshire during the Roman period. In many ways the copper alloys of Roman Yorkshire are directly comparable to those of the wider Roman world. In some cases Yorkshire alloys may differ from the broad traditions of the Roman world because of local historical circumstances.

The production of copper alloys in the Roman World

During prehistory there were a large number of changes in the kinds of copper alloys produced. At any one time in prehistory, however, only one or occasionally two alloys tended to be used. Where more than one alloy was used this can often be explained simply in technological terms. For the Roman period, however, there is abundant evidence for the use of a range of copper alloys. Copper, bronze, brass were used, with and without the addition of lead, and complex alloys were also formed by mixing these different alloys. The variations in alloy composition reflected in part the availability of raw materials and the technology needed to produce particular alloys, but were also influenced by social and economic factors. Different traditions of alloying had grown up in various parts of what were to become the Roman Empire.

The alloys used in the Mediterranean period during the Republican period were usually formed through the addition of tin and lead to copper to produce leaded and unleaded bronze. The amounts of tin and lead added seems to have been determined in many cases by the nature of the object being cast (Craddock 1975; 1988). Most objects were made of tin bronze with between 5 and 15% tin. Mirrors were produced using higher than normal tin levels (20% or more) which produced a silvery-white alloy known as *speculum*. Statues were always made from bronze with the addition of substantial amounts of lead, while smaller statuettes had more varied lead contents. The use of leaded bronze can also be seen in the production of some vessels (den Boesterd and Hoekstra 1965). The consistent alloy composition of many Roman artefacts suggests the use of recipes. Craddock has had considerable success in relating the composition of many artefacts to recipes mentioned by Pliny (Craddock 1988).

The Augustan period saw the first widespread use of brass in the Ancient World when it was used in the production of *sestertii* and *dupondii* (Dungworth 1996b). Brass was also used in the production of many of the military fittings of the Principate (such as *lorica segmentata* hinges — Bishop and Coulston 1993, 183). Prior to this brass was almost unknown due to the difficulties involved in its production (Craddock 1978; Bayley 1984). Zinc is a highly volatile element which boils before copper melts. Thus when copper is smelted any zinc present as an impurity in the ore will usually be lost as a vapour. The systematic production of brass in the Ancient World only began once the cementation process had been developed (Craddock 1978; Bayley 1984). In the cementation process zinc ore and pure copper were placed together in a sealed crucible and heated. The heat would reduce the zinc ore to zinc but the escape of the zinc vapour was prevented as the crucible was sealed. The reduced zinc vapour would then diffuse into the copper to produce brass. The zinc content of cementation brass would usually be between 23 and 28% (Werner 1970). The relative obscurity of this method of production, and the fact that many of the earliest artefacts made of brass were 'official' (coins and military equipment), has led some to suggest that production was controlled through a 'state monopoly' (Grant 1946; Caley 1964; Bishop and Coulston 1993, 183). The monopoly thesis seems somewhat unlikely, however, as brass was quickly used for the production of a range of 'civilian' artefacts, such as Nauheim and Colchester A brooches (Bayley 1992). A detailed examination of the incidence of brass on sites in Roman Yorkshire (see below) also suggests that brass was widely available.

The nature of the cementation process produced a fairly distinctive brass while the use of alloying recipes for bronzes resulted in a limited number of 'stock' alloys in common usage in the Roman Empire. Many everyday copper alloy objects, however, were composed of alloys of a mixed nature, and contain copper, zinc, tin and/or lead (Craddock 1975; Bayley 1992; Dungworth 1997; Riederer and Briese 1974). Copper alloys which contain zinc and tin are usually referred to by the modern term gunmetal. It is possible that such alloys were deliberately created (such as by adding tin to brass) but it has been observed that the zinc and tin contents of Roman alloys are negatively correlated (Craddock 1975, fig 4; Dungworth 1997, fig 2) which suggests that ancient gunmetals were usually formed by mixing scrap bronze and brass together. This does not mean that the smiths did not care about the composition of their alloys. The brass used to manufacture some first-century brooches (such as the Hod Hill and Aucissa types) has a relatively low zinc content (around 20%) and small quantities of lead and tin are also present. This alloy was probably formed by carefully mixing bronze and brass in the ratio 1 to 4. The consistent composition of these (and other) brooches suggests that this mixing often followed strict recipes (Bayley 1992). One of the areas of Roman copper metallurgy in need of further exploration is the relationship between the use of recipes for some artefacts and the use of scrap for others. This curious situation may have arisen due to variations in the modes of production or due to differing attitudes to the use of artefacts.

The production of copper alloy artefacts in Roman Yorkshire

The majority of Roman period copper alloy artefacts found in

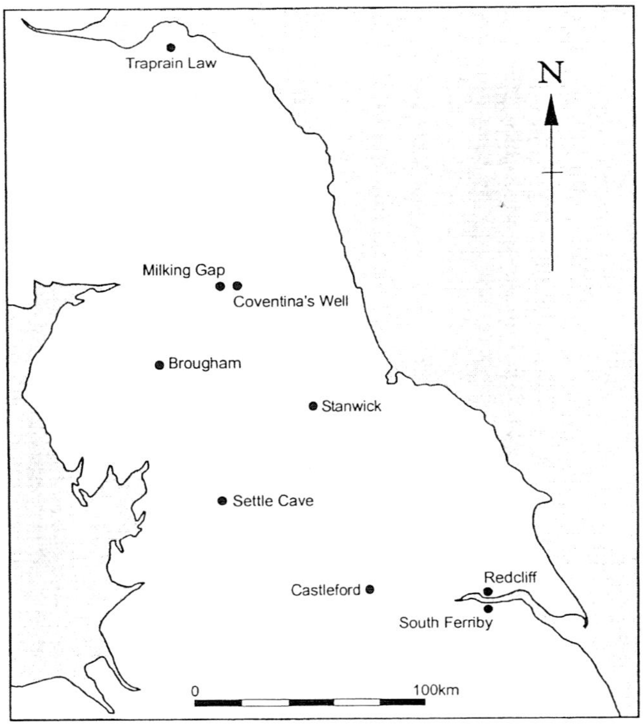

Figure 1. Map of northern Roman Britain showing the sites mentioned in the text.

Yorkshire (Figure 1) have compositions similar to those found in objects analyzed from other parts of the Roman world, such as the *lorica segmentata* fittings from the recent excavations at Castleford are made of brass (Dungworth 1998; cf Bishop and Coulston 1993, 183). These similarities may arise because some items were manufactured in a single workshop or factory and then traded throughout the Empire. Alternatively these artefacts may have been produced in a number of workshops throughout the Empire but to a standard 'recipe' (cf. Welbourn 1985). A few artefact types have distributions limited to Yorkshire or northern Britain and so are likely to be local products. This section will examine the use of copper alloys in the production of two specifically Yorkshire artefacts — Dragonesque and Fantail brooches (Figure 2).

Dragonesque Brooches

Dragonesque brooches have been studied in a series of short articles by Bulmer (1938) and Feachem (1951; 1968). The Bulmer/Feachem typology divides brooches according to the nature of the enamel decoration. The earliest type (here labelled P) is plain and consists largely of an S-shape. From this was derived the two simplest forms of enamel decorated brooches with central zones of decoration (those with a circular motif are labelled R and those with a series of squares are labelled S). The transformation of the squares into lozenges leads to the last type (labelled L). Bulmer envisaged a chronological sequence from type P to types S and R. Finally type L developed out of type S (see Bulmer 1938, figs 3 and 4). This typology is now widely used (the section of Dragonesque brooches in Hull's widely used but unpublished typology is essentially the same as the Bulmer/Feachem scheme outlined above). The relative rarity of Dragonesque brooches has made dating difficult but they appear roughly at the same time as the Roman Conquest and continue in use until the middle or end of the second century AD.

A fuller understanding of Dragonesque brooches can now be gained due to the recent analysis of the alloys used in their manufacture (Table 1). The types of brooches as defined by the Bulmer/Feachem typology do not show the clear association of type with metal content seen in some first-century bow brooches (such as Colchester A, Hod Hill, etc) (see Bayley 1995). A distinction can be made, however, between the early P and R types (plain and circular decorative motif) which generally have modest zinc contents and relatively high tin contents (mean values are 3.3% and 7.9%, respectively), and the later S and L types (square and lattice decoration) which have higher zinc and lower tin contents (mean values of 8.7% and 3.4%, respectively). Some unenamelled Dragonesque brooches have intricate moulded decoration (usually including a central roundel) and are referred to in Table 1 as P/R. All that remains of one of the brooches from Kelco cave (Feachem 38) is the head and so it cannot be assigned to a type with certainty. The lack of enamel on the surviving fragment and the composition (high tin and low zinc) suggest that it belongs to and early P or R type.

The generally modest zinc levels (usually less thn 15%) in all Dragonesque brooches suggests a limited use of scrap metal derived from the Roman World (although some was used as only one of the brooches had no zinc present). The trend from low zinc, high tin alloy to moderate zinc, low tin alloy may indicate a gradual increase in the use of Roman scrap metal (especially brass).

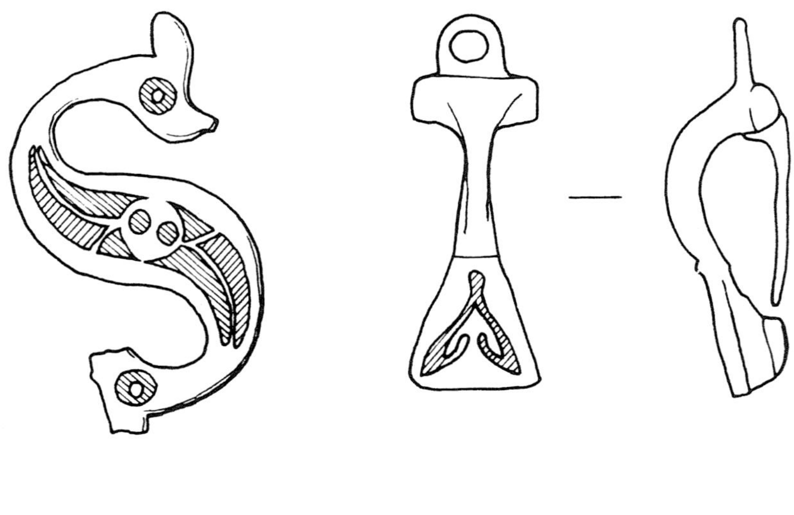

Figure 2. Dragonesque and Fantail brooches. (Scale 1:1)

One of the Dragonesque brooches which has been analyzed, that from Milking Gap, has a very low zinc content. In fact the alloy used for this brooch is very similar to that used for Iron Age artefacts in northern Britain (Dungworth 1996a). From this it may be deduced that Dragonesque brooches were first produced before the Roman Conquest (it is possible, however, that this alloy type continued in occasional use after the Conquest). If the Milking Gap brooch is one the earliest surviving examples of this type then the Bulmer/Feachem typological scheme (ie. P before R) may be incorrect.

Fantail brooches

Fantail brooches are another of the well-known types of bow brooch from Roman Britain. They all have catch-plate ends which expand but they come in various shapes and sizes: some have discs or plates added to the bows, while others have trumpet heads. There is one type, however, which is small, plain, and bears a triangular zone of curvilinear decoration on the tail usually filled with enamel (Figure 2). These Fantail brooches vary little in their size or decoration and have been found in considerable numbers in southern Yorkshire and the northern parts of Lincolnshire and Nottinghamshire (notably Castleford and South Ferriby). The restricted distribution and form of these brooches suggests that they may be a local product. Four of these brooches have been analyzed (Dungworth 1995) and found to be leaded bronze although the tin levels are relatively modest (3–6.5%). A similar alloy was used for the production of the cosmetic and mortar pestles (Jackson 1985). The addition of such a low proportion of tin to copper is curious as the vast majority of Roman period bronzes have around 10% tin (as do Iron Age bronzes). Where

Provenance	Reference	Bulmer / Feachem Type	Cu	Zn	Pb	Sn	Fe	As
Attermire Cave (Settle)	Feachem 39	P	84.6	1.38	5.30	8.53	0.16	nd
Bainesse Farm, Catterick	cf Feachem i/iii	P	88.9	4.37	1.23	5.47	0.06	nd
Kelco Caves (Settle)	Feachem 38 (head only)	?	83.0	0.60	3.37	13.0	nd	nd
Kelco Cave (Settle)	Feachem 40 Bulmer H	P/R	79.6	6.03	3.23	10.9	0.16	nd
Edinburgh Castle	Bulmer H cf Feachem 40	P/R	83.3	12.2	0.81	3.22	0.39	0.11
Attermire Cave (Settle)	Feachem i	R	86.0	0.62	3.75	9.45	0.22	nd
Victoria Cave (Settle)	Feachem 24 Bulmer H	R	86.8	1.91	2.01	9.17	0.09	nd
Victoria Cave (Settle)	Feachem 42 cf Bulmer H	R	87.8	0.32	7.64	4.28	nd	nd
Milking Gap	Bulmer H2	R	89.6	0.12	0.39	9.10	0.37	0.38
Watercrook	cf Bulmer H1	R	80.2	3.12	4.99	11.0	0.16	nd
Victoria Cave (Settle)	Bulmer D3	S	84.9	4.61	6.11	4.26	0.15	nd
Old Wintringham	cf Bulmer D2	S	87.7	1.82	0.65	9.21	0.14	0.43
Castleford	cf Bulmer C	S	84.1	6.30	6.26	3.19	0.19	nd
Watercrook	cf Bulmer D3	S	85.5	11.7	0.32	2.11	0.29	nd
Traprain Law	Feachem 39	S	87.0	5.67	3.18	3.70	0.28	0.17
Sewells Cave (Settle)	Feachem ii	L	88.1	8.57	0.69	1.71	0.18	0.68
Britain (unprovenanced)	Bulmer G1	L	80.0	15.2	2.50	1.40	0.17	ND
Richborough	Bulmer 17	L	62.1	15.3	0.20	1.20	ND	ND

Table 1. Alloy Composition of Dragonesque brooches from northern Britain (the last two were analyzed by Craddock (1975) and Bayley (1992)). [nd = not detected; ND = not determined]

Roman alloys have such low levels of tin they usually have some zinc present (the inverse correlation between zinc and tin mentioned earlier). The reasons for the low level of tin in these bronzes are unclear. It is unlikely that tin was expensive or scarce as it is regularly found in a range of alloys from northern Britain at moderate (5–15%) and even high levels (up to 25%). There are no obvious metallurgical reasons for using a low tin bronze for the production of small decorative objects — the alloy would not be particularly easier to cast or work than a higher tin bronze. A low-tin bronze would, however, have been slightly redder than the normal bronze (ie. 8–12% tin). It would seem, therefore, that the low level of tin was a deliberate choice designed to make the object appear distinctive. This alloy may have been most distinctive simply to the metalworkers who produced and used it (rather than those who used and wore the ornaments made from it). The encoding of technological activity in ritual and the role of cosmological criteria in craft activities are well known in the ethnographic literature (such as van der Merwe and Avery 1988).

Copper alloys from Roman sites in Yorkshire

The alloys used in Roman Yorkshire reflect the modes of production (both local and empire-wide). One of the most significant aspects of alloy choice is the relationship between artefact typology and alloy composition. Artefacts (such as specific brooch types) found in northern and southern Britain as well as the continent often have identical compositions. Many other artefacts, however, do not seem to be made to a 'recipe'.

An alternative approach (ie. non-typological) explores the relationship between settlements and the alloys used (Dungworth 1995; 1997). In particular, this information can be used to examine the relationship between site status and the use of brass. As mentioned above, brass has traditionally been seen as a Roman metal and saw little or no use in northern Europe before the first century AD. If brass was produced and supplied by a 'state monopoly' then the largest proportion of brass would be expected on highly Romanised settlements such as towns and forts. The range of alloys (including brass) used on different types of Roman-period sites in northern Britain is shown in Figure 3. It can be seen that the proportion of brass on a range of Romanised sites (forts, *vici*, towns, and villas) is broadly similar (about 20% of alloys from these sites are brasses). This suggests that there was little differentiation between these types of sites in terms of the availability of brass. The proportion of brass on these sites does not seem to reflect any military–civil or urban–rural settlement hierarchy. In fact, the category of site with the highest incidence of brass is the small isolated rural farmstead. The high proportion of brass on such sites is the opposite of what might be expected from traditional explanations of the Roman occupation of Britain. The indigenous inhabitants of northern Britain should not be seen as passively receiving those elements of Roman culture offered them: they apparently had more access to a 'Roman' resource than many more Romanised communities. The high proportion of brass on small rural sites seen here is similar to that seen in late Iron Age and 'Celtic' metalwork (Dungworth 1996a). Brass objects are common on Yorkshire sites which were occupied immediately prior to the Conquest (such as Stanwick and Redcliff). The widespread use of brass in the early Empire suggests that there was not an official monopoly on the production. Brass was widely available throughout northern Britain, even for indigenous manufacture (for a broadly similar picture from a Polish site, see Stos-Gale 1993).

There are a few classes of sites, especially hillforts and caves, where much lower proportions of brass (ie. less than 20%) are found. Given the above discussion of indigenous metalwork, it would be difficult to argue that the low proportion of brass on these sites reflects a lack of Romanization. This may be resolved by arguing that material on these sites was carefully selected before deposition (Hill 1994).

It has become increasingly accepted that the material culture of the Iron Age which survives in the archaeological record is highly influenced by active selections made by the depositors and so 'cannot be taken at face value' (Hill 1994). The low levels of brasses from some Roman-period sites in northern Britain may reflect similar structured deposition. Sites which have produced low levels of brass include Traprain Law, the Settle Caves, Coventina's Well, and the Brougham cemetery.

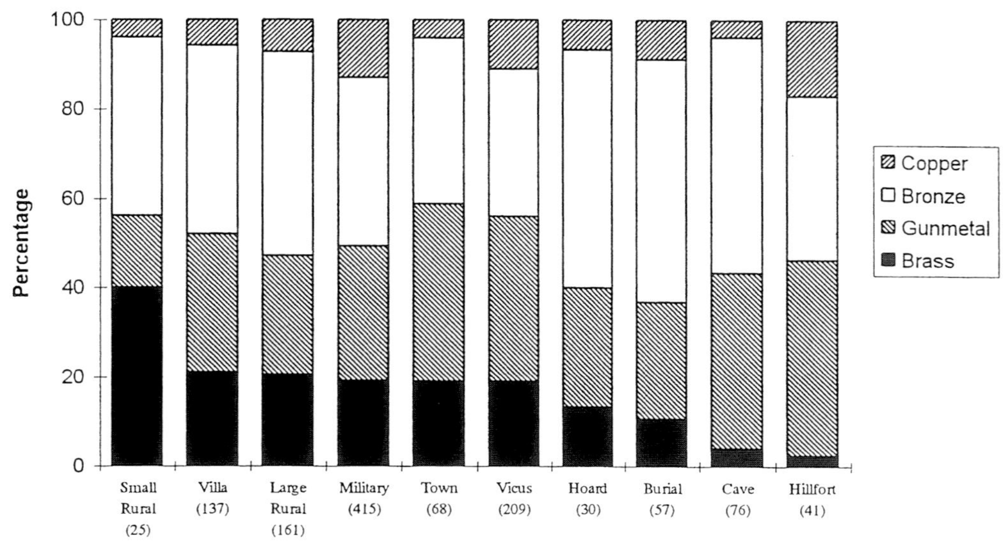

Figure 3. Alloys used on a range of Roman sites in northern Britain

All four of these sites can be easily interpreted as arenas for ritual activities. Millett (1993) has demonstrated that the artefacts deposited in an early Roman cemetery were carefully selected and did not reflect the full range of artefacts available. A similar process would seem to be operating in the deposition of copper alloys on a range of Roman-period 'ritual' sites. Brass may have been seen as an inappropriate metal for deposition in some circumstances. Alternatively, some artefacts (which were only incidentally made of brass) may have been regarded as inappropriate.

Conclusions

Copper alloys were used to produce a wide range of Roman decorative metalwork. The composition of the alloys used (ie. the levels of zinc, tin and lead) varied considerably. The traditional Iron Age copper alloy in northern Britain was a tin bronze. While tin continues to be the most ubiquitous alloying element in Roman alloys it was almost always accompanied by at least small quantities of zinc. The traditional alloy used for the production of cast objects in the Mediterranean world (from the Hellenistic period onwards) was a leaded bronze. The expansion of the Roman Empire in to north-western Europe occurred at a time when brass first became widely used. All of these copper alloys (and mixtures of these) were used in Roman Yorkshire. The few types of copper alloy object which are concentrated in Roman Yorkshire (such as Dragonesque and Fantail brooches) tend to have rather mixed compositions (containing appreciable levels of tin, zinc and lead). This probably reflects the widespread use of scrap metal.

Bibliography

Bayley, J, 1984 Roman brass-making in Britain. *J Hist Metallurgy Soc* **18**, 42–43

----, 1992 *Non-Ferrous Metalworking in England Late Iron Age to Early Medieval.* Unpublished Univ London PhD thesis

Bishop, M C, and Coulston, J C N, 1993 *Roman Military Equipment.* London: Batsford

den Boesterd, M H P, and Hoekstra, E, 1965 Spectrochemical analyses of the bronze vessels in the Rijksmuseum G M Kam at Nijmegen. *Oudheidkundige Mededelingen* **46**, 100–27

Bulmer, W, 1938 Dragonesque brooches and their development. *Antiq J* **18**, 146–53

Caley, E R, 1964 *Orichalcum and Related Ancient lloys.* Numismatic Notes and Mono **151**. New York: American Numismatic Soc

Craddock, P T, 1975 *The Composition of Copper Alloys in the Classical World.* Unpublished Univ London PhD thesis

----, 1978 The composition of the copper alloys used by the Greek, Etruscan and Roman Civilisations 3: The origins and early use of brass. *J Archaeol Sci* **5**, 1–16

----, 1988 Copper alloys of the Hellenistic and Roman World: new analyses and old authors. In J E Jones (ed) (1998) *Aspects of Ancient Mining and Metallurgy. Acta of a British School at Athens Centenary Conference, 1986.* Bangor: Univ College Wales, Bangor, 55–65

Dungworth, D B, 1995 *Iron Age and Roman Copper Alloys from Northern Britain.* Unpublished Univ Durham PhD thesis

----, 1996a The production of bronzes in Iron Age Britain. *Proc Prehist Soc* **26**, 1–23

----, 1996b Caley's 'zinc decline' reconsidered. *Numismatic Chron* **156**, 228–34

----, 1997 Roman copper metallurgy: results from the analysis of artefacts. *J Archaeol Sci* **24**, 901–10

----, 1998 EDXRF analysis of copper-alloy samples. In H E M Cool and C Philo (eds) (1998) *Roman Castleford Excavations 1974–85. Volume 1: The Small Finds.* Wakefield: Yorks Archaeology **4**, 117–20

Feachem, R W, 1951 Dragonesque fibulae. *Antiq J* **31**, 32–44

----, 1968 Notes — Dragonesque fibulae. *Antiq J* **48**, 100–02

Grant, M, 1946 *From Imperium to Auctoritas: A Historical Study of Aes Coinage in the Roman Empire, 49BC–AD14.* Cambridge: Cambridge Univ Press

Hill, J D, 1994 Why we should not take the data from Iron Age settlements for granted: recent studies of inter-settlement patterning. In A P Fitzpatrick and E L Morris (eds) (1994) *The Iron Age in Wessex: Recent Work.* Salisbury : Trust for Wessex Archaeology, 4-8

Jackson, R, 1985 Cosmetic sets from Late Iron Age and Roman Britain. *Britannia* **16**, 165-92

Merwe, N J, van der, and Avery, D H, 1988 Science and magic in African technology: traditional iron smelting in Malawi. In R Maddin (ed) (1988) *The Beginning of the Use of Metals and Alloys.* Cambridge, MA: MIT Press, 245–60

Millett, M J, 1993 A cemetery in an age of transition: King Harry Lane reconsidered. In M Struck (ed) (1993) *Römerzeitliche Gräber als Quellen zu Religion, Bevölkerungsstruktur und Sozialgeschichte.* Mainz: Institut für Vor- und Frühgeschichte der Johannes Gutenberg-Universität, 255–82

Riederer, J, and Briese, E, 1974 Metallanalysen römischer Gebrauchsgegenstände. *Jahrbuch des Römisch-Germanischen Zentral Museums Mainz* **19**, 83–88

Stos-Gale, Z A, 1993 The origin of metals from the Roman period levels of a site in southern Poland. *J European Archaeol* **1.2**, 101–31

Welbourn, D A, 1985 Craft specialization and complex societies: a critique. In T C Champion and J V S Megaw (eds) (1985) *Settlement and Society: Aspects of West European Prehistory in the First Millennium BC.* Leicester: Leicester Univ Press, 123–31

Werner, O, 1970 Über das Vorkommen von Zink und Messing im Altertum und im Mittelalter. *Erzmetall* **23**, 259–69

NON-FERROUS METALWORKING IN ROMAN YORKSHIRE

JUSTINE BAYLEY

Abstract

The paper reviews the evidence for non-ferrous metalworking in Yorkshire and the surrounding areas in Roman times, comparing the sites and finds with those from other parts of Britain. The various metalworking processes are outlined and the evidence for each presented. There is evidence for primary extraction of lead as well as secondary working of pewter, lead, silver, gold, copper, and its alloys.

Introduction

When we look at metalworking, Yorkshire is in many ways no different from the rest of Roman Britain. There are metalworking finds from forts, vici, towns, villas and other small settlements, a dispersal of activity that is typical of the Roman period (Bayley 1998a).

The map (Figure 1) shows the sites which are mentioned below. Doubtless there are others that have produced similar material, but those illustrated are a representative cross-section and include all the major finds.

Metalworking is not a homogeneous activity but a series of individual processes which are carried out sequentially, either on a single site or at a number of different locations. Figure 2 shows the different processes for which archaeological evidence has been found, and how the product of one process is the raw material for the next; when eventually discarded or broken, objects were recycled, completing the cycle.

Metal extraction

The first group of processes are those concerned with the primary extraction of metal from mineral ores. First the minerals had to be located, removed from the ground and extraneous rock separated from them. The ores may have been roasted before they were smelted in one or more stages, giving metals of variable purity. Unlike many parts of

Findspot	*Date*	*Reference*
Hurst, Swaledale	117-138	Raistrick 1927; 1930, 181; *RIB* II.1, 2404.64
Crackpot, Swaledale	possibly Roman	Hartley and Ingilby 1956, 251; Clough 1962, 106
South Cave, Brough-on-Humber	2nd-3rd C	Gowland 1901, 402 table 3; *RIB* II.1, 2404.47
Brough-on-Humber (5 pigs)	2nd-3rd C	Smythe 1939-40; Corder and Richmond 1942, 28, nos 2-6; *RIB* II.1, 2404.48-50, 2404.52, 2404.55
Heyshaw Moor, Nidderdale (2 pigs)	81	Gowland 1901, 382, pl lvii.12, 402 table 3; *RIB* II.1, 2404.61-62
Grassington	98-117	Raistrick 1927, 81; *RIB* II, 2404.63
Ellerker/Brantingham, Brough-on-Humber	-	Tylecote 1986a, 65 table 39; *RIB* II.1, 20404.58
Belby, Howden	-	Webster 1952-3, 22, no 18
Faxfleet	-	Corder and Richmond, 1942, 28 no 8; *RIB* II.1, 2404.54

Table 1: Lead pigs found in Yorkshire

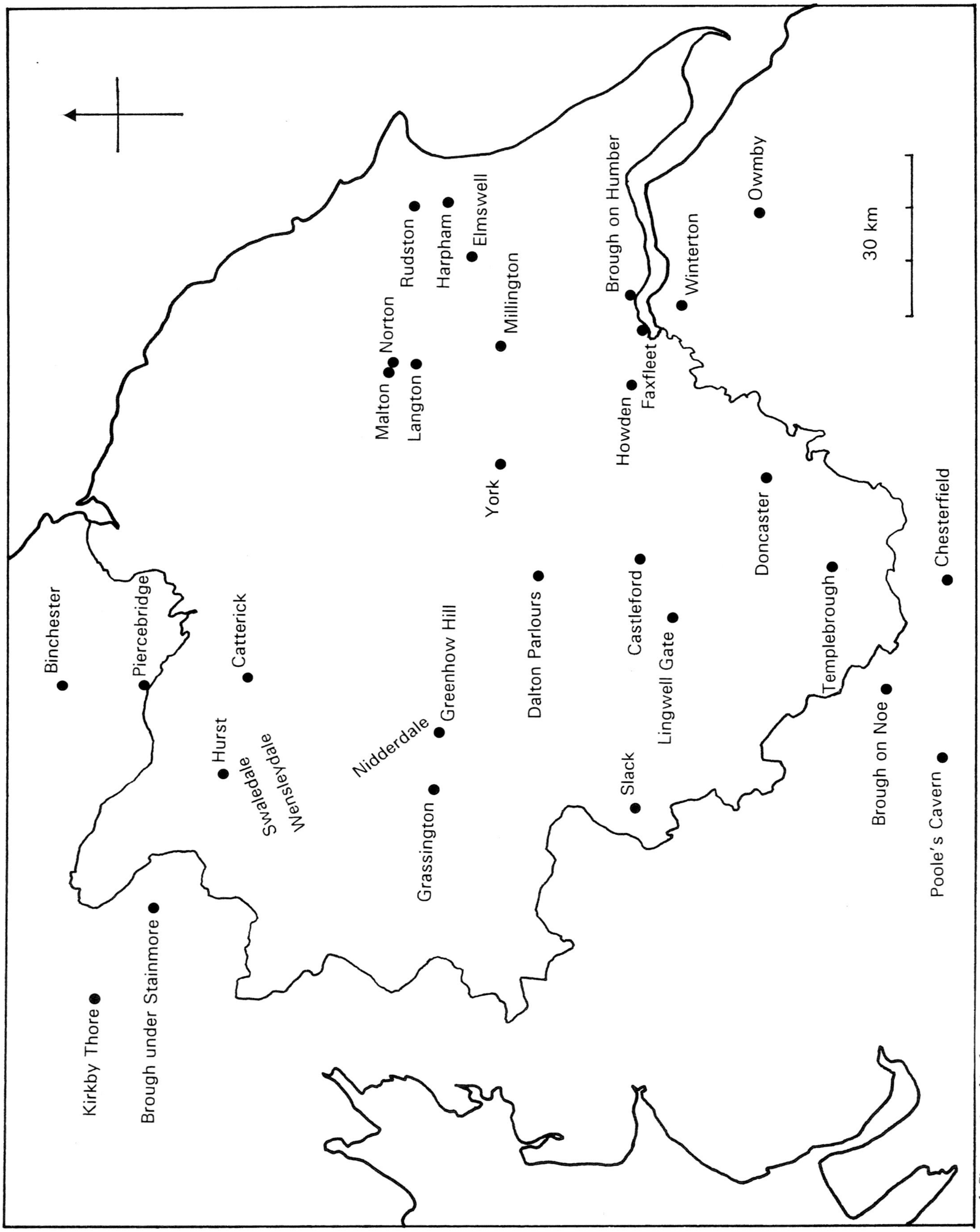

Figure 1. Sites mentioned in the text.

England, Yorkshire does have deposits of metal ores in the Pennines so mining and primary metal smelting can be expected here. In the south of the county there are finds of galena (the main lead ore) at Slack (Barber 1867, 297; 1870, 9; Davies 1935, 164), and evidence for lead smelting at Templeborough (West 1931, 40); lead extraction was also widespread immediately to the south in Derbyshire, at sites such as Brough-on-Noe (Dearne 1993, 157-61). There are also extensive lead deposits further north, in Wensleydale and Swaledale for instance, and there is some evidence that these areas were exploited in the Roman period (Clough 1962, 106) though White (1988, 204) considers the evidence poor. Willies and Cranstone (1992) present recent work on lead smelting sites in the Pennines. However, none of the sites they include are dated before the twelfth century AD. Fleming has more recently suggested the Marrick Moor to Ulshaw Bridge road in Swaledale could be Roman and thus suggests that Hurst, Marrick Moor and Fell End are areas where Roman mines should be sought (Fleming 1996, 100). Future fieldwork may yet demonstrate a Roman origin for the exploitation of lead in the Yorkshire Dales.

Newly smelted lead was cast into rectangular ingots, typically 500 x 140 x 120mm and weighing about 80 kg (250 Roman pounds). Most of the known ingots can be dated to the first and second centuries AD (Tylecote 1986a, 62-65) so it is possible that newly smelted lead was moved around in other forms at later dates, though there is no evidence for it. The ingots were carried considerable distances as finds from Brough-on-Humber (Smythe 1939-40; Wacher 1969, 25) and Faxfleet (*RIB* II.1, 2404.54) show; these had most probably been transported down the Trent and its tributaries from the Derbyshire/South Yorkshire area, though they could have come from further north.

Secondary metalworking

All other metalworking (the processes below 'smelting' in Figure 2) can be considered as secondary, and it is evidence for these processes that is so common throughout Roman Britain. The metals that were worked included gold, silver, copper and its alloys, lead and pewter (an alloy of lead and tin) and each of these is considered in turn.

Pewter

Pewter was used for making vessels from the third century onwards. Stone moulds that stacked, allowing one or more plates to be cast at once, are relatively common finds, though rare in northern Britain. There are three Yorkshire examples: from Langton villa (Goodall 1972, 34-36), Parliament Street, York (Tweddle 1986, 199) and Catterick (Blagg, in Wilson 2002, 300). It is not possible to be sure why pewter working was uncommon in northern Britain, but the distance from the tin deposits of Cornwall must have been a factor.

Lead

Lead was more widely used, for plumbing in buildings (roof flashings, pipes and for seating iron fittings in stonework) as well as for small portable objects such as weights. The problem with leadworking is that it leaves very little by way of specifically identifiable finds. Lead can be melted in any ordinary pot and may leave no trace in it. Because it is so soft no specialised tools are needed to work it so there is little evidence for lead working. However, the prevalence of stone buildings mean that there would have been a need for plumbing, so lead use, and hence lead working, must have been widespread. The finds of lead ingots have already been mentioned and to add to these is Ramm's (1978, 121) suggestion that lead found at Langton may have been cast there.

Silver

Lead did however have another use which can leave behind identifiable debris. When silver was recycled it was first purified (refined) and then alloyed with a known amount of copper. The refining process that was practised is known as cupellation and was carried out in the following way: the silver was melted together with a larger amount of lead on an absorbent hearth, usually lined with bone ash (burnt crushed bone). The molten metal had air blown across its surface, oxidising the lead to litharge (lead oxide) which then dissolved any base metal such as copper that had been present in the silver. The litharge was absorbed by the hearth lining, producing what are known as litharge cakes, and the purified silver was left sitting on the surface of the hearth. The litharge

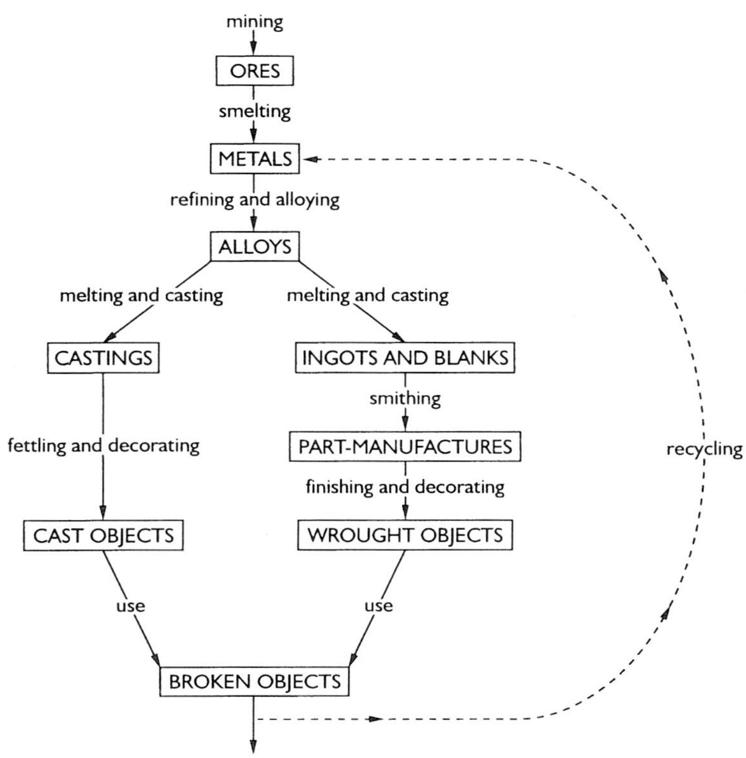

Figure 2. Flow chart showing metalworking processes.

cakes formed as a by-product of silver refining are different from litharge that is sometimes found on lead smelting sites as the former always contain a significant copper content. Litharge cakes are typically 100-200mm across and up to 40mm thick; one has been found in Doncaster (Tylecote 1986b).

Cupellation was also carried out on a far smaller scale when it was necessary to test the purity of a batch of silver. The same process was used but instead of a hearth, a shallow ceramic vessel 30-50mm across was used; one has been identified at Piercebridge (Bayley 1988). The other evidence for silver working is crucibles that were used to melt the metal which have been found at Castleford (Bayley 1998b) and Doncaster (Bayley 1986).

Gold

Gold scrap was re-melted at Tanner Row, York (Budd 1987a; 1987b) and there is the well known inscription from Norton which refers to a slave working for a goldsmith (*RIB* I, 712). There is a further suggestion of goldworking at Millington but no specific evidence is cited (West 1931, 46).

Copper and its alloys

By far the largest group of metalworking finds are those that provide evidence of wrought working and casting of copper and its alloys. As Figure 2 shows, we must expect to find evidence of both wrought and cast manufacturing, though the nature of the processes means that the latter will always be better represented in the archaeological record. A total of 14 Yorkshire sites have produced finds of this sort (see Gazetteer) and in several cases finds have come from more than one location within the same settlement.

What is immediately obvious from the gazetteer is that the commonest finds are crucibles. This is partly because they are durable, and also they are readily recognisable, even in small fragments, as their vitrified surfaces are often brightly coloured. Crucibles would also have been used in quantities as there is little evidence that they were used more than once as the cost of losing metal into the fire would have been high compared with the cost of the crucible. Crucibles may be found where wrought metalworking was being carried out as metal was melted and cast into small bar or disc-shaped ingots which were then hammered out into bars, rods or sheet metal. The ingots could have been cast in depressions in the workshop floor though there are some stone ingot moulds, such as those from Vindolanda (Bidwell 1985, 152-54), which could also have been used to make blanks for wrought metalworking.

Clay moulds should also be common finds as they were not used more than once, whether they were investment moulds or piece moulds. The former were made by encasing a wax model of the desired object in soft clay, drying and firing it to melt out the wax. Piece moulds were made by pressing a permanent model made of wood or metal onto one lump of clay and then pressing another piece of clay onto its other side; the two valves of the mould were then taken apart, the model removed and the mould reassembled, dried and fired. The advantage of a piece mould was that the model could be reused many times over, allowing mass production of a sort; most small Roman castings were made in piece moulds. However, clay moulds were normally made of relatively soft fabrics which are not durable so they often only survive in recognisable form where they were buried soon after they had been used, and so avoided the abrasion and disintegration which must have been their usual fate. Large numbers of clay moulds must have been produced and used on many sites. The quantities found in excavations depend not on the scale of metalworking activity but on deposition processes; only where the moulds were dumped soon after use in a pit or ditch that was not subsequently disturbed, as at Castleford (Bayley and Budd 1998), do more than a few fragments survive.

The spoon moulds from Castleford (Bayley and Budd 1998), and the moulds from Harpham (Goodall 1972, 37; Ramm 1978, fig 46), Malton (in Malton Museum) and Rudston (Stead 1980, 107, fig 71) are all for objects that were cast in something close to their final form; little cold working of the castings other than removing the sprue (casting waste) would have been needed to turn them into usable artefacts. The vessel parts cast in multi-piece moulds at Castleford (Bayley and Budd 1998) would likewise have needed cleaning up before being enamelled and assembled into complete flasks. Other moulds from Lingwell Gate and York were for coins, and represent the practice of counterfeiting which was common in the later Roman period (Boon 1974; see also Cool, this volume; Swan this volume).

The rest of the finds are scrap metal, often offcuts from wrought metalworking, though the finds from Templeborough included sprues (May 1922, 75-76, pl xvi nos 10, 16, 19). Among the finds from Catterick were the remains of two wooden trays that had been filled with brass filings, doubtless collected in a workshop and destined for remelting (Heyworth and Wilthew 1988; Bayley et al, in Wilson 2002,). Similar boxes containing scrap metal came from a late second-century workshop at St Mary Bishophill Senior in York (Ramm 1976, 39) and larger metal fragments would also have been collected for remelting.

Conclusions

Yorkshire is one of only a few English counties where non-ferrous metal ores were exploited in Roman times. This is not because of the selectivity of the miners and smelters, but because the ores are not widely distributed. The Pennines contain a number of discrete lead deposits so it not surprising that there is evidence for primary lead working here, as ores were normally smelted near to where they were found. There is much uncertainty as to the scale and extent of this industry, but the tantalising scraps of evidence do prove it existed.

In the Roman period most types of secondary non-ferrous metalworking were carried out all over the country in settlements of all sizes and types (Bayley 1998a). The bulk of the evidence from Yorkshire comes from forts or from romanised settlements such as towns and villas; little has been found on native sites. A similar pattern is found for metal objects, but both distributions reflect the frequency of urban excavations as much as the real distribution of finds. Thus, until more native settlements are excavated, one can only speculate whether metalworking was concentrated in romanised settlements, perhaps because here there was a large and continuing demand for its products (which were necessary to those who had a Romanised lifestyle), or whether its practice was truly widespread.

Almost all the secondary non-ferrous metalworking processes that have been identified in Roman Britain are represented among the finds from Yorkshire. Those that are missing are the least common or least easily recognised so it is possible that further scientific examination of old finds will discover examples; new excavations, where small and apparently insignificant finds are retained, are another possible source of finds to fill these gaps.

Metalworking in Roman Yorkshire can thus be seen to comprise the expected mix of secondary processes and the unusual, though here not unexpected, evidence for primary extraction of lead.

Gazetteer of metalworking sites

Note: Where specific copper alloy names are given, the material has been chemically analysed. The exception to this is information from older reports where "bronze" is used as a conventional description for all copper alloys.

Binchester, Durham (NZ 210313)
Fort with workshop.
Refs: Wilson 1972, 309

Brough-on-Humber, Yorks (SE 9326)
2nd-4th century
Finds include a rim sherd, possibly from a conical crucible, and copper alloy scrap and sheet clippings. Five lead pigs have also been found. They may be accidental losses during transhipment or stock retained for local use.
Refs: Ramm 1978, 60, 119; Biek and Wacher, in Wacher 1969, 227-31

Brough-on-Noe, Derbys (SK 184825)
later 2nd-3rd century
The fort acted as a control point for the Roman lead mines in Derbyshire. Finds of a furnace, hearths, slag, galena, lead offcuts and objects suggest lead working on site.
Refs: Frere 1984, 289; 1985, 282; Dearne 1993, 157-61

Brough-under-Stainmore, Cumbria (NY 7914)
Objects said to be flawed and unfinished castings, and a lead model of a brooch were found.
Refs: Bateson 1981, 104; Collingwood 1931c

Castleford, Yorks (SE 4225)
late 1st-4th century
About 25 crucible fragments have been found on a number of different excavations. Several different fabrics and types, all handmade, are represented. Most appear to have been used for copper alloy working though silver was detected on three. Nearly 1000 fragments of late 1st century clay piece moulds for casting vessels were found. Relief decoration on the moulds produced champleve fields in the castings to take enamel. At least 20 different designs have been identified.

Excavations have produced 833 fragments of late 3rd-4th century clay piece moulds for casting purse-shaped spoons of two very similar patterns. The minimum number of spoons cast was 55. The two-piece moulds were luted together with clay and assembled into cone shaped multiple units with a single sprue cup so about 16 spoons were cast at a time by running molten metal down the handles towards the bowls.
Refs: Bayley and Sherlock 1986; Bayley 1998b; Bayley and Budd 1998

Catterick, Yorks (SE 2397)
late 1st-3rd century
Two wooden 'trays' 0.5m square, containing brass scrap and at least filings, a stone mould for a pewter vessel and a fragment of a crucible used for melting gunmetal were found.
Refs:Wacher 1978, 91; Heyworth and Wilthew 1988; Blagg, in Wilson 2002, 300; Bayley *et al*, in Wilson 2002, 164-66

Catterick: Bainesse, Yorks (SE 241973)
Finds include a shallow handmade crucible used to melt leaded gunmetal.
Refs: Wilthew 1986; Bayley *et al*, in Wilson 2002, 164-66

Chesterfield, Derbys (SK 385711)
early- mid 2nd century
Final phase of occupation of fort was "... marked by a number of ovens and furnaces connected with metal working". Finds include parts of four crucibles, two about 40mm diameter with traces of silver and one conical in form.
Refs: Hart 1981, 87; Ellis 1989, 93-94

Dalton Parlours, Yorks (SE 402445)
mainly 3rd-4th century
Finds from this villa include crucibles with "bronze" residues.
Refs: Goodburn 1978, 428

Doncaster, Yorks (SE 5703)
Excavations produced fragments of five crucibles of a variety of fabrics. Three (one possibly from a beaker form) were used to melt copper alloys, probably brass, and two (one probably from a shallow hemispherical crucible) to melt silver.
Refs: Bayley 1984a

Doncaster: Frenchgate and St Sepulchre Gate, Yorks (SE 5703)
probably 2nd century
Excavations outside the fort produced a hemispherical crucible which had been used to melt silver. Also found was a sample of litharge with "some copper but little silver" suggesting silver may have been recovered from coins by lead soaking.
Refs: Bayley 1986; Tylecote 1986b

Elmswell, Humberside (SE 9958)
Crucible fragment found in pit 2.
Refs: Hartley 1954, 13

Faxfleet, Yorks (SE 8624)
Lead pig found.
Refs: Corder and Richmond 1942, 28; Ramm 1978; *RIB* II.1, 2404.54

Greenhow Hill, Yorks (SE 1164)
Lead mine. Finds include old slag.
Refs: Davies 1935, 164

Harpham, Yorks (TA 0961)
An ashy layer in a ditch at this villa site produced copper alloy scrap (sheet, wire and parts of objects), four complete crucibles with an added outer layer of less refractory clay and a number of clay piece moulds for a variety of objects including a ring and three "composite moulds" for casting studs. The crucibles are illustrated by Ramm as hemispherical

or thimble-shaped but the extra outer layers may, in some cases, be hiding flat-based beaker-form crucibles.
Refs: Goodall 1972, 37; Ramm 1978, fig 46

Kirkby Thore, Cumbria (NY 6325)
2nd century ?
The site has been described as "... a centre of the local bronzeworking industry" and traces of the manufacture of trumpet brooches are said to have been found here. The evidence for brooch manufacture has been queried by Boon and Savory (1975, 43).
Refs: Collingwood 1931a; 1931b; Hartley 1954, 13

Langton, Yorks (SE 7967)
3rd-4th century
Finds from this villa succeeding a native farmstead included a stacking limestone mould for pewter plates and five conical crucibles, probably used to melt copper or one of its alloys. Lead may also have been cast here.
Refs: Goodall 1972, 34-36; Ramm 1978, 121

Lingwell Gate, Yorks (SE 3320)
c 200 AD
A narrow-necked conical crucible and moulds for casting coins are thought to have come from here.
Refs: Brailsford 1964, 78; Faull and Moorhouse 1981, 168

Malton, Yorks (SE 7871)
later 1st-4th century
Finds from the settlement outside the fort include two complete small conical crucibles and fragments of other larger conical ones as well as part of a piece mould for casting a ring with a diameter of about 80mm and copper alloy scrap, both sheet metal and large blobs and lumps.
Refs: [Finds in Malton Museum] Ramm 1978; Robinson 1978

Millington, Yorks (SE 8352)
late 3rd century
Goldsmith.
Refs: West 1931, 46

Norton, Yorks (SE 7971)
An inscription refers to a goldsmith's shop.
Refs: *RIB* 712; Wacher 1978, pl 29

Owmby, Lincs (TA 0704)
Unfinished Roman brooch found.
Refs: Whitwell 1982, 133

Piercebridge, Durham (NZ 213157)
mostly 4th century
Finds from workshops to the east of Dere Street include a fragment of a clay piece mould and 10 fragments of crucibles, mostly handmade thumbpots, which had been used to melt a range of copper alloys. There was also one small cupel with traces of silver in it and two fragments from a multipart, possibly brass-making crucible.
Refs: Wilson 1975, 235; Bayley 1988

Poole's Cavern, Derbys (SK 0572)
The metal finds from the site include evidence for both casting and smithing of copper alloys; a casting jet and other metal waste, scrap metal sheet and bars and rods, two doming punches and a block. An unfinished brooch-casting is recorded as well as several lead brooches. These were not used as finished objects but as patterns to make clay piece moulds. No crucibles and/or clay moulds have been noted from the site but the pottery has yet to be catalogued.
Refs: Mackreth 1983; Bayley and Branigan 1989

Rudston, Yorks (TA 0967)
Probably 2nd century
Finds from this villa include three clay mould fragments, one an ingot mould, a possible *tuyere*, parts of eight crucibles used for melting copper alloys and scrap and spilt metal.
Refs: Stead 1980, 107, fig 71

Slack, Yorks (SD 9728)
Much galena was found at this fort site.
Refs: Barber 1867, 297; 1870, 9; Davies 1935, 164

Templeborough, Yorks (SK 4191)
mid 1st century onwards
A lead smelter is recorded as well as several crucibles (some thumbpots), scrap "bronze", slag and casting waste (i.e. sprues).
Refs: May 1922, 80-81, pl xv.c; Hartley 1954, 12; West 1931 40

Winterton, Lincs (SE 911182)
Copper alloy working.
Refs: Wilson 1971, 258

York: Aldwark, Yorks (SE 607521)
c 200 AD
Counterfeit coin moulds found.
Refs: Schofield *et al* 1981, 111; Wilson 1975, 236

York: 9 Blake Street (SE 6051)
late 1st century
Finds from the *praetentura* area of the fortress include two sherds from a crucible used to melt impure silver. Copper alloy offcuts and lead offcuts and spillages came from later Roman contexts on the site.
Refs: Cool *et al* 1995, 1592

York: Church Street, Yorks (SE 6051)
Roman: 2nd - early 3rd century
Finds include a crucible sherd.
Refs: MacGregor 1976, 21

York: Parliament Street, Yorks (SE 6051)
Fragment of a limestone mould for a pewter dish or bowl found.
Refs: Tweddle 1986, 199

York: Rougier Street, Yorks (SE 6051)
4th century ?
Sherds of three crucibles used to melt copper alloys (?leaded gunmetal) found.
Refs: Budd 1987b

York: St Mary Bishophill Senior, Yorks (SE 6051)
late 2nd century
Metal workshops contained two superimposed boxes containing 'bronze' working scrap and a crucible.
Refs: Wacher 1975, 171; Ramm 1976, 39; 1978, 119; Biek, in Ramm 1976, 68;

York: Tanner Row, Yorks (SE 6051)
late 2nd-early 3rd century
Finds include the base of a beaker-form crucible used to melt copper alloy and a possible clay mould fragment as well as a

reused sherd, half a small beaker-form crucible and two fragments of a shallow dish, all with traces of gold and apparently used as cupels.
Refs: Budd 1987a; 1987b

York: Yorkshire Museum, Yorks (SE 6051)
Four complete crucibles used to melt brass and gunmetal are among the collection. All are from old excavations in York and are undated. The two large ones (360 and 245ml) are flat-based beakers which are a typical Roman form but the smaller ones (17 and 19ml) are thumb pots which are not a diagnostic form and could equally be post-Roman.
Refs: Bayley 1984b

Bibliography

Barber, F, 1867 The Roman station at Slack. *Archaeol J*, **24**, 289-99

----, 1870 On the Roman station at Slack. *Yorks Archaeol J*, **1**, 1-11

Bateson, J D, 1981 *Enamel-working in Iron Age, Roman and Sub-Roman Britain: the Products and Techniques*. Oxford: Brit Archaeol Rep, Brit Ser **93**

Bayley, J, 1984a *Crucibles from Doncaster, Yorks*. London: Ancient Monuments Lab Rep **4185**

---- , 1984b *Some Crucibles from the Yorkshire Museum*. London: Ancient Monuments Lab Rep **4432**

---- , 1986 A crucible from St Sepulchre Gate (Site DEH). In Buckland and Magilton (1986), 196

---- , 1988 Crucibles from Piercebridge, Co Durham. London: Ancient Monuments Lab Rep **118/88**

----, 1998a Metals and metalworking in the first millennium AD. In J Bayley (ed) (1998) *Science in Archaeology: an Agenda for the Future*. London: English Heritage, 161-68

----, 1998b The crucibles. In Cool and Philo (1998), 223-24

--- , and Branigan, K, 1989 The Romano-British metalwork from Poole's Cavern, Buxton. *Derbys Archaeol J* **109**, 34-50

----, and Budd, P, 1998 The clay moulds. In Cool and Philo (1998), 195-222

----, and Sherlock, D, 1986 Roman spoon moulds from Castleford, West Yorkshire. *Antiq J* **66**, 382-84

Bidwell, P T, 1985 *The Roman fort of Vindolanda at Chesterholm, Northumberland*. Archaeol Rep **1**. London: HBMC

Boon, G, 1974 Counterfeit coins in Roman Britain. In J Casey and R Reece (eds) (1974) *Coins and the Archaeologist*. Oxford: Brit Archaeol Rep, Brit Ser **1**, 95-171

---- , and Savory, H N, 1975 A silver trumpet-brooch with relief decoration, parcel gilt, from Carmarthen, and a note on the development of the type. *Antiq J* **55**, 41-61

Brailsford, J W, 1964 *Guide to the antiquities of Roman Britain* (3rd edn). London: Trustees of the British Museum

Buckland, P C, and Magilton, J R, 1986 *The Archaeology of Doncaster. 1, The Roman Civil Settlement*. Oxford: Brit Archaeol Rep, Brit Ser **148**

Budd, P, 1987a *Crucible Fragments from Tanner Row, York*. London: Ancient Monuments Lab Rep **18/87**

----, 1987b *Examination of Some Crucible Fragments from Tanner Row and Rougier Street, York*. London: Ancient Monuments Lab Rep **82/87**

Clough, R T, 1962 *The Lead Smelting Mills of the Yorkshire Dales*. Leeds: [privately published]

Collingwood, R G 1931a Roman objects from Stanwix and Thatcham. *Antiq J* **11**, 37-45

---- , 1931b Roman objects from Stanwix. *Trans Cumberland Westmorland Antiq Archaeol Soc* (new ser) **31**, 69-80

---- , 1931c Objects from Brough-under-Stainmore in the Craven Museum, Skipton. *Trans Cumberland Westmorland Antiq Archaeol Soc* (new ser) **31**, 81-86

Cool, H E M, Lloyd-Morgan, G, and Hooley, A D 1995 *Finds from the Fortress*. Archaeol York **17/10**. York: CBA

---- , and Philo, C, (eds) 1998 *Roman Castleford: Excavations 1974-85, Vol 1 The small finds*. Wakefield: Yorks Archaeol **4**

Corder, P, and Richmond, I A, 1942 Petuaria. *J British Archaeol Assoc* (ser 3) **7**, 1-30

Davies, O, 1935 *Roman Mines in Europe*. Oxford: Clarendon Press.

Dearne, M, (ed) 1993 *Navio: The Fort and Vicus at Brough-on-Noe, Derbyshire*. Oxford: Brit Archaeol Rep, Brit Ser **234**

Ellis, P, 1989 Roman Chesterfield: excavations by T Courtney 1974-78. *Derbys Archaeol J* **109**, 51-130

Faull, M L, and Moorhouse, S A, 1981 *West Yorkshire: An Archaeological Survey to AD 1500. Vol I*. Wakefield: West Yorks Metro County Council

Fleming, A, 1996 Early roads to the Swaledale lead mines. *Yorks Archaeol J* **68**, 89-100

Frere, S S, (ed) 1984 Roman Britain in 1983; I Sites explored. *Britannia* **15**, 266-332

---- , (ed) 1985 Roman Britain in 1984; I Sites explored. *Britannia* **16**, 252-316

Goodall, I H, 1972 Industrial evidence from the villa at Langton, East Yorkshire. *Yorks Archaeol J* **44**, 32-37

Goodburn, R, (ed) 1978 Roman Britain in 1977; I Sites explored. *Britannia* **9**, 404-72

Gowland, W, 1901 The early metallurgy of silver and lead: Part I, Lead. *Archaeologia* **57(2)**, 359-422

Hart, C R, 1981 *The North Derbyshire Archaeological Survey to AD 1500*. Chesterfield: North Derbys Archaeol Trust

Hartley, B R, 1954 Bronze-worker's hearth [Heronbridge]. *Chester Archaeol Soc J* **41**, 1-14

Hartley, M, and Ingilby, J, 1956 *The Yorkshire Dales*. London: Dent

Heyworth, M, and Wilthew, P T, 1988 *Technological Material from Catterick, North Yorkshire*. London: Ancient Monuments Lab Rep **215/88**

MacGregor, A, 1976 *Finds from a Roman sewer system and an adjacent building in Church Street*. Archaeol York **17/1**. London: CBA

Mackreth, D, 1983 The Roman brooches. In D Bramwell, K Dalton, J F Drinkwater, M Hassell, K L Lorimer and D M Mackreth (1983), Excavations at Poole's Cavern, Buxton: an interim report. *Derbys Archaeol J* **103**, 52-61

May, T, 1922 *The Roman Forts of Templebrough near Rotherham*. Rotherham: Henry Garnett and Co

Raistrick, A, 1927 Notes on lead mining and smelting in West Yorkshire. *Trans Newcomen Soc* **7**, 81-96

---- , 1930 A pig of lead, with Roman inscription, in the Craven Museum. *Yorks Archaeol J* **30**, 181-82

Ramm, H G, 1976 Excavations on the site of St Mary's Church, Bishophill Senior. *Yorks Archaeol J* **48**, 3568

----, 1978 *The Parisi*. London: Duckworth

RIB I = Collingwood, R G, and Wright, R P, 1965 *The Roman inscriptions of Britain. I. Inscriptions on Stone*. Oxford: Clarendon Press

RIB II.1 = Frere, S S, Roxan, M, and Tomlin, R S O, 1990 *The*

Roman inscriptions of Britain. II. Instrumentum domesticum. Fasc 1. The military diplomata; metal ingots; tesserae; dies; labels; and lead sealings (RIB 2401-2411). Gloucester: Sutton

Robinson, J F 1978 *The Archaeology of Malton and Norton.* Leeds: Yorks Archaeol Soc

Schofield, J, Palliser, D, and Harding, C, 1981 *Recent Archaeological Research in English towns.* London: CBA

Smythe, J A, 1939-40 Roman pigs of lead from Brough. *Trans Newcomen Soc* **20**, 139-45

Stead, I, 1980 *Rudston Roman Villa.* Leeds: Yorkshire Archaeological Society

Tweddle, D, 1986 *Finds from Parliament Street and Other Sites in the City Centre.* Archaeol York **17/4**. London: CBA

Tylecote, R F, 1986a *The Prehistory of Metallurgy in the British Isles.* London: Inst Metals

---- , 1986b Litharge from a second century pit (Site DG, Pit 21) in Frenchgate, Doncaster. In Buckland and Magilton (1986), 196

Wacher, J S, 1969 *Excavations at Brough-on-Humber 19581961.* London: Rep Res Comm Soc Antiq London **25**

---- , 1975 *The Towns of Roman Britain.* London: Batsford

---- , 1978 *Roman Britain.* London: Dent

Webster, G, 1952-3 The lead mining industry in North Wales in Roman times. *Flints Hist Soc Pub* **13**, 5-33

West, L C, 1931 *Roman Britain: The Objects of Trade.* Oxford: Blackwell

White, R, 1988 A Pennine Gap? The Roman period in the North Yorkshire Dales. In: J Price and P R Wilson (eds) (1988), *Recent Research in Roman Yorkshire.* Oxford: Brit Archaeol Rep, Brit Ser **193**, 197-217

Whitwell, J B, 1982 *The Coritani: Some aspects of the Iron Age Tribe and the Roman Civitas.* Oxford: Brit Archaeol Rep, British Ser **99**

Willies, L, and Cranstone, D, (eds) 1992 *Boles and smeltmills: Report of a Seminar on the History and Archaeology of Lead Smelting.* Matlock Bath: Hist Metallurgy Soc

Wilson, D R (ed) 1971 Roman Britain in 1971; I Sites explored. *Britannia* **2**, 243-88

---- , 1972 Roman Britain in 1971; I Sites explored. *Britannia* **3**, 298-351

---- , 1975 Roman Britain in 1974; I Sites explored. *Britannia* **6**, 221-83

Wilson, P R, 2002, *Cataractonium. Roman Catterick and its Hinterland. Excavations and Research 1958-1997.* CBA Res Rep 129

Wilthew, P, 1986 *Examination of Slag and Other Material from Various Sites Excavated by the CEU.* London: Ancient Monuments Lab Rep **72/86**

THE LEATHER TRADES IN ROMAN YORKSHIRE AND BEYOND

CAROL VAN DRIEL-MURRAY

Dedicated to Prof Dr Willy Groenman-van Waateringe on her 65th birthday: 05.02.1998.

When the Roman troops crossed into Britain, they were accompanied by impressive technological support, in which leather occupied a prominent role: saddles and harnessing, bags, covers for helmets and shields, pouches, dispatch cases and sheeting. But most unfamiliar and wondrous to the native Britons must have been the complex hobnailed footwear and the huge expanses of waterproof leather tenting (Groenman van Waateringe 1967; van Driel-Murray 1990). Leather is the most versatile of materials: durable, flexible, waterproof and easily worked, it has only been partially surpassed in recent times by plastic. It is with the coming of Rome that leather suddenly enters the archaeological record, marking a distinct break in tradition with the introduction of a new technological package, undoubtedly accompanied by more complex organisation of manufacture, supply and distribution. It is not only the military sites, with their obvious and massive demands, which produce clear evidence for the importance of leather in everyday life in the Roman period, but also the civilian settlements, though here evidence is for the most part confined to footwear.

North Britain is, in many respects, exceptionally well suited to a regional approach. Geographically distinct, almost entirely within the military zone, with supply routes - the point of entry of new influences - directed south, the region has, above all, been remarkably well-served by a long history of research, resulting in a body of material evidence unparalleled in its depth and intensity. This allows potential sources of bias to be identified and eliminated and sequences, laboriously constructed from sites of different dates and character, and scattered throughout the Roman empire, can here be tested and assessed. Though the all-pervading military influence might be considered as a source of bias in itself, evidence from London and the Continent confirms the strongly international and essentially homogenous character of Roman culture as far as footwear and other leather goods are concerned. The main distorting factor is that of unequal survival. Vegetable tanned leather only survives in waterlogged conditions, while leather cured by other means is not even water resistant and will therefore not survive at all, save under very exceptional circumstances. Consequently, the contribution of the leather trades to the Roman economy has been grossly underestimated, though on analogy with the situation in Late Medieval Europe it should have occupied third place, ranking just behind food production and the textile industry in importance. Recent excavations are beginning to redress the situation: just taking *Vindolanda* (Chesterholm) Period IV (*c* AD 106-115) as an example, leather goods account for 47% of the finds, pottery a mere 10% (Figure 1, based on Birley 1994).[1]

Raw Material

The leather trades encompass two distinct branches of endeavour: the processing of raw or green hides into leather and the manufacture of the leather goods themselves. Hides and skins are a by-product of butchery and the leather trades are closely associated with the supply of animals for food. While there is little problem concerning the local origin of the cattle hides employed in footwear manufacture, the source of

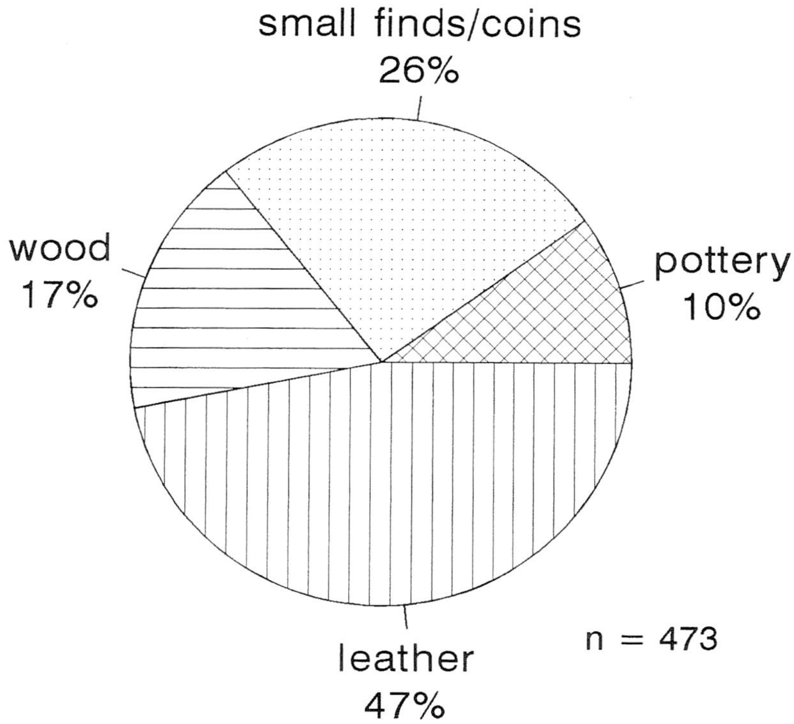

Figure 1. Vindolanda Period IV: percentages of finds by material.

the thinner skins used for sheet leather - in particular the vast quantities necessary for military purposes - is a matter of debate (van Driel-Murray 1985). Sheet leather which I have personally examined from various sites in North Britain (e.g.. Castlecary, Castleford, Carlisle, Newstead, Ribchester, *Vindolanda*) is predominantly of goatskin, as it is on Continental sites. Previous analyses frequently refer to `calfskin', but most of these identifications are erroneous, having failed to take account of the worn condition of the leather, which may obscure the secondary follicles characteristic of goat and sheepskins: the same applies to the occasional mention of `horsehide' or `pigskin'.[2] True wooled sheepskin gives poor quality leather and is rare, but may, like doeskin, in fact have been used more commonly for soft, luxury leathers which tend to be cured, using alum or oil, and which would not usually survive in waterlogged conditions. At Welzheim (Baden-Württemberg, Germany), doeskin was used exclusively to make slippers of the type shown on Figure 5, and here it had probably been preserved by a sort of natural `secondary tannage' in the oak lined well. At *Vindolanda*, such slipper tops, where preserved, are of goatskin, but it is noteworthy that soles far outnumber the surviving tops. Hair samples from textiles and, more rarely leather, also point to the continued presence of Iron Age sheep breeds, with a generalised medium fleece, which may be considered as a source for the rather inferior leather used for the smaller back panels of tents at *Vindolanda*.[3]

There is no indication as to the origin of the six goatskins mentioned in one of the *Vindolanda* texts (Bowman and Thomas 1994, *Tab Vindol* 309), but as goat bones are rare in archaeological contexts, it is frequently assumed that the large numbers of goatskins required by the army must have been imported from the Mediterranean. This begs the question whether the army could afford to remain dependent on such extremely long lines of communication or whether, in view of the general pattern of self- sufficiency, there would be an attempt to stimulate local production.[4] Interestingly, post-medieval records of Scottish border raids reveal that while wealthy farmers owned large numbers of cattle, smallholders tended to concentrate on sheep and - particularly - goats. Noddle (1994, 122) cites one raid of 1521 in which 4000 goats were lifted, against 2000 sheep and only 200 cattle. Though conditions were very different in Roman times, these records do reveal that the north of Britain is a potential source of goatskins, and also shed some light on possible reasons for the archaeological invisibility of goats in general. If the goat is the smallholder's stock, the evidence for slaughter and consumption would occur on rural sites, where bone assemblages are usually small, making it even more difficult to separate the bones of sheep and goat. A similar lack of goat bones in medieval contexts is occasionally, and perhaps significantly, compensated by marked concentrations at known (urban) tannery or horn working locations, such as several sites in York (O'Connor 1989, 17). At similar locations in the Netherlands (e.g.. Prummel 1978), import from the Mediterranean is automatically assumed, though without any supporting evidence. On consideration, however, the risk of transport damage, the additional weight and general awkwardness make it far more likely that the presence of horns and hoofs in fact indicate local supply, drawn direct from butchers and smallholders (Serjeantson 1989). This is indeed the context in which Schmid (1973) made her now classic observations on the association between skin processing and such refuse: hill farmers curing just one or two skins which were to be sold in town at a later date. Similarly, the early Roman pits in Augsburg, filled with black organic debris and goat and sheep horns and metapodia, are more likely to represent local supplies: it is almost inconceivable that heavy and useless bones would be transported over the Alps, only to be discarded in Augsburg (Schmid 1969, 105). Both O'Connor (1989, 13) and Maltby (1989, 75) draw attention to highly specialised, large-scale carcass processing in Roman towns, including York, in which efficient processing of the by-products - hides, horn and bone - was an integral part. O'Connor (1989, 15) further points out that it is the secondary products which tend to determine the source of the meat supply to towns, not the independent choice exerted by the consumer. In such circumstances, it is highly probable that the army would have actively stimulated goat breeding: the absence of bone evidence must be regarded as an archaeological blind spot, perhaps caused by the lack of excavation (or conservation) at appropriate sites.

Tanneries

A common misconception in the archaeological literature is the equation of products with the location of manufacture, when just the converse is true: only offcuts will remain in a workshop as the products are sold and dispersed. Worn out old shoes and cut-up tents are evidence of refuse disposal, not of manufacture and are even further removed from the process of tanning itself.[5]

In a meticulous reassessment of the evidence, Hooley (2002) has effectively disposed of the much quoted "tannery" at Catterick: neither the leather finds nor the structures can be linked in any way to tanning or to any form of skin processing. Similarly, the squalid conditions at *Vindolanda*, taken together with the vast quantity of worn-out leather goods, which were long regarded as clear evidence for a tannery can now be seen as the backfill of abandoned structures with refuse, dung and stable litter.

To locate tanneries, a combination of structural and osteological evidence must be sought, preferably in combination with chemical analysis of soil samples. Structural evidence alone is rarely sufficient since vats, access to water, drainage and rinsing facilities and substances such as ash and urine are common to a number of trades. In medieval towns such industries were often located in the same areas and trades such as butchers, tanners, horn and bone workers, glue makers and extractors of neatsfoot oil are, furthermore, so interdependent that it may be difficult to isolate the manufacturing waste of a single discrete branch (Maltby 1989, 75). At the *vicus Vitudurum* (Oberwinterthur, Switzerland), two houses have been claimed as Roman tanneries: both contained arrangements of three vats made of halved wine barrels, set in clay and waterproofed on the inside with pitch or resin, and were associated with marked concentrations of cattle horn cores (Clerici 1983; Fellmann-Brogli 1991; Morel 1991, 117, Abb 81). In a subsequent phase, leather offcuts around one of the yards seemed to confirm the link with the leather trade. However, the evidence for an extensive industry in horn and bone products and the fact that the vats were only 0.55m deep, suggests rather that the vats were used for soaking horns to facilitate the removal of the outer cover (and thus explaining the presence of both horns and hair). MacGregor (1989, 117) mentions a medieval horn soaking pit (appropriately enough) in Hornpot Lane, York, where similar clay and timber lined pits contained some 200 cattle and goat horn cores. Thus, despite the fact that horns are often considered to be the classic tannery indicator, the absence of metapodia, the lack of pits big enough to take cattle hides and the fact that the offcuts are of *goat*skin, not cowhide, are significant factors in rejecting *Vitudurum* as a candidate for a Roman tannery. With their greater regard for hygiene, the Romans may have been more rigorous in banning noisome industries from towns if space allowed, and it may be that tanneries - and with them the missing goat bones - should be sought in the extra-mural settlements.

Military Supply

As leather production was initially set up to meet military demands and the army remained the largest consumer of leather goods, the organisation of military supplies will be the dominant feature in North Britain. At the conquest, the Roman forces were faced with a major logistical problem. It is becoming increasingly clear that tanning with vegetable extracts was unknown to the Iron Age peoples of North-west Europe. Native skin processing was confined to various methods of curing, such as smoking or oiling, for small scale leather production, probably on an individual and household basis. There was, therefore, no native source of supply and no possibility of expanding native production, even though livestock - much of it used to feed the army - was readily available. The entire process had to be set up from scratch, using either army or immigrant personnel and owing nothing to native British traditions. Since tanning is a long and complex process, demanding considerable investment in materials and installations, this newly introduced industry could only have been located in the more secure areas of the province, and the campaigning armies would have been dependent for some time on transport over considerable distances.

Signs of stress and uncertainty of supply, discernible in the leather complexes from early forts such as Castleford and *Vindolanda* (Figure 2), point to severe problems in maintaining production right into the early years of the second century (van Driel-Murray 1993; 1998, 291, 333-34). Though recycling sheet leather is standard military procedure, at Castleford especially the extent of salvage reaches extreme proportions, with carefully pared-off lengths of seams being redeployed as binding cord, and bundles of cut-off caliga loops retained to repair other old shoes (Figure 2d); this is also a feature of Augustan Mainz (Göpfrich 1986, Abb 37 nos 25 and 26). At *Vindolanda*, serviceable sections of various cut-up old boots were even cobbled together to eke out their life. With a new sole underneath one of these strange, composite boots they lasted long enough for both the replacement nailing and the new outer sole to wear completely through, proof of the solidity of Roman shoe construction and the heavy wear they were expected to endure (Figure 2a-c). *Vindolanda* is, of course, likely to experience transport difficulties at the best of times, but it is noticeable that the stress observed in the first three periods (*c* AD 85-106) is largely absent from Period IV onwards, nor are these indicators observable on the Antonine Wall sites

It has been suggested that supply bases are a feature of the conquest phase, while later, forts may have been more independent in the organisation of their supplies. This is certainly the case in Egypt (Alston 1995, 110) where officers approached local officials direct for the necessary goods, but this was a highly regulated society and in Britain the use of middlemen (such as the documented Octavius, see below) seems more likely. Moreover, in the case of hides, which were 'produced' by the units themselves as a by-product of the food supply, different systems may well have operated. In the earlier periods of *Vindolanda* (until *c* AD 106), cow hide offcuts stamped with VIIIIB or IXB certainly suggest that tanned hides were being claimed by, or allocated to, individual units (here the *Cohors IX Batavorum*), from regional supply points. An alternative which needs to be considered is, however, that the unit retained control over the hides taken from its own food animals, even though tanning took place elsewhere. Whether such unit stamps were being applied before or after tanning is clearly crucial here: the *Vindolanda* stamps were certainly applied before finishing and probably before tanning.[6] Vegetable tanning is most efficiently organised as a continuous process, and it is unlikely that individual Wall forts would have had the facilities. The presence of tanneries at suitable locations, serving the forts of both Walls, seems logical, but it must be emphasised that there is as yet, no evidence as to their location, or even whether they were actually in military or civilian hands.

The well-known document *Tab Vindol* 343 (Bowman and Thomas 1994, 321-29 no 343; Birley 1993, 60) concerning large consignments of hides indicates the scale of military demand. The writer, Octavius, is a middleman, involved in processing raw materials: he purchases grain in the ear and has it threshed, collects hides from various sources - some are military allocation, others perhaps private - and has them tanned (he has 'completed 170 hides'). He works partly on contract but could also be approached by other military personnel for supplies and there seems to be ample opportunity for private gain (all this closely paralleled in the Egyptian documentation, Alston 1995, 110-12). We may suspect a mix of officially regulated military contracts and private speculation, exploiting the shortcomings of the official system, for which there is such clear evidence in the earlier phases at *Vindolanda*. Particularly interesting in this text is that Candidus at *Vindolanda* could, apparently, order (military) personnel at Catterick to release hides to Octavius for processing: though this may illustrate the context for the VIIIIB/IXB stamps on offcuts, it must be stressed that these belong to Periods II and III, while the Octavius letter comes from the (end?) of Period IV and, if Anthony Birley is correct in his identification of the legionary centurion in the letter, the

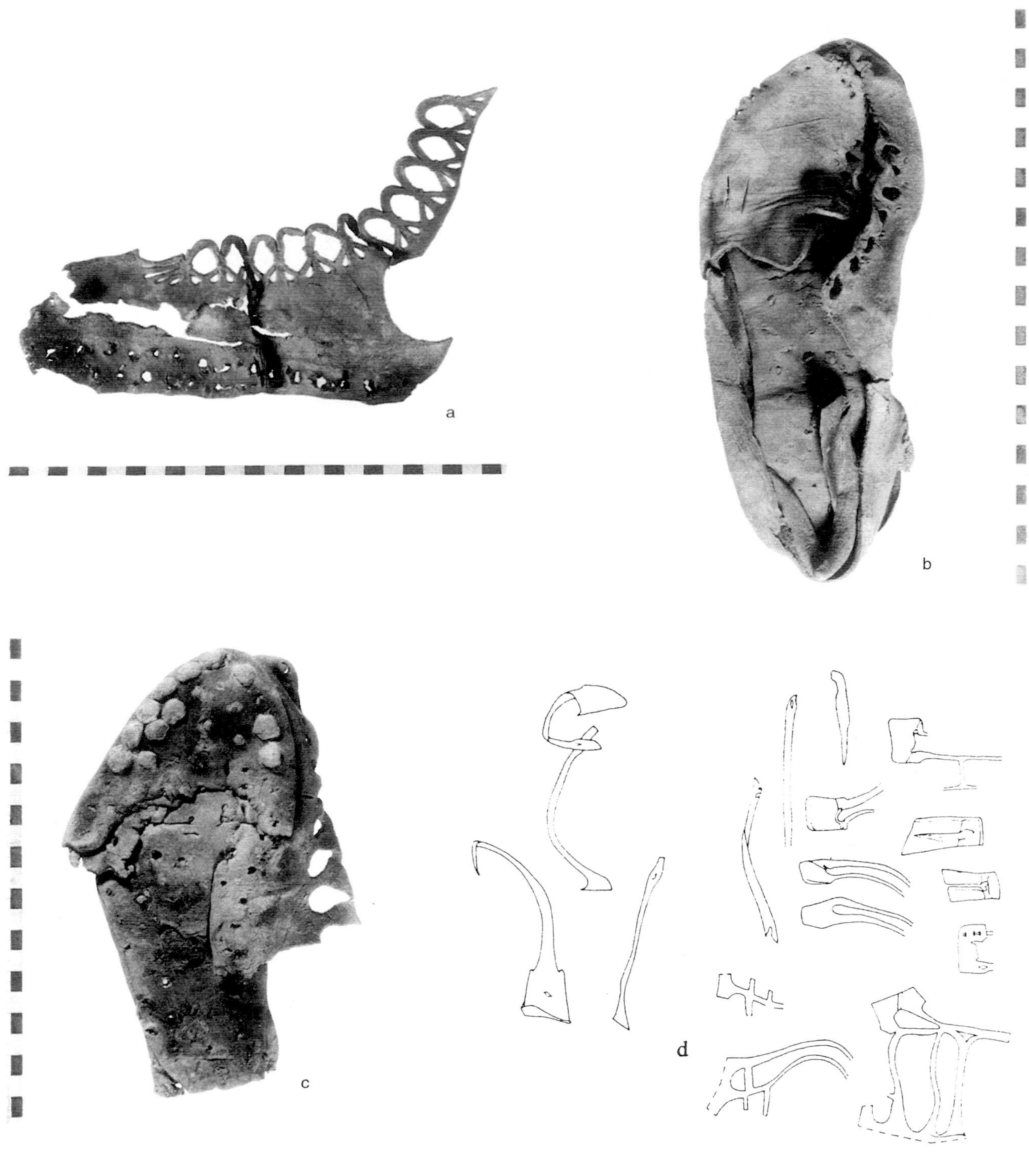

Figure 2. a) Cut-up boot upper of the style Croglin (Vindolanda L 1600); b and c) cut down and modified boots of the style Fell (Vindolanda L 1052, 1357); d) collection of cut caliga loops from Castleford. (Scale 1:3)

exceptionally large consignment involved might be directly related to the start of building activities on Hadrian's Wall. Catterick, as Hooley (2002) points out, was well-suited as a collection and processing point (temporary curing if not tanning, cf. van Driel-Murray 1985, 62) for hides from the Vales of Mowbray and York, but again it should be noted that the leather deposits there appear to date to the later Hadrianic/Antonine period, and cannot therefore be brought into association with the activities described in letters from *Vindolanda* Periods III or IV. The text and the finds may illustrate the processes, but are not actually part of the same pattern.

In the hinterland, with the establishment of towns and *vici*, different systems probably developed quite quickly, with the military relinquishing even more routine activities to civilian contractors (such as Octavius). The sheer number and lack of duplication of stamps on offcuts which probably identify tanneries (van Driel-Murray 1977; Rhodes 1987) suggests numerous individual owners, many of them citizens of substance in view of the frequent use of the *tria nomina*. Publication of the leather from York may well clarify the relationship of military and civilian production and whether there was any development through time (see note in Cool, this volume).

Firmly tied into the sequence of military annexation are several large and relatively well-dated leather complexes from North Britain composed predominantly of recycled military equipment from unit work places. Complexes from Castleford, *Vindolanda* (van Driel-Murray 1993; 1998), Carlisle (Winterbottom 1991, 244-5) and Catterick (Hooley 2002) are, despite variations in dating, broadly comparable and fit the general model of military production already established (van Driel-Murray 1985). Footwear was a basic necessity and evidence for its production turns up at virtually every fort, though sometimes discrete from other leather waste. Otherwise, leather working at auxiliary forts was confined to maintenance of equipment and careful recycling: these leather complexes contain virtually no unused goatskin offcuts, which would point to the manufacture of new equipment, but only old, cut-up, reused and often unrecognisable material. An exception is perhaps the production of highly individual equipment, specific to a particular unit, such as the horse chamfrons which were being made in one of the cubicles in the *Vindolanda* Period III *praetorium* (van Driel-Murray 1993, 57 and fig 25). In auxiliary forts, small rooms around both the *principia* and the *praetorium* may have fulfilled the role of the *tabernae* along the main roads of legionary fortresses (Speidel 1996, 193-4) and other workshop debris should be sought in these areas. At Valkenburg (Netherlands) for instance, the camp shoemaker clearly worked in a room of the *principia* (van Driel-Murray 1985, fig 5). In *Vindonissa*, texts mention a shield maker working in the *taberna* number 8, while number 30 was occupied by an *armorum custos* (and 13 by a *vinarius*: Speidel 1996, 193). It is significant that large numbers of goatskin primary offcuts only come from legionary fortresses such as *Vindonissa* (Windisch), Bonn and Mainz (Gansser-Burckhardt 1942, 99-104; Göpfrich 1986, nos 185-96; van Driel-Murray and Gechter 1983, 55: it remains to be seen whether future analysis of material from York will confirm this pattern for Britain). It is likely, therefore, that the manufacture of complex but standardised equipment like tents and shields was the preserve of the legionary *fabricae*. This would explain the extreme standardisation of the equipment, but does not, of course exclude recycling and general maintenance taking place in the same workplaces. If this model is valid, production of tents and such like would, at least initially, have been concentrated in York and Chester, and the analyses of leather complexes should take account of the possibility of different spheres of influence. In theory, one might expect a western axis supplied from Chester, and an eastern axis running from York to *Vindolanda* and Newstead, but very much more work needs to be done and to a level of detail that unfortunately surpasses the resources usually available. It is in any case likely that once Hadrian's Wall was built, supply lines would be shortened by moving facilities up nearer the location of the troops themselves, and the easing of supply problems at *Vindolanda* may show the results of this.[7] The supply of the Antonine Wall forts is a case apart, but so little sheet leather is available that speculation is fruitless.

These dumps of manufacturing waste should not be regarded as normal ongoing workshop refuse, but are generally to be linked to particular circumstances, in particular the redeployment of troops in the first and early second century.[8] The greater stability of occupation and better organisation of refuse disposal are probably the major factors influencing the notable decline of sheet leather in complexes dating to the mid second century and later (Winterbottom 1991, 244). Some of the sheet leather from the later ditches at *Vindolanda* (end of the second century onwards) looks distinctly mouldy and insect damaged, and it may be that periods of peace resulted in a falling off of standards in maintenance and unwelcome surprises when tents had to be got out in emergencies.

Footwear

Despite its position at the uttermost limits of the Roman empire, there is little to suggest that northern Britain lagged behind metropolitan fashions as far as footwear was concerned, nor that regional idiosyncrasies were strongly developed at any period. The constant movement of military personnel and especially the rotation of offices amongst the urban elite - men who travelled with their families and household in attendance - served to link frontier communities to the urban heartland, while assured income provided the lower ranks with the resources for social emulation. The documented movement of detachments between London and *Vindolanda* (Bowman and Thomas 1994, *Tab Vindol* 154) provide the context for the remarkable similarities in footwear at both settlements at the turn of the second century (in particular Period III at *Vindolanda* and the collection of footwear from the Bank of England excavations in London). Stamps, especially on sandal soles, may give an insight into more specific contacts, but given that footwear obviously travelled with its owners, such links are open to numerous interpretations. A distinctive and beautifully cut stamp depicting a most official-looking eagle (Figure 3a) occurs on identical sandals at *Vindolanda* and London, while at the end of the second century a complex and unusual stamp at *Vindolanda* (Hassall and Tomlin 1989, 341) is exactly paralleled at York (*RIB II.4* 2445.9; Hassall and Tomlin 1987, 374: QFIMCV | ESV | ORM), both placed on the outer soles of the same type of elegant cork slippers (Figure 3b). We can only speculate about shopping trips to London or York, or on the origin of the wardrobe brought by new arrivals to *Vindolanda*. With so few finds to go on, the direction of contacts and the location of putative `centres' remains elusive.

Each unit was evidently able to supply its own footwear needs with soldiers set to work presumably under supervision of a skilled shoemaker (Bowman and Thomas 1994, *Tab Vindol* 155). On discharge, army shoemakers would play an important role in the dissemination of Roman technology and styles and increasingly blurring the distinction between military and civilian. If existing apprenticeship contracts are anything to go by, sons of army shoemakers would

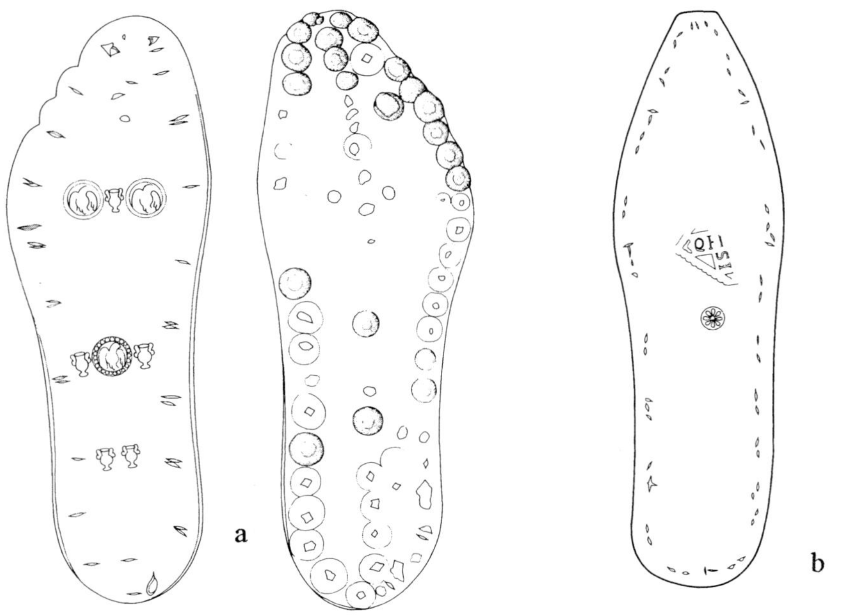

Figure 3. a) Stamped sandal insole (Vindolanda L 319); b) stamped slipper outer sole

presumably learn their trade young from their fathers, and if they entered the army in their late teens they would already be eligible to enrol as *immunes* (Bradley 1991, 107). North Britain was conquered too late for the classic army boot, the *caliga*, beloved of military re-enactment societies, to be much in evidence. Castleford has yielded several extremely worn specimens (van Driel-Murray 1998, 291-98), but none occur in the earliest phases of *Vindolanda* or Carlisle, suggesting that by about AD 85 the *caliga* with the distinctive high ankle loops had disappeared entirely (Figure 4).[9] With the abandonment of this boot, it becomes impossible to distinguish military from civilian footwear on style alone, but the continued occurrence of offcuts within the forts at both *Vindolanda* and Carlisle (Winterbottom 1991, 318-28; 1992, 95), as well as references in the *Vindolanda* texts (Bowman and Thomas 1994, *Tab Vindol* 155) indicate that military shoemakers were still active, even if they were now making different styles of footwear, including single piece shoes. Civilian involvement outside the forts cannot be excluded, and a too rigid distinction between makers and users should probably not be drawn at any time. It would seem just as feasible for the resident military shoemaker to supply the needs of soldier's families as for soldiers to obtain their shoes from a civilian shoemaker outside the camp, especially if, as the *Vindolanda* texts reveal all too clearly, the continuous official deductions from soldiers' pay for `replacements' became too heavy a burden. True *caligae*, with four thongs drawn up between the toes (uncomfortable for the modern wearer and therefore left out of most reconstructions) could not have been worn with socks and the rather curious toeless boots depicted on military tombstones may be intended to represent textile foot wrappings with knobs indicating the positions of the upstanding tabs of painted *caliga* loops. The abandonment of the caliga has been regarded as a concession to comfort, and though indeed the standard Period III boot at *Vindolanda* (Figure 8.1) is suitably sensible, one of the other main military boots in use (Figure 9.2) is as open and lacy as the *caliga*, and a lot more stylish - though it could be - and in view of the lavish decoration, presumably was - worn with (coloured) socks (Bowman and Thomas 1983, *Tab Vindol* 38).

In contrast to prehistoric and indeed, medieval, footwear, Roman shoemakers employed several distinct technologies to manufacture a wide variety of shoe styles. Most closely associated with the Roman presence are the complex hobnailed shoes and sandals, but equally characteristic are sewn constructions (Göpfrich 1986, 22, fig 42), cork slippers (Figure 5), and wooden over shoes (van Driel-Murray 1993, pl. V), all of which are well represented in Northern Britain. Single piece shoes (*carbatinae*) are technologically similar to native forms on the Continent (there are no surviving Iron Age shoes from Britain), and these might be expected to be the vehicle for the expression of any native influences. However, even

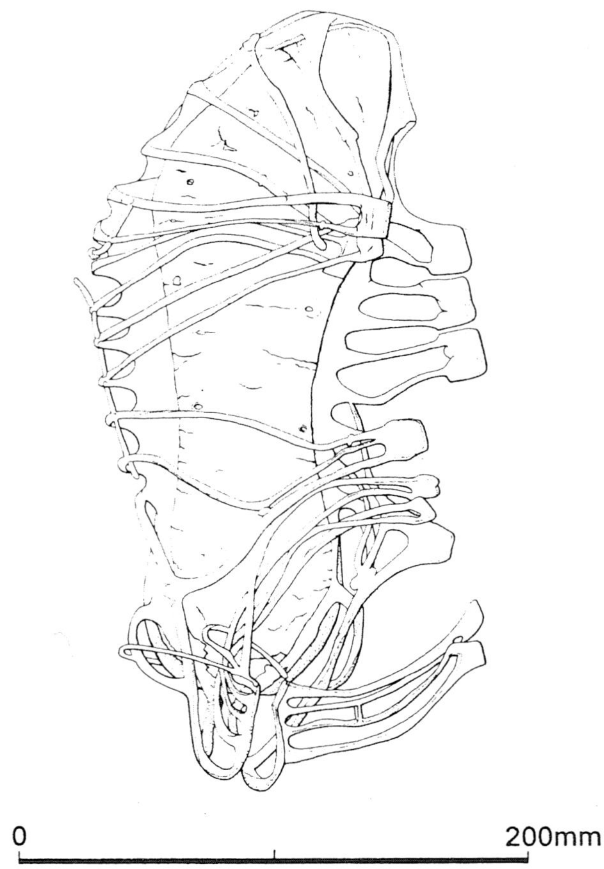

Figure 4. Caliga from Castleford. (Scale 1:3)
(Illustration courtesy of West Yorkshire Archaeology Service)

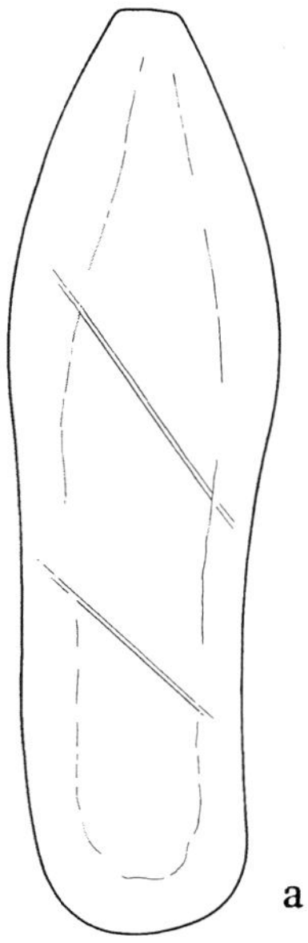

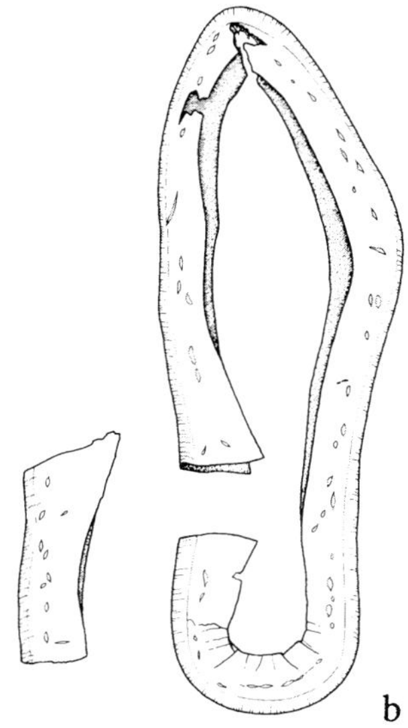

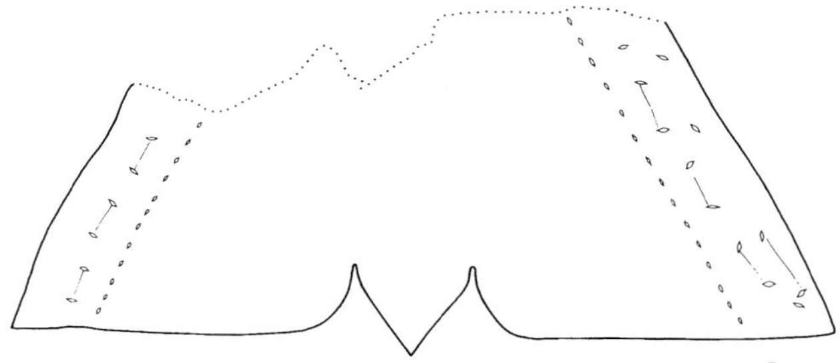

Figure 5a Slipper, Vindolanda L1945

a Inside of outer sole showing bracing impression
b Two pieces of sewn leather envelope
c Fragment of the deerskin upper

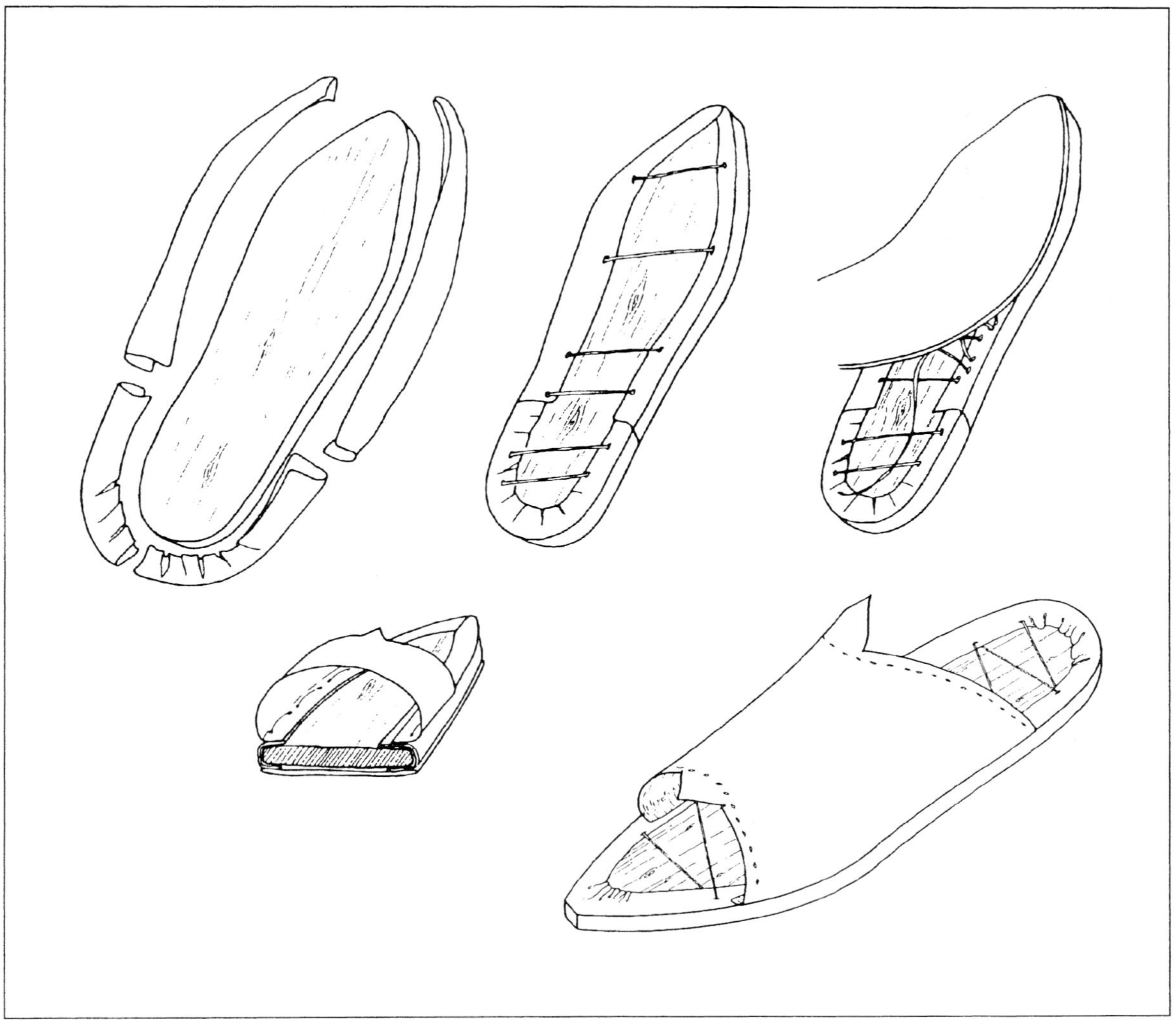

Figure 5b. Manufacturing stages of leather-covered cork slipper as reconstructed from the finds at Welzheim. (© B Brouwenstijn)

Figure 6. Carbatina dated to c AD 80-85 from Castleford. (Scale 1:3) (Photo F Gijbels, IPP)

the earliest examples in the North, such as an elegant example from Castleford (Figure 6), are all directly related to the civilian styles found in London, which seem to have their roots in Gaul, thereby, perhaps, indicating the source of the immigrants who introduced both tanning and the mixed Gallo-Roman clothing traditions to Britain. The most striking feature of Roman footwear is the rapidity of change, affecting the style of the upper, the shape of the sole, nailing patterns and decorative elements and at a rate consistent throughout the Western Empire (Figure 7). Though the lack of published comparative data from York is particularly regrettable, parallels for virtually every kind of footwear used in the frontier forts can be found in contemporary complexes in London and even further afield. On account of this international aspect and to emphasise that these are very real fashion changes, rather than the simple evolution of designs, I prefer to use `brand names' for shoe styles rather than the more usual archaeological terminology of `type 1, 2 or 3' which implies some kind of logical development. Styles of footwear changed radically within a generation or so, and people made a conscious choice between the products on offer. The exceptionally large quantity of footwear from the four earliest periods at *Vindolanda* illustrates the trend of continuous change remarkably well (Figure 7). These developments and their dating can be paralleled all over northern Britain and, indeed, the Continent, thus allowing poorly dated or smaller collections to be viewed within a broader context. Because of the quantities involved, the significance of even quite minor variants becomes apparent: the Batavians switched massively from the oval lace holes of the Chesterholm to the round holes of the Fell (Figure 8.1) in Period III, while the Tungrians abandoned the decorative Croglin entirely (Figure 2a). Other than this, there is no suggestion of unit preferences or ethnic differences except in so far as the Balmoral (Figure 8.2) is clearly related to the presence of the prefect and his family in the *praetorium* of Periods II and III. Unfortunately, Period V is much disturbed as a result of the cutting of the two later ditches and the odd `tails' of some of the styles clearly represent redeposited material. Nevertheless, the hiatus in dumping between Period IV and the contents of the `Antonine Ditch' stands out clearly, with few styles spanning the mid second century. Fitting into this gap is a remarkably homogenous association of three main styles which dominate the Antonine Wall sites (Robertson *et al* 1975, 68-75, types A, B and C) and also represented at Hardknott (Charlesworth and Thornton 1973), Catterick and, to some extent, Newstead. The virtual absence of these styles in the huge complex from *Vindolanda* (Figure 8.3 illustrates one of the few examples) suggest a very much reduced occupation there throughout the 140's and into the 170's. On the other hand, several of these shoes are present in the earlier ditches at Birdoswald (Mould 1997, figs 238 no 5, 239 no 8), suggesting that not all garrisons from Hadrian's Wall were withdrawn.

The close dating of certain types of footwear sheds an interesting additional light on the history of fort occupation. Some of the single piece shoes (*carbatinae*) at Bar Hill can, for instance, now be paralleled in several well-dated Flavian/Trajanic contexts in London, as can one of the few nailing patterns to be illustrated (Robertson *et al* 1975, fig 26 no 59), thus perhaps supporting the slight evidence for a Flavian phase (Robertson *et al* 1975, 6). Later occupation is suggested by the occurrence of shoes which appear in the `Antonine Ditch' deposits at *Vindolanda* and at Welzheim (in the first decade of the third century) (Figure 8.6; Robertson *et al* 1975, fig 24 no 47, fig 25 no 53): clearly a re-examination of all the nailing patterns here is required to identify any more later styles. Similarly, at Newstead, the Flavian occupation is clearly marked by footwear styles and nailing patterns very similar to those at *Vindolanda*, but there are also a number of shoes which definitely belong to the late second century, and these include at least one in a woman's size, thus reinforcing the evidence for the presence of soldier's families at all times (van Driel-Murray 1995). Although the exact find locations of the leather in these old collections are unknown, the combination of stylistic dating and analysis of sizes may yet give further insights into the nature of military activity beyond Hadrian's Wall and its intensity.

Occupation at *Vindolanda* resumes with dumping in the `Antonine Ditch' covering the period *c* 170 into the first decades of the third century, a period well represented in Continental leather complexes (Busch 1965). Characteristic are huge, shapely, pointed heavily nailed soles, mostly in male sizes, the Roman equivalent of cowboy boots, which must have transformed the gait of the wearers in much the same manner (Figure 10b and 12). At *Vindolanda* and Bewcastle (Thornton 1991, fig 112) these belonged to soldiers, but elsewhere, including York (Skeldergate, MacGregor 1978, fig 27 no 350) and London, they were equally popular with civilians. Thus although a similar sole, belonging to a body found in Grewelthorpe Moor, near Ripon, and recovered with its textile insole as well as yellow stockings and brightly coloured clothing, preserves sufficient characteristics for a late second-century date to be beyond dispute, it cannot, unfortunately, tell us whether the owner

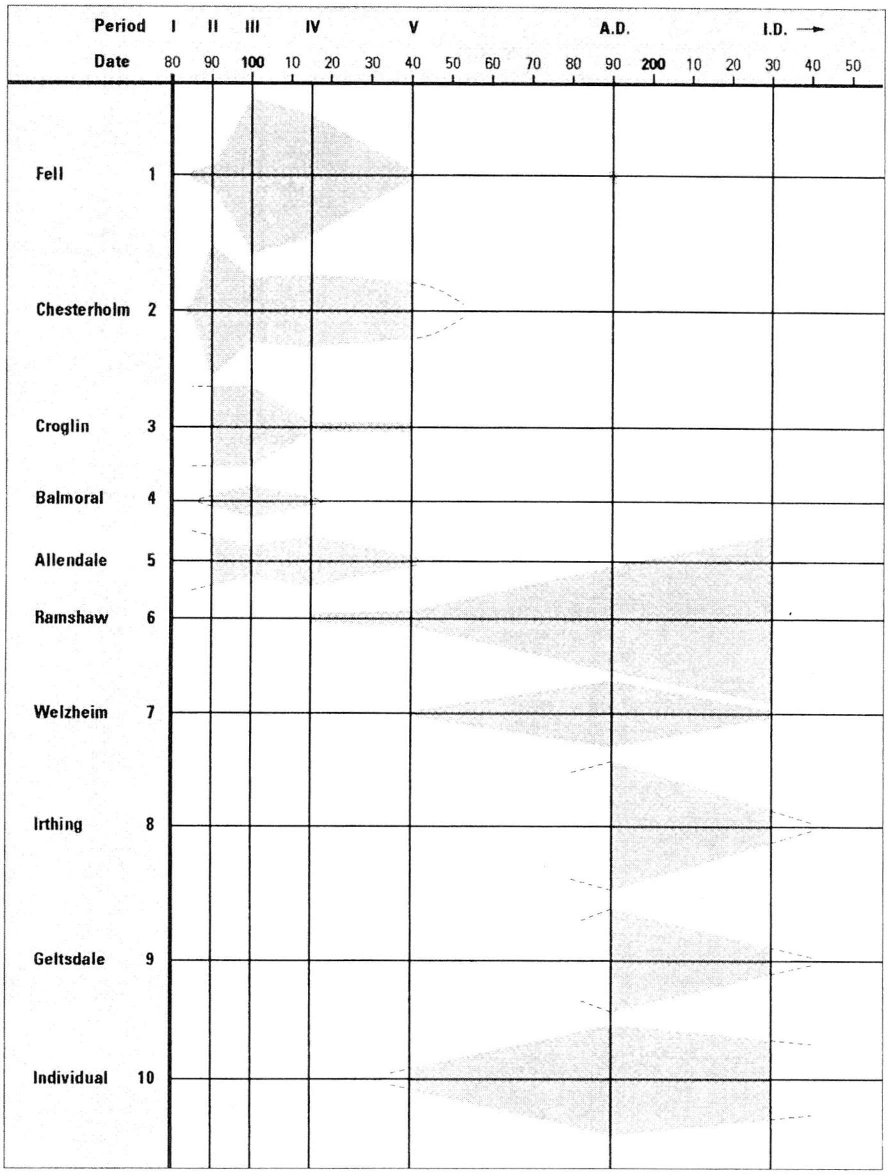

Notes to the chart:

AD = Antonine Ditch, centred on AD 190 for convenience; ID = Inner ditch centred on AD 230 for convenience, though continuing into 4th century.

No. 2, `Chesterholm', not illustrated, a variant of the Fell. Under no. 10, `individual' are grouped styles with no more than two occurrences.

Figure 7. Changing fashion between c AD 90-250 based on a selection from Vindolanda.

was a soldier or civilian (Turner *et al* 1991). From the beginning of the third century, there is a marked increase in individuality, with an explosion of different styles, increasingly showy decoration and much experiment with cutting patterns and construction methods (Figure 7 no 10). The so-called `Inner Ditch' contains some of the latest material at *Vindolanda*, including some paralleled at Birdoswald (Mould 1997, figs 240 nos 17-21, 241 nos 25, 28), in the Skeldergate well in York (MacGregor 1978, fig 28), as well as in late third- and fourth-century sites in London (Figure 11a and b). Noticeable is that the almost invariable toe seam of the earlier periods gives way to side seams, sometimes in combination with a toe seam, but more often without. In consequence, the vamp begins to be decorated with cut-outs, tiny brass studs (worth looking out for in burials, Figures 8.4, 11b), sewn appliqués or stitched imitations of the old toe seam. Though related to earlier styles, new sewn and single piece shoes begin to appear, and the multiplicity of unnailed footwear at this period means that burials which lack hobnails may certainly not be described as `unshod'. This was unfortunately not recognised by Clark (1979, 180) and the role of footwear in this and other late cemeteries is probably very much more complex. Indeed, the characteristic military shoe (though by no means exclusive to soldiers) of the period, as depicted on tombstones, occurs in both a sewn and a nailed version, which might equate with either the `*caligae militares sine clabo*' or the `*campagi militares*' of Diocletian's Edict, priced at 100 and 75 denarii respectively. (For example, Figure 11b would be recognisable in a burial due to the hobnails, while the sewn Figure 11a would not). The unnailed shoes from Deurne (Braat 1973) are associated with a spur, making it tempting to associate similar examples such as Figure 11a from *Vindolanda* with the presence of cavalrymen. The nailed construction does continue into the fourth century, as finds from wells at Dalton Parlours (Mould 1990), Rothwell Colliery (unpublished) as well as York (MacGregor 1978), attest, but the evidence becomes increasingly patchy and uneven, making it difficult to disentangle truly new influences from developments on traditional themes.

Rural Settlements

Rural areas are poorly represented in the available complexes of leatherwork, and it is, consequently difficult to assess the uptake of Roman fashions outside urban and military settlements. The key to the dissemination of footwear styles into the countryside lies in the marketing networks, for if footwear was hawked around local fairs, the rural population would have had access to exactly the same styles as everyone else. Shoemaking offcuts are so ubiquitous that it is safe to assume that almost any conglomeration could support one or

1. The Fell
2. The Balmoral
3. The Hardknott
4. The Ramshaw (unusual decorated example)
5. The Ilkley
6. The Welzheim with fastening method

Figure 8. Selected footwear styles (© B Brouwenstijn)

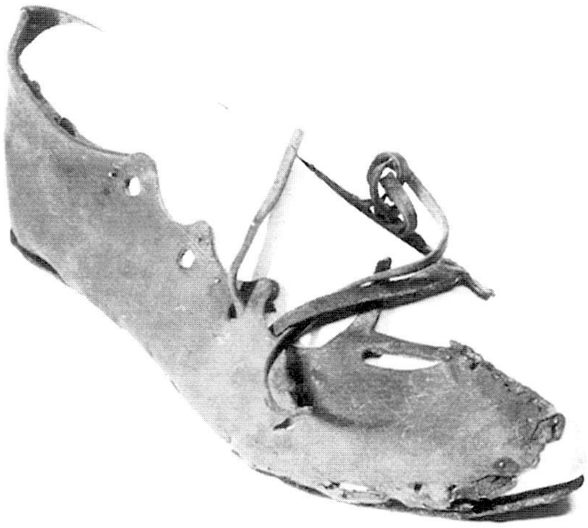

1

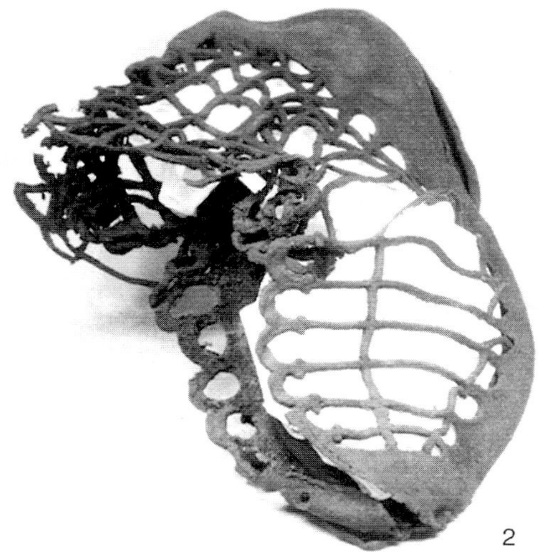

2

Figure 9 Selected footwear styles

1. The Geltsdale (example from Welzheim). (Photo F Gijbels IPP)
2. The Allendale in condition as excavated

more shoemakers. Post-medieval footwear approaches the Roman in complexity and longevity so it may not be entirely inappropriate to refer to sixteenth-century Leiden where there was a shoemaker or cobbler in virtually every house block (van Oerle 1975, maps 7-11). If York is in any way comparable, we could expect a shoemaker in every *insula*. Rural markets, fairs and itinerant peddlers would ensure wide distribution and regional homogenisation (de Ligt 1993; MacMullen 1970), but the possibility of itinerant shoemakers cannot be excluded. The skill required to make even the simplest of Roman footwear should not be underestimated - as anyone who has attempted to make replicas will know. The unknown factor is the extent of the survival of native, household, production. On ethnographic parallels, both curing and leather working are female tasks, so any major shift towards Roman styles of clothing would affect women's economic roles, as well as signalling a conscious acceptance of the new order. Furthermore, the change to rigid close-fitting shoes would affect a person's stance and gait, thus forming a highly visible statement of that person's position in society, and in view of the bunions which Molleson begins to find on Roman feet, one exacted at the price of some discomfort (Molleson 1993, 199). Note however that the Romans did not use stirrups, (cf. van Driel-Murray and Connolly 1991). Other than the wells on villa sites, rural sites tend to lack suitable waterlogged conditions, so neither the Roman tanned products nor the native cured skins stand much chance of survival. On the other hand, there is one handy indicator for the most characteristically 'Roman' of all introduced clothing, in the form of the iron hobnails, whose survival is probably more widespread than excavation reports admit. Occasionally, careful excavation of hobnail concentrations can reveal the shape of the sole and the nailing pattern, both of which are useful indicators for dating in their own right. Thus a decayed sole in a rural enclosure at Pontefract is one of the late second/early third-century pointed type mentioned above (Figure 12), with other concentrations of small-headed, neat nails (diameter *c* 5mm) of a type often used for the elaborate nailing patterns on third-century footwear. To the excavators, the lack of finds suggest the site was no more than a stock yard, but if so, the herdsmen were wearing the fashionable 'cowboy boots'. The evidence may be thin, but it does seem that by the late second century, all levels of the population had access to the latest styles and there was little difference between soldier and civilian, urban or rural.

Conclusion

It is the massive find complexes from waterlogged contexts in towns and forts which provide the basic framework for analysis and interpretation, but these are, at the same time costly and time consuming to research and the state of publication leaves much to be desired. Almost virgin territory when Groenman-van Waateringe defined the parameters for presentation, illustration and quantification in her thesis of 1967, archaeological leather research has seen rapid progress and it is particularly regrettable that the long delay between submission of reports and their publication (8-10 years seems to be the norm) mean that new insights and interpretations permeate the major reports but slowly. The presentation of the primary data is, therefore, essential, for only then can new ideas be tested against existing material. In particular, the relatively rapid presentation of find associations from Carlisle (Winterbottom 1991; 1992) will compel a reassessment of several major complexes of military leatherwork still 'in press'.

For the army, leather was an important raw material, and, on account of the very large quantities involved, models developed to explain the organisation of supply, and the different manufacturing procedures can be applied more generally to clarify the logistics of the Roman army in Northern Britain. In addition, the identification and reconstruction of individual items of equipment - a topic not tackled here - contribute to our understanding of the army, its material legacy and changing appearance through time. The ability to recognise newly identified items in old catalogue reports is clearly of considerable importance to students of military equipment (van Driel-Murray 1988; 1989a; 1989b;

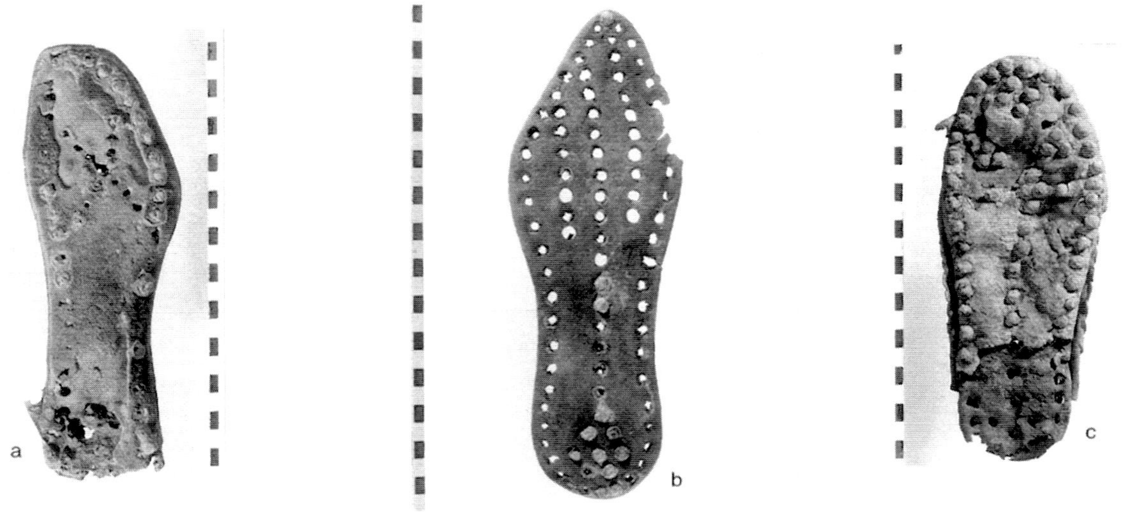

Figure 10. Changing shape of soles: a) late first century; b) late second/early third century; c) late third/fourth century. (Scale 1:4)

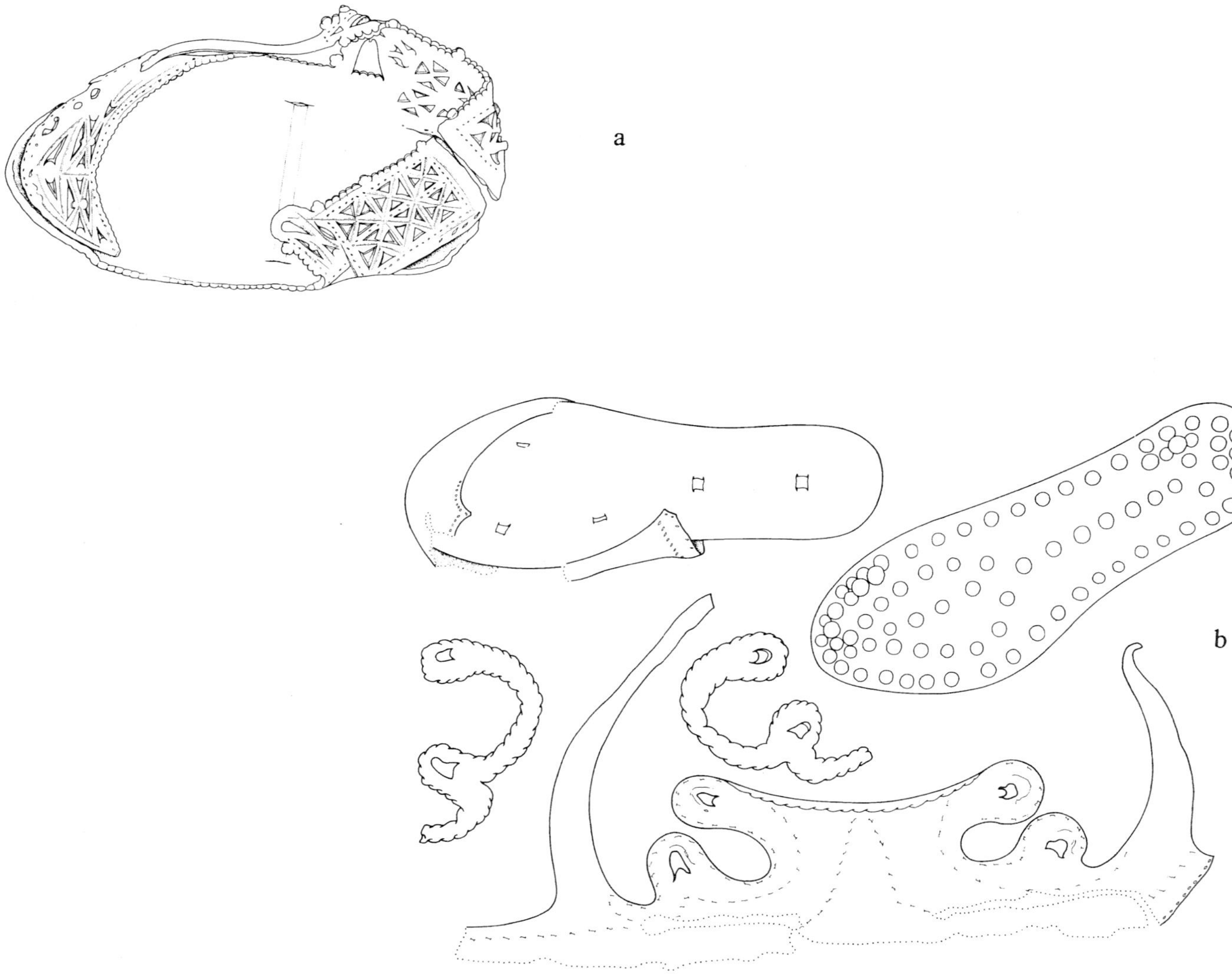

Figure 11. Late third/fourth Century footwear from Vindolanda. Vamp of b) decorated with brass studs. (Scale 1:3) (© B Brouwenstijn)

Figure 12. In situ hobnails at Apple Tree Close, Pontefract. (Illustration courtesy of West Yorkshire Archaeology

1990; van Driel-Murray and Connolly 1991). Questions which still need to be elucidated are the origin of hides and skins and the location of tanneries. Rigorous criteria for the identification of tanneries need to be developed by leather specialists, archaeozoologists and soil scientists in co-operation to ensure that likely sites are given the priority they require in archaeological assessments.

Because of the sheer quantities involved, the analysis of footwear is a useful tool with which to clarify the role of the cities (London, York, Chester) in the transmission of technology and style. As chronology becomes finer, regional differentiation and variations in the take up of particular styles may emerge, but as yet the overwhelming impression is one of the rapid spread of a remarkably homogenous, internationally oriented package of footwear styles, the social implications of which call into question the now common opinion that Roman culture was no more than a thin veneer over essentially native society. A major problem at the moment is the decline in the quantity and quality of the evidence from the late third century onwards. This is a critical period in which the established patterns may be expected to undergo profound transformations. Leather footwear, that most sensitive indicator of social change, may be expected to record these transformations with greater clarity than most other find categories.

Acknowledgements

I would like to thank B Brouwenstijn for producing the line illustrations and particularly for his reconstructions, and M Ydo for the photographs. Throughout, reference is made to much unpublished and ongoing research, both my own and that of others, who have kindly allowed me access to their results. Rather than give cumbersome and pointless references to unpublished material, the sites can be listed here: Castlecary, London (Museum of London holdings), Newstead, Pontefract and Rothwell Colliery, all unpublished or ongoing research, and Welzheim, Germany, since published (van Driel-Murray 1999). Ribchester was kindly shown to me by Ms Howard-Davis, Lancaster Archaeological Unit. I am especially grateful to Quita Mould for sharing her information on Dalton Parlours and Birdoswald in advance of publication, to Sue Winterbottom for giving me access to her immense archive of leather from Carlisle, to D Hooley for allowing me to refer to the Catterick report prior to publication and the West Yorkshire Archaeology Service for permission to reproduce illustrations.

Endnote

Manuscript completed 1997, publication details revised 1999.

Footnotes

1. Even if the pottery numbers were assigned as bulk, compensation is offered by the exclusion of offcuts and featureless, though registered, pieces as well as the fact that much leather was also registered as bulk - including an almost complete tent.

2. Due to its fat content and deep hair implant, pigskin only comes into regular use after the Middle Ages.

3. Ryder 1991, an analysis carried out on hair taken from leather from *Vindolanda*, which confirmed that the *Vindolanda* tent panels were predominantly of goatskin, as microscopic examination had already suggested. About 18% of the *Vindolanda* textiles are of generalised medium wools from `primitive' sheep breeds. This research was funded by the Leathersellers' Company Charitable Fund, to whom we are grateful for the support offered.

4. The origin of the large numbers of male goats offered to Mercury in the Temple at Uley also needs to be considered in this context (Woodward 1992, 75).

5. The `tannery' at Lullingstone villa (Meates 1979, 105-10) is a good illustration of the total misunderstanding of the relationship between refuse and manufacture.

6. In van Driel-Murray 1993, 63, I considered that these stamps had been applied after tanning, but in view of the differences with stamps and incisions on finished items (such as footwear), I am now more inclined to regard them as marks applied to the raw hides. Experiment would resolve the difficulty.

7. It must be pointed out that shield-makers are mentioned at least twice in the *Vindolanda* tablets (Bowman and Thomas 1994 nos 160, 182), though in view of the repeated references to work gangs being employed elsewhere, including

Corbridge, this need not mean that shield were being produced at *Vindolanda* itself. Indeed, only a single fragment of a shield cover has turned up there.

8. The centurion Sollius Iulianus known from a building stone on Hadrian's Wall is also mentioned in a scratched graffito on leather, probably a tent door, from York (Hooley 1988/9, fig 6; Hassall and Tomlin 1987, 373-4, no. 31): was the detachment moved up to the Wall from here, or did they come to York after their building stint was completed? Both moves are occasions for extensive checking and refurbishing equipment and for turning out old stores (cf. Castleford and Vindolanda, van Driel-Murray 1993, 27).

9.One is also reported from Catterick, there are a few in early contexts in London, but Rhodes' claim for `caligae' at the Billingsgate site (1980, 113-14) must be discounted, as the soles bear no resemblance to caligae, and in fact display the normal second century construction for shoes with separately cut, foot-enclosing uppers where the laminae or middle soles are attached to the insole by thonging.

BIBLIOGRAPHY

Alston, R, 1995 *Soldier and Society in Roman Egypt*. London: Routledge

Birley, A R, 1993 Review of the texts. *The Early Wooden Forts. Reports on the Auxiliaries, the Writing Tablets, Inscriptions, Brands and Graffiti*. Hexham: *Vindolanda* Res Rep **II**

Birley, R, 1994 *The Early Wooden Forts*. Hexham: *Vindolanda* Res Rep **I**

Bowman, A, and Thomas, J D, 1983 *Vindolanda: The Latin Writing-Tablets*. London: Britannia Mono **4**

---- , and ---- , 1994 *The Vindolanda Writing-Tablets (Tabulae Vindolandenses II)*. London: British Museum Press

Braat, W C, 1973 Der Funde von Deurne. In H Klumbach (1973) *Spätrömische Gardehelme*. München, 51-83

Bradley, K R, 1991 *Discovering the Roman Family*. Oxford: Oxford University Press

Busch, A L, 1965 Die römerzeitlichen Schuh- und Lederfunde der Kastelle Saalburg, Zugmantel und Kleiner Feldberg. *Saalburg-Jahrbuch* **22**, 158-210

Charlesworth, D, and Thornton, J H, 1973 Leather found in *Mediobogdum*, the Roman fort of Hardknott. *Britannia* **3**, 141-52

Clark, G, 1979 *The Roman Cemetery at Lankhills. Pre-Roman and Roman Winchester, Part II*. Oxford: Oxford University Press

Clerici, R, 1983 Römische Holzfässer aus Vitudurum. *Helvetia Archaeologica* **14**, 14-24

Driel-Murray, C van, 1977 Stamped leatherwork from Zwammerdam. In B L van Beek, R W Brandt and W Groenman-van Waateringe (1977) *Ex Horreo*. Amsterdam: Cingula, 151-64

----, 1985 The production and supply of military leatherwork in the first and second centuries AD: a review of the archaeological evidence. In M C Bishop (ed) (1985) *The Production and Distribution of Roman Military Equipment*. Oxford: Brit Archaeol Rep, Int Ser **275**, 43-81

---- , 1988 A fragmentary shield cover from Caerleon. In J C Coulston (ed) (1988) *Military Equipment and the Identity of Roman Soldiers. Proceedings of the Fourth Roman Military Equipment Conference*. Oxford: Brit Archaeol Rep, Int Ser **394**, 51-66

---- , 1989a The *Vindolanda* chamfrons and miscellaneous items of leather horse gear. In C van Driel-Murray (ed) (1989b) *Roman Military Equipment: the Sources of Evidence*. Oxford: Brit Archaeol Rep, Int Ser **476**, 281-318

---- , 1989b A circular shield cover and the reconstruction of the accompanying shield. *Exercitus* **2.7**, 132-34

---- , 1990 New light on old tents. *J Roman Military Equipment Stud* **1**, 109-38

---- , 1993 The leatherwork. In C van Driel-Murray, J P Wild, M Seawood, J Hillam and R Birley (1993) *The Early wooden forts. Preliminary reports on the leather, textiles, environmental evidence and dendrochronology*. Hexham: *Vindolanda* Res Rep **III**

---- , 1995 Gender in Question. In P Rush (ed) (1995) *Theoretical Roman Archaeology: Second Conference Proceedings*. Aldershot: Avebury, 1-21

---- , 1998 The leatherwork from the fort. In H E M Cool and C Philo (eds) (1998) *Roman Castleford Excavations 1974–85. Volume 1: The Small Finds*. Wakefield: Yorks Archaeology **4**, 285-334

---- , 1999 *Das Ostkastell von Welzheim, Rems-Murr-Kreis: Die römischen Lederfunde*. Stuttgart: Thiess, Forschungen und Berichte zur Vor- und Frühgeschichte in Baden-Wüttemberg **42**

---- , and Gechter, M, 1983 Funde aus der Fabrica der Legio I Minervia am Bonner Berg. *Rheinische Ausgrabungen* **23**, 1-83

---- , and Connolly, P, 1991 The Roman cavalry saddle. *Britannia* **22**, 33-50

Fellmann-Brogli, R, 1991 Leder aus Vicus Vitudurum-Oberwinterthur. *Vitudurum* **5**. Zürich: Berichte der Zürcher Denkmalpflege Archäologische Monographien **10**, 43-58

Gansser-Burckhardt, A, 1942 *Das Leder und Seine Verarbeitung im römischen Legionslager Vindonissa*. Basel: E Birkhäuser and Cie

Göpfrich, J 1986 Römische Lederfunde aus Mainz. *Saalburg-Jahrbuch* **42**, 5-67

Groenman-van Waateringe, W, 1967 *Romeins lederwerk uit Valkenburg Z.H.* Groningen: Wolters

Hassall, M W C, and Tomlin, R S O, 1987 Roman Britain 1986. II. Inscriptions. *Britannia* **18**, 360-437

---- , and ---- , 1989 Roman Britain 1988. II. Inscriptions. *Britannia* **20**, 327-45

Hooley, A D, 1988/89 Roman leather from the General Accident Site, York and Catterick, North Yorkshire. *Yorks Archaeol Soc Roman Antiqs Sec Bull* **6**, 16-25

---- , 2002 Leather from the 1958-9 By-pass excavations (Site 433). In P R Wilson (2002) *Cataractonium. Roman Catterick and its hinterland. Excavations and Research 1958-97* Part II. CBA Res Rep **129**, 318-80.

Ligt, L, de, 1993 *Fairs and Markets in the Roman Empire*. Unpublished PhD dissertation, Free Univ Amsterdam

MacGregor, A, 1978 *Roman Finds from Skeldergate and Bishophill*. Archaeol York. The Small Finds **17/2**. London: CBA, 50-54

---- , 1989 Bone, antler and horn industries in the urban context. In Serjeantson and Waldron (1989), 107-28

MacMullen, R, 1970 Market days in the Roman Empire. *Phoenix* **24**, 333-41

Maltby, M, 1989 Urban-rural variations in the butchery of cattle in Romano-British Hampshire. In Serjeantson and Waldron (1989), 75-106

Meates, G W, 1979 *The Lullingstone Roman Villa*. Maidstone: Kent Archaeol Soc Mono **1**

Molleson, T I, 1993 Part 2. The human remains. In D E Farwell and T I Molleson (1993) *Poundbury: Volume 2 The Cemeteries*. Dorchester: Dorset Nat Hist Archaeol Soc Mono **11**, 142-214

Morel, P, 1991 Die osteologischen und anthropologischen

Untersuchungen. *Vitudurum* **5**, Zürich: Berichte der Zürcher Denkmalpflege Archäologische Monographien **10**, 79-176.

Mould, Q, 1990 The leather objects. In S Wrathmell and A Nicholson (eds) (1990), *Dalton Parlours. Iron Age Settlement and Roman Villa.* Wakefield: Yorks Archaeol **3**, 231-35

----, 1997 Leather. In T Wilmott (1997) *Birdoswald. Excavation of a Roman Fort on Hadrian's Wall and its Successor Settlements: 1987-92.* London: English Heritage Archaeol Rep **14**, 326-41

Noddle, B, 1994, The under-rated goat. In A R Hall and H K Kenwood (eds) (1994) *Urban-Rural Connections: Perspectives from Environmental Archaeology.* Oxford: Oxbow Mono **47**, 117-27

O'Connor, T, 1989 What shall we have for dinner: food remains from urban sites. In Serjeantson and Waldron (1989), 13-23

Oerle, H A van, 1975 *Leiden binnen en buiten de stadsvesten.* Leiden: Brill

Prummel, W, 1978 Animal bones from tannery pits. In `s-Hertogenbosch (1978) *Berichten Rijksdienst voor het Oudheidkundig Bodemonderzoek* **20**, 399-422

Rhodes, M, 1980 Leather footwear. In D M Jones and M Rhodes (eds) (1980) *Excavations at Billingsgate Buildings, Lower Thames Street, London 1974.* London: London and Middlesex Archaeol Soc Spec Pap **4**, 99-128

---- , 1987 Inscriptions on leather waste from Roman London. *Britannia* **23**, 173-81

RIB II.4 = Frere, S S, and Tomlin, R S O, 1992 *The Roman Inscriptions of Britain. Vol II. Instrumentum Domesticum, fasc 4.* Stroud: Sutton

Robertson, A, Scott, M, and Keppie, L, 1975 *Bar Hill: A Roman Fort and its Finds.* Oxford: Brit Archaeol Rep, Brit Ser **16**

Ryder, M L, 1991 *Report on hair and wool from Roman leather from Vindolanda.* Unpubl rep, 16.12.1991

Schmid, E, 1969 Knochenfunde als Archäologische Quellen. *Archëologie und Biologie, Forschungsberichte* **15**, 100-11

---- , 1973 Ziegenhörner als Gerberei-Abfall. *Schweiz. Volkskunde* **63, h 5/6**, Basel, 65-66.

Serjeantson, D, 1989 Animal remains and the tanning trade. In Serjeantson and Waldron (1989), 129-46.

---- , and Waldron, T, 1989 *Diet and Crafts in Towns.* Oxford: Brit Archaeol Rep, Brit Ser **199**

Speidel, M A, 1996 Das römische Heer als Kulturträger. In R Frei-Stolba and H E Herzig (1996) *La politique édilitaire dans les provinces de l'Empire romain IIème -IVème siècles après J-C.* Bern, 187-209

Thornton J, 1991 Shoe leather. In P S Austen (1991), *Bewcastle and Old Penrith.* Kendal: Cumberland Westmorland Antiq Archaeol Soc Res Ser **6**. 219-34

Turner, R C, Rhodes, M, and Wild, J P, 1991 The Roman body found on Grewelthorpe Moor in 1850: a reappraisal. *Britannia* **22**, 191-201

Winterbottom, S, 1991 The sheet leather objects. In T G Padley and S Winterbottom (1991) *The Wooden, Leather and Bone Objects from Castle Street, Carlisle: Excavations 1981-2.* Kendal: Cumberland Westmorland Antiq Archaeol Soc Res Ser **5** fasc **3**, 244-328

---- , 1992 Leather. In I Caruana (1992), Carlisle: excavation of a section of the Annexe Ditch of the first Flavian fort, 1990. *Britannia* **23**, 45-109

Woodward, A, 1992 *Shrines and Sacrifice.* London: Batsford/English Heritage

THE JET INDUSTRY AND ALLIED TRADES IN ROMAN BRITAIN

LINDSAY ALLASON-JONES

One industry which is generally associated with both Roman and nineteenth century Yorkshire is the carving of jet for the production of jewellery and domestic artefacts. One of the best collections of such artefacts from Romano-British contexts is housed in The Yorkshire Museum and this has been studied as part of a long-term research project at the University of Newcastle upon Tyne (Figure 1). The Yorkshire Museum collection was recently published in detail in the Museum's catalogue series (Allason-Jones 1996); the present paper seeks to set this Yorkshire material into its context as part of what is beginning to be recognised as an extremely complex, country wide industry.

During the early years of the Roman occupation of Britain black jewellery does not seem to have been highly rated yet, for some reason as yet unexplained, by the end of the second century AD this jewellery was becoming more and more fashionable until a clearly identifiable industry had been established by the beginning of the third century (Allason-Jones 1996, 15-17). Traditionally all this black jewellery has been identified as jet with a presumed source on the Yorkshire coast at Whitby, but the recent research has shown that jet was only one of the materials being used and that the immediate vicinity of Whitby was not the only source being exploited (Allason-Jones and Jones 1994). Cannel coal, detrital coals, shales (including torbanite), and on occasions domestic coals and burnt bones, were also being used to make small personal objects with a black shiny finish.

What all these materials have in common is their black colour and their ability to be polished to a shine and both criteria are due to the fact that the stones are rich in organic material.

Figure 1. Grave group found in 1874 under the new Railway Station, York. The artefacts were found in a wooden box with bronze fittings. The illustrated artefacts consist of three cabled jet armlets, a large plain armlet, two small armlets or rings, the jet head from a bone pin, a ring-shaped glass flask and a small phial of green glass. When found the group also included eight small beads and three pins. © Yorkshire Museum.

This may be from terrestrial plant material which divides into three groups: vitrinite, inertinite and liptinite. A normal UK coal consists of 60% vitrinite, 20% inertinite and 20% liptinite; the rocks selected by the Roman craftsmen for carving, on the other hand, contain a higher organic content rich in liptinite and amorphinite. Liptinite has its origins in resin, spores, pollen and cuticle (leaf coating) as well as algae; amorphinite requires the burial of planktonic material in anaerobic mud on the sea floor, otherwise it will oxidise away. It is the waxy, fatty, oily faction of the liptinite and amorphinite which gives the good polish, whilst the high organic content also makes the rock soft and easy to carve.

'Jet' as a geological term is restricted to a form of vitrinite derived from logs of wood which were washed into the sea where they sank into the airless sediment on the sea floor. The mud of the sea floor was rich in the remains of planktonic material and algae which generated hydrocarbons as they decomposed. These hydrocarbons impregnated the wood preventing the vitrinite from developing its brittle character. `Cannel coal' is formed from plant debris accumulating on the bottom of a lake in a coal-forming swamp. Although it can include clay and pyrite in small quantities it is its wide range of plant components, including spores, resin and algae, all of which are waxy in character, which produces the polishing characteristics of cannel. This material derives its name from `candle' because it will burn with a flame like a candle due to its high liptinite content. `Shale' is a common sediment which consists almost wholly of clay minerals. Some shales are black due to a high content of phytoclasts or hydrocarbons and capable of being polished either because the phytoclasts are dominantly liptinites or because of the presence of the hydrocarbons. The more hydrocarbon-rich shales are the oil source rocks containing abundant amorphinite, the best known of which are the Kimmeridge Clay, which outcrops from Dorset to Teesside, and the Jet Rock shale in which the logs of true jet are found. The phytoclast-rich shales are so common that identifying a specific source is impossible.

The Roman authors seem to have referred to all these materials as `jet' (*gagates*) without fully comprehending the differences between the materials. Pliny, for example, described jet as being `black, smooth, porous, light, not very different from wood, and brittle, and has an unpleasant smell when rubbed. Anything inscribed in it on earthenware is indelible. When it is burnt it gives off a smell like that of sulphur. What is remarkable is that it is ignited with water and quenched by oil' (*Nat Hist*, 36). Its source, he claimed, was `a district and a river in Lycia known as Gages', and this was how the material got its name. He further informed his readers that jet `is found also to be washed up by the sea on the promontory of Leucolla (in Pamphylia) and to be gathered at places up to a distance of a mile and a half'. Much of Pliny's account is an accurate description of jet but, regrettably, it would appear that what he was actually describing was asphalt, which shares many of the characteristics of jet, is also called *gagates* in Latin, and is to be found in Lycia. More recent attempts to find jet in Turkey have failed (Spratt and Forbes 1847).

Writing a century after Pliny, Solinus clearly derived much of his information about `jet' from Pliny, or from a source common to both authors, but apparently had additional information to draw upon as he was the first to state that jet was found in Britain, `in great abundance and of very fine quality' (*Collectanea rerum memorabilium* 22,II (19)). Neither Pliny nor Solinus refer to shale; indeed, no word for shale was recorded until the Middle Ages when the term derived from the Germanic *shala*. Both authors' comments about coal were largely confined to its use as a fuel (see Dearne and Branigan 1995).

Today, geologists are aware that jet can be found throughout the world in Russia, Germany, Portugal, France, and North America, and analysis has shown that sources in the Rhineland and Spain (particularly Asturias on the north coast: Suárez-Ruiz *et al* 1994) were being exploited between the second and fourth centuries AD. Shales and coals, on the other hand, although more widely available, appear to have been rarely used during the Roman period for any other purpose than as fuel except in the British Isles and Germany. Modern geologists can differentiate between these materials by identifying their botanical constituents and their rank. Rank can be determined chemically but also optically under a microscope. The higher the rank the brighter the polished surface of the coal will appear under the microscope. Identification involves measuring the reflectance of the wood (vitrinite), that is the percentage of light reflected back from the polished surface. The Roman craftsmen, of course, did not have modern analytical techniques available to them and the recent research at the University of Newcastle upon Tyne has shown that it is not possible to differentiate between the materials by eye alone (Allason-Jones and Jones 1994; Allason-Jones 1996). If the craftsmen were simply using the material closest to hand they would, no doubt, have learnt quickly by experience that their local coal and shale measures were or were not suitable for carving. It might also be expected that the bulk of the jet/shale/coal artefacts would be found closest to the most conveniently available sources of the most easily carved materials but this is not so. Only at Halton Chesters, on Hadrian's Wall, have two armlets been found which appear to have been carved from shale recovered from within a mile of the findspot (Allason-Jones and Dore *forthcoming*), whilst one of the largest assemblages of black jewellery - at South Shields fort - appears to spurn the natural deposits within walking distance of the fort, the craftsmen apparently preferring to obtain jet from Whitby, torbanite from Midlothian, and shales and cannels from Northumberland and the Midlands (Allason-Jones and Jones 1994; Allason-Jones and Miket 1984). Even in the famous jet-working city of Roman York half-finished artefacts of Kimmeridge clay and cannel coals have been found (Allason-Jones 1996). Clearly there was a lively trade in black stones throughout Britain and possibly Europe. The question remains, however, did the Roman craftsmen differentiate between the materials and how did they know which materials would be most suitable for which artefacts?

The first question may be answered by an analogy to modern woodworkers. An experienced wood carver will know by touch alone that a particular piece of wood is suitable for his purpose and will confidently reject other pieces which, to the untutored eye, look perfectly reasonable. The importance of this empirical knowledge, built on years of experience and skill, should not be underestimated; such knowledge enables a craftsman to make accurate judgements that often defy modern analytical techniques. In the case of jet and associated materials, subtle indicators such as weight, shine, and tiny imperfections suggesting inclusions would all be noted, probably subconsciously, leading to the decision to use a piece. Even today the jetworkers of Whitby know instinctively whether the pieces they gather from the beaches are workable or not. Such skills leave no trace in the archaeological record.

It is also possible that the craftworkers used geographical data to aid their decision making. It would be known that the black material from Dorset came in large slabs which could be carved into table legs, table tops, trays and armlets whilst the smaller fragments found on the Yorkshire coast were better suited for objects requiring more delicate carving. Although the table and tray factories seem to have been confined to Dorset, the discovery of blocks of unworked jet and shale and half-finished objects made from these materials at both York and South Shields, points to the craftshops having access to a wide variety of sources and that they were either able to order what was required or had a regular supply of their preferred material. Whilst a regular supply of jet from Whitby inland to York or up the coast to South Shields, or even of shale and cannel coals from the Tyne Valley along the military roads to South Shields, might not occasion surprise, the discovery of jet and shale artefacts in quantity in Silchester, Wroxeter or London clearly points to a more complex trade pattern, whilst the indications that raw material was being obtained from Midlothian, beyond the frontiers of the Roman Empire, as well as from the ambiguous area immediately to the north of Hadrian's Wall, raises issues of cross-frontier trade (Allason-Jones and Jones 1994).

Occasionally it would seem that the craftsmen were not too concerned as to the source or type of material they worked in, but there does appear to have been a bias towards making pins from jet, armlets, tables and trays from shale, and finger-rings and beads from cannel coal, but this is not an invariable rule and certain artefacts, such as pendants, can be found in any material. Cannel coal, with its predisposition to shatter when carved, may have been deemed more appropriate for smaller objects whilst shale, being stronger, could take the heat and stresses of lathe-turning and was thus more suitable for armlets, but the evidence suggests that the jewellers were willing to make any object out of any material as long as the final product could be polished to a black shine; unpublished examples in the Yorkshire Museum suggest that, if the more usual materials were not available, then substitutes could be used, such as burnt bone.

The gathering and carving of jet and shale was not a complicated process and required little equipment. Much of the raw material could be gathered by beach-combing or by searching along the base of cliffs after storms or cliff falls. It is possible that the beachcombing method supplied enough to satisfy the market but if demand outstripped supply a 'drift' or tunnel could have been cut into a hillside or a pit dug down from the surface using simple tools. In the nineteenth century a jet or shale miner relied solely on a pick or shovel, often with the point of the pick drawn out by a blacksmith so that it was sharp enough to bite into the rock. Obviously, it is difficult to identify with any certainty picks or shovels used for Roman jet or shale mining as opposed to the tools used for other activities, but the type of pick found at Hod Hill, Dorset, amongst other sites, with a spike blade on one side and a chisel edge on the other would have been a suitable tool for the purpose (Manning 1985, C1). Roman shovels, like nineteenth century miners' shovels, tended to be of wood and surviving examples are therefore rare.

Cliff falls and the activities of medieval and later miners have removed all identifiable traces of Roman mining for jet or shale, whether from surface digging or from the method called 'dessing' whereby a person was lowered down from the cliff top on a rope in order to chip out the desired layers. The First Edition 6 inch Ordnance Survey maps of Roseberry Topping in Cleveland, however, show a number of shallow pits, labelled as 'Ancient British Settlements', which some authorities have interpreted as either Roman or medieval jet workings (Owen 1975, 18).

As jet is a comparatively light material, which is more often found in small lumps than in large blocks, it is likely that the first part of its journey to the workshops involved people carrying the loads of raw material on their backs in sacks or baskets, particularly as most of the jet beaches are at the base of steep cliffs necessitating a stiff climb to get the harvest to a road. Transporting the heavier raw materials, such as shale, which was often used in large slabs for the manufacture of tables and trays, may have involved the use of wagons drawn by oxen or mules. Evidence is lacking to indicate whether the people who gleaned the material were responsible for getting their material to the workshops or if middlemen were involved at a central collection point. A cache of unworked jet pebbles found at Goldsborough Signal Station may represent one person hoarding casually found pieces until they could be passed on for carving (Hornsby and Laverick 1932).

Carving the material also required simple tools and in the nineteenth century the first stage involved removing the outer skin or 'spar' with a heavy chisel of iron or steel with a lead-weighted handle. The block was then chopped into pieces suitable in size for the intended product using a chisel or saw. A sandstone wheel was then used to grind off the rough edges and the piece handed to the specialist craftsmen who would shape the article before it was finished off by polishing with jeweller's rouge, paraffin, linseed oil and lamp-black. The process involved several people working in a production line and most of the heavier equipment, such as the grinding wheels, were bought from wholesale suppliers, although the smaller tools used by the specialist craftsmen were often handmade for specific purposes and handed down from father to son (McMullen 1992; Muller 1980).

In contrast, the Roman industry does not appear to have been arranged along factory lines but each individual seems to have handled each stage of the process personally. The discovery at York of several roughouts for beads and armlets (Figure 2) may suggest that the first stage was carried out by apprentices with the master craftsman completing the object,

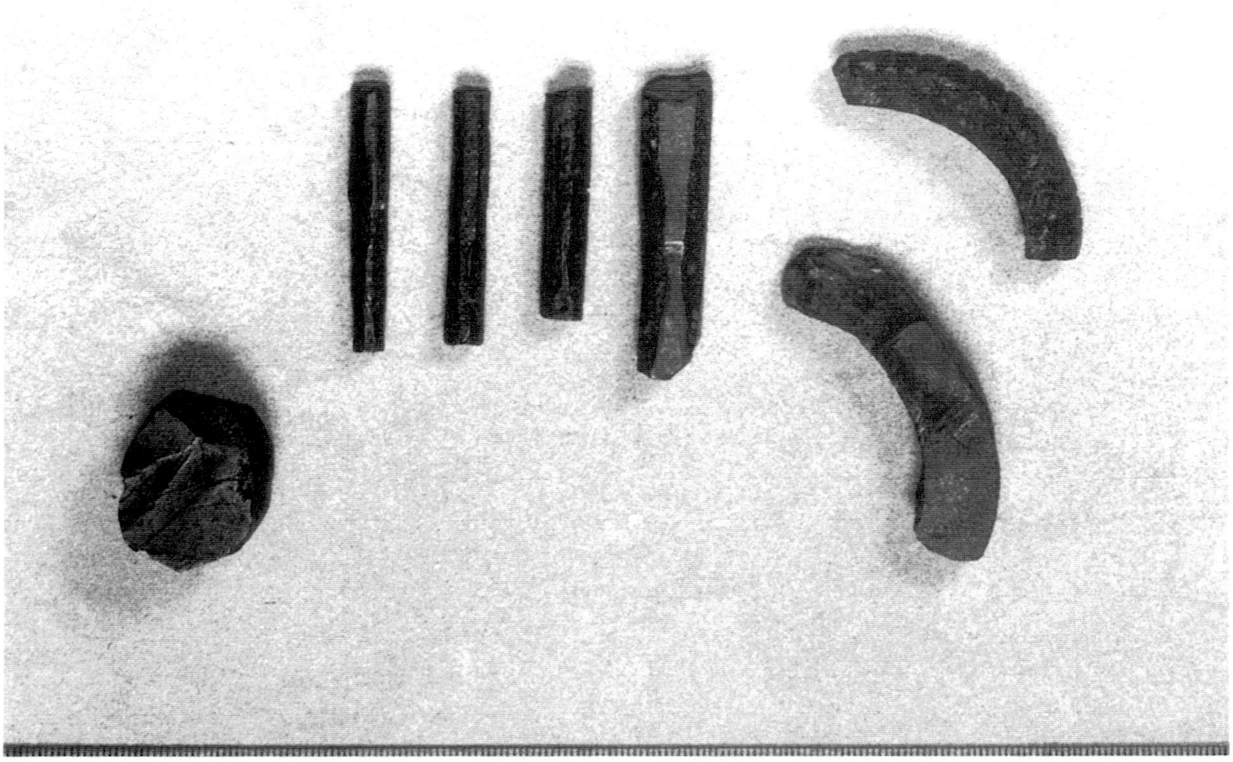

Figure 2. Rough-outs for armlets and cylindrical beads, from York. © Yorkshire Museum.

but may equally indicate that some workers preferred to make artefacts stage by stage in batches. The largest piece of equipment required by a jet or shale carver would have been a lathe, probably worked by a treadle: such lathes were introduced into shale armlet cutting in the Glastonbury region in the Iron Age (Gray 1953). To prepare a lathe-turned armlet the turner first cut a roughly circular slab, about 20mm thick, from the source block using a saw or a chisel. Two circular holes were then drilled or pecked into both faces so that the disc could be firmly secured to the lathe. Having prepared and secured the blank the turner would shape the edge of the disc before cutting a V-shaped circular groove into both faces; the armlet ring could then be snapped from its central core. At the Kimmeridge shale centres of Dorset a distinctive narrow flint blade was used on the lathe (Davies 1936, 208, fig 3), but these appear to be confined to the area and it is likely that most craftsmen used iron blades on their lathes. This method of cutting armlets requires a high degree of skill and implies that a long apprenticeship had to be served before a young craftsman could be considered competent (David Sim *pers comm*).

The central discs which are left after an armlet has been cut on a lathe are very distinctive and caused much curiosity in the early nineteenth century when antiquarians first took notice of them. Initially they were interpreted as some sort of trading token and called `Kimmeridge shale money' (Austen 1857; Calkin 1955) but it soon became clear that they were, in fact, the waste products of armlet manufacture. The amount of armlet cores found during the investigations of the nineteenth century indicates that the Kimmeridge factories were extraordinarily productive: in 1846 a Mr Pennie found 4 or 5 bushels during drainage operations, Mr Groves claimed to have dug up nearly 300 pieces in an hour, and in 1859 Mr Austin discovered 600-800 examples from two excavation trenches three feet square. The evidence of the ridges left around the inner faces of armlets cut on a lathe indicates that jet armlets were produced in the same way but no armlet cores of jet have been found. This may be because even the smallest fragment of jet can be used for making beads or finger rings, so the cores would provide valuable recycling material. Shale was not so easy to work in small pieces and, as it was more freely available, there would be little need to utilise every scrap. It is also possible that the heat created during the lathe-turning process would have made the shale cores very brittle and less easy to recarve; in the nineteenth century the objects being lathe-turned had to be kept cool by constant application of water to stop them splitting.

The rough lathe scars on both the jet and shale armlets would have been filed smooth with an iron file - several of the York pieces still show traces of file marks (Allason-Jones 1996, 29-36). A number of iron files have been found in third- and fourth-century contexts in Roman Britain which are invariably of a tapering square-section with teeth on all four faces (Manning 1985, 11); once again it is impossible to tell a jet-worker's file from one used by a metalworker, a carpenter or a blacksmith. The stage at which the decoration was added would have depended on the type of armlet being made. Those with dot-and-ring motifs (Allason-Jones 1996, nos 78-79), or shallow notches confined to the face (Allason-Jones 1996, nos 63, 67), would have had the decoration added at a very late stage, whilst those with cable decoration (Allason-Jones 1996, nos 71-77), or deeply cut chevrons which impinged on the sides (Allason-Jones 1996, 52-56), would have had to have the decoration roughed out at an early stage. Two unfinished armlets from York show shallow grooving on the outer face which seems to suggest a planning stage in the decoration (Allason-Jones 1996, nos 316-17).

It was in the Iron Age that decoration began to be applied to shale armlets and many of the decorative motifs used in the third and fourth centuries AD can be traced back to these origins. At Maiden Castle in Dorset, for example, a circular armlet decorated with horizontal grooves may be a precursor of the octagonal armlets known both in Germany and Britain in the third century AD (Wheeler 1943, fig 109, no 9; cf. York: Allason-Jones 1996, nos 69-70), whilst at the Meare Lake Village in Somerset several shale armlets have been found with motifs which may have paved the way for the cabled armlets of the Roman period as well as the more elaborately decorated examples (Gray 1953, pl LIII, K39, K47-55, K79). One of the Meare armlets appears to have been a two-piece armlet of a similar type to one found at York (Gray 1953, pl LIII.K58; Allason-Jones 1996, no 82).

For the other products, the main cutting seems to have been carried out with saws, chisels, drills and knives. Some of the roughouts for beads and finger-rings found during the Railway Station excavations at York show clear signs of having been chiselled or pared to shape (Allason-Jones 1996, nos 327-35). Some Roman chisels had a wooden handle with a mushroom-shaped end to make them comfortable for use (Manning 1985, 22), whilst others had solid iron shanks which could take a hammer blow (Manning 1985, 23). The latter type was probably used when splitting the large blocks of raw material prior to shaping; the former for more delicate work. A two-part knife handle from York (Allason-Jones 1996, no 299) (Figure 3) has clear marks on its flat back which indicate that it was sawn from a block; as this face would not be seen when the handle was complete there was no need for the saw marks to be filed away.

Modern jetworkers are divided as to whether it is simpler to drill the hole for a bead first and then shape the face or *vice versa*. The second-century bead blanks from York Railway Station indicate that when making a cylindrical bead a Roman craftsman first cut a rod of approximately the correct diameter using a lathe (Figure 2) - the dimples at both ends of a number of examples are clearly centre marks from a lathe. A line was then scored around the shank to indicate where each bead should be snapped off. It is not clear if the hole was then drilled before or after the outer face was decorated; the pillar beads from one necklace at York (Allason-Jones 1996, no 13) show that at least in one case the holes were drilled after the decoration was applied. This procedure would seem sensible as the lathe would hold the bead firmly whilst the decoration was applied; if the hole was drilled first then the bead might slip when chisel pressure was applied. Some of the decoration, such as the incised transverse grooving on cylinder beads, was clearly cut on a lathe and it would, no doubt, have been more convenient to do all the lathe-work together.

The holes in the beads could have been drilled either with a bow-drill or a strap-drill. Both work on the same principle, the only difference being that whilst the bow-drill was worked by the craftsman on his own, holding the drill with one hand and moving the bow with the other, the strap-drill

Figure 3. Knife handle showing saw marks on its inner face. © Yorkshire Museum.

required a second pair of hands to pull alternately on the ends of the strap which drove the drill while the first person guided the bit. The strap-drill may have been unnecessarily powerful for delicate jewellery work and the bow-drill is the more likely piece of equipment. Experiments by David Sim, University of Reading, have shown that it will take 30 minutes to drill a hole through a 20mm shale bead using a reciprocating drill. Shale is easy to saw but can split when drilled if care is not taken. It will also take the edge off a drill very quickly requiring constant resharpening. Alternatively, a sharply pointed hand-held awl or augur could have been used, although it is difficult to make a long straight hole using a hand-held tool. Most beads in cross-section show the 'hourglass' shape to their holes which indicates that they have been drilled first from one end and then from the other. This would have been an easier method, even if it depended on considerable accuracy, and would also have stopped heat or a vacuum building up inside the hole as the drill worked, problems which often shatter beads made today with electric drills.

Pins would have been the simplest objects to make. A square-sectioned rod was first cut from the source block using a saw. The head end was then clamped firmly and the shank of the pin was either lathe-turned or whittled to shape: both methods have been noticed amongst the York products (Allason-Jones 1996). The pin could then be removed from the clamp and the head cut to shape and decorated. The method may explain why pins with faceted cube heads appear to have been the most popular - starting from a square-sectioned rod it was a simple matter to cut the corners off the clamped block to produce an attractive head, the facets of which would catch the light when worn in the hair.

The finer decoration found on beads, armlets or finger-rings, and the carving of elaborate pendants (Figure 4), would have been cut with small chisels and points of the type used by craftsmen working in metal or semiprecious stones. Most of these would have been of iron or bronze but small chips of flint may have been used when available. These smaller tools would probably have been used for cutting finger-rings, although the basic method would have been the same as that used for cutting armlets.

Whatever the product, it would be the final appearance which would have ensured sales and this meant that each piece needed to be polished to a smooth, shiny, black finish. For objects of jet this would entail the application of oil mixed with jet dust, which would both smooth and polish at the same time as it was rubbed into the surface. Olive oil may well have been used but lanolin-rich sheep's wool would also have been effective. The final buffing with a clean cloth or piece of fine leather was probably done when the object was placed on display at point-of-sale. Cannel coal could also be finished off by simply buffing to a shine, but shale required more effort as it might need colouring as well as polishing and might absorb a large amount of oil if it was very dry. In the nineteenth century lamp-black mixed with paraffin and linseed oil was used; in the Roman period lamp-black, burnt wood or bone mixed with olive oil were the likely methods of colouring shale.

Other than the shale armlet cores found in Dorset, the largest group of roughly blocked-out or half-finished artefacts of black jewellery from Roman Britain comes from York (Figure 2). Elsewhere, a few half-finished pieces, such as the fox's head from South Shields (Allason-Jones and Miket 1984, no 7.170), and armlets from Wroxeter (*unpubl*), are known but in these cases it is still unclear whether the objects were being carved completely at these centres or were started elsewhere and simply finished-off at point-of-sale. Analysis of some of the rectangular blocks of jet at South Shields have shown them to be of Whitby jet rather than a more locally available material, which suggests that craftsmen were acquiring blocks in unworked form and then carving them in their workshops (Allason-Jones and Jones 1994). Silchester, which has produced a large number of jet and shale artefacts, has not yet revealed any evidence for production on site but, as the tools required would have been equally suitable for other crafts and every scrap of jet or cannel coal can be used for small beads or as polishing dust, jet or cannel coal workshops are difficult to identify unless substantial numbers of blocks or half-finished artefacts are found.

Figure 4. Pendant with the head of Medusa. © Yorkshire Museum.

Comparison between the major assemblages of jet and shale objects in Britain shows that while some objects, such as the pins with faceted cube heads, appear throughout the country, each assemblage has its own particular objects and styles. At Silchester, for example, there is a strong bias towards dishes, trays, and tables as well as armlets decorated with lathe-turned grooves and ridges, presumably because a high proportion of the Silchester objects are made from Dorset shales and look back to the Iron Age for their inspiration. Silchester has produced no finger-rings, only one cantharus-headed pin, and no pillar or barrel beads (Lawson 1975). South Shields, on the other hand, has a particular type of finger-ring with an S-motif carved in high relief on the panel (Allason-Jones and Miket 1984, nos 7.98-102), which has not been found at York, as well as a large number of disc spindlewhorls and gaming pieces which are also missing at York (Allason-Jones and Miket 1984, nos 7.171-90, 164-68). Wroxeter shares notched armlets and cylinder beads with York (Allason-Jones 1996, nos 57-68), ridged armlets and globular spindlewhorls with Silchester (Lawson 1975, nos 40-3, 107-08), and a number of bead types with South Shields, but also has a small, loaf-shaped type of bead which appears to be unique to the site. The material being excavated in London has yet to produce any unusual styles but seems to have examples of most of the artefact types from the other sites. The collections from Bonn and Cologne in Germany have also produced some interesting comparative material (Hagen 1937). Whilst many of the armlets, beads and pins can be compared with those from British sites, one type of finger-ring with grotesque heads carved on the shoulders is confined to Bonn, as are finger-rings with rivet holes drilled in the panels. Cologne is particularly notable for its elaborately carved armlets, only a few of which have been found in Britain, mostly at York (Allason-Jones 1996, nos 52-53).

Every now and again a distinctive piece can be matched with an exact parallel at another site. Bead No 40 at York (Allason-Jones 1996), for example, has parallels only at Silchester (Lawson 1975, fig 2, no 17) and Merten in Germany (Hagen 1937, C29), while a York necklace (Allason-Jones 1996, no 7) has an unusual type of disc bead which is also to be seen at Chelmsford (Drury 1973), Colchester (Crummy 1983, nos 807-09), Wroxeter (*unpubl*) and Cologne (Hagen 1937, D8). However, no pattern has emerged by plotting these exotic items. Such individual pieces appearing on different sites might be the result of travellers moving around the empire, the families of the military taking up new postings, or relatives and friends sending each other gifts. The plainer artefacts, such as the undecorated armlets or the faceted-cube pins, may well have been made at more than one centre as they are the logical products of the material given the technology available. The more complicated items suggest a lively trade in black jewellery.

Although there are strong similarities between the objects found in the Rhineland and those found in Britain, analysis has revealed anomalous pyrite in the German material which suggests that the German craftsmen were using local sources of jet and shale for their products. The precise significance of these anomalies is not yet clear and further work is required before the precise source of the Bonn and Cologne material can be identified. There are no signs, so far, that cannel coal was being exploited for jewellery by the German craftsmen. However, some of the similarities of decorative detail between the German and British objects, particularly those from York, are too close to be explained by coincidence or technology, and must suggest some communication, even if this was just individual craftsmen travelling between the two communities. Precisely why jet and shale carving became popular in Britain and the Rhineland in the second to fourth centuries AD when it failed to do so in the other provinces cannot be explained by the availability of the raw material alone and it is possible that the movement of troops or traders between the two provinces may have played a part. Evidence is beginning to emerge from France and Spain of a small trade in jet goods in the third century; a few artefacts made of jet from an unlocated source have been discovered at Bordeaux (Christophe Sireix *pers comm*), whilst three objects from York have produced a reflectance measurement of 0.35/37 in analysis, which is consistent with the measurements recorded for jet from Asturias on the north coast of Spain (Suárez-Ruiz *et al* 1994). This is the first time that any evidence has emerged to suggest that jet, either as a raw material or as finished objects, might have been brought from the Roman province of Spain to Britain. No analysis has been done so far to reveal whether British jet or shale was being exported in any quantity to Roman Spain.

The amount of jet jewellery found in Britain is now large enough for gaps in type and styles to be revealed. For example, there are no ear-rings of jet. Whilst it is possible that some of the loose beads might have hung originally on metal hoop ear-rings, a study of Romano-British ear-rings has failed to find any examples, although metal ear-rings with stone, glass and bone beads are well attested (Allason-Jones 1989). Nor are there any small pieces of inlay which might have fitted into finger rings or brooches; even the commonly found disc brooches with black conical insets rely solely on glass for their effect. The lack of elasticity in the black materials, which means that they break easily under stress, particularly if allowed to become dry, may explain why they were not used as ear-ring hoops or carved into bow brooches, but its non-appearance as a secondary element in jewellery may suggest that the electrostatic properties noticed by Pliny were important (*Nat Hist*, 36). It may have been considered essential for jet to have contact with the body if it was to have any effect, that is, either worn next to the skin or in the hair.

Other than objects related to spinning, such as spindlewhorls and spindles (see Allason-Jones and Miket 1984), the only other significant group of practical artefacts found in Britain is that of tubular handles. Few handles of jet or shale have been found other than in York and it would appear that the people of Roman York only shared their taste for black handles with the German population on the Rhineland (Hagen 1937). Handles require an appreciable amount of raw material, enough in some instances to make several necklaces; as a result their production may have appealed only to craftsmen working near the sources of supply with large blocks available. None of the handles found in York have retained any blades, so it can only be presumed from the evidence of the German examples that they were intended for knives rather than mirrors. The smaller handles might have been suitable for scalpels or other medical instruments but jet does not seem to have been popular with the medical profession as a handle material, although it was used as a medicinal ingredient (see Pliny *Nat Hist*, 36). The lack of blades is to be regretted as the Cologne and Bonn knives have a distinctive leaf shape, which is not commonly found in Britain, and the survival of the blades might have added to the discussion as to the possible trade links between the Rhineland and York.

The recent use of reflectance measurements as an accurate, non-destructive method of identifying black materials has revealed a more complex industry than had been appreciated previously. This work is in its infancy and much is still to be done before the jet, shale and cannel coal industry of Britain, and Yorkshire's place in that industry, can be fully understood and put into its Empire wide context.

Bibliography

Allason-Jones, L, 1989 *Ear-rings of Roman Britain*. Oxford: Brit Archaeol Rep, Brit Ser, **201**

---- , 1996 *Roman Jet in the Yorkshire Museum*. York: Yorkshire Museum

---- , and Dore, J N, *forthcoming* Excavations at Halton Chesters, 1959-61. *Archaeol Aeliana*

---- , and Jones, J M, 1994 Jet and other materials in Roman artefact studies, in *Archaeol Aeliana* 5 ser **22**, 265-72

---- , and Miket, R F, 1984 *Catalogue of Small Finds from South Shields Roman Fort*. Newcastle upon Tyne: Soc Antiq Newcastle Mono **2**

Austen, J H, 1857 On the Kimmeridge coal money. *Purbeck Soc Papers* **1**, 82-97

Calkin, J B, 1955 Kimmeridge coal money. *Proc Dorset Nat Hist Archaeol Soc* **75**, 45-71

Crummy, N, 1983 *The Roman Small Finds from Excavations in Colchester, 1971-9*. Colchester: Colchester Archaeol Rep **2**

Davies, H F, 1936 The shale industries at Kimmeridge, Dorset. *Archaeol J* **93**, 200-13

Dearne, M J, and Branigan, K, 1995 The use of coal in Roman Britain. *Antiq J* **75**, 71-105

Drury, P J, 1973 Romano-British jet objects from Chelmsford. *Antiq J* **53**, 272-73

Gray, H St G, 1953 Objects of Kimmeridge shale. In H St G Gray and A Bulleid (1953) *The Meare Lake Village*, vol 2. Taunton: privately printed, 259-68

Hagen, W, 1937 Kaiserzeitliche Gagatarbeiten aus dem rheinischen Germanien. *Bonner Jahrbücher* **142**, 77-144

Hornsby, W, and Laverick, J D, 1932 The Roman signal station at Goldsborough, Whitby. *Archaeol J* **90**, 203-53

Lawson, A J, 1975 Shale and jet objects from Silchester. *Archaeologia* **105**, 241-75

Manning, W H, 1985 *Catalogue of the Romano-British Iron Tools, Fittings, and Weapons in the British Museum*. London: British Museum

McMullen, M, 1992 *Whitby Jet through the Years*. Hull

Muller, H, 1980 *Jet Jewellery and Ornaments*. Princes Risborough: Shire

Owen, J S, 1975 Jet mining in North East Yorkshire. *Cleveland Indust Archaeol* **3**, 13-22

Pliny *Natural History*

Solinus *Collectanea rerum memorabilium*

Spratt, T A B, and Forbes, E, 1847 *Travels in Lycia*. London

Suárez-Ruiz, I, Iglesias, M J, Jiménez, A, Laggoun-Défarge, F, and Prado, J G, 1994 Petrographic and geochemical anomalies detected in Spanish Jurassic jet. *American Chemical Soc* **1994**, 76-78

Wheeler, R E M, 1943 *Maiden Castle, Dorset*. Oxford: Soc Antiq London Res Rep **12**

SOURCES OF BUILDING MATERIALS IN ROMAN YORK

G D GAUNT and P C BUCKLAND

Introduction

The Roman fortress of *Eboracum*, founded in or about AD 71 as a base for the Ninth Legion, lies in the angle of the confluence of the little River Foss and the Ouse, at the point where the southward flowing River Ouse breaches the York Moraine, an arcuate but essentially east-west aligned glacial ridge across the otherwise flat and low-lying Vale of York. The site lies not far above the conjectured natural tidal limit in Roman times (P Jones *pers comm*). These geographical features gave the site obvious local strategic advantages, especially in terms of communication and transport, during the initial conquest and occupation of the region. The Vale of York provides the only entirely lowland route between the southern and northern parts of Britain, and *Eboracum* became the military administrative centre for the north. The fortress was rebuilt in stone from the early second century onwards, and was surrounded by extensive civilian occupation (Ottaway 1994), which was granted the status of a *colonia* during the early third century. Twice in the later empire, under Septimius Severus and Constantius Chlorus, it provided a base for the emperor engaged in military activities in Scotland, and Constantine, the son of Constantius Chlorus, was proclaimed emperor at York in 306. It is perhaps a measure of the regional advantages of the site that it was perpetuated, since, compared with other Roman sites to the east and west, it has one major disadvantage, there is no suitable source of building stone for at least 15 km in every direction. Along the rivers, the preferred means of moving heavy materials, the distances are considerably greater.

There are two reasons for this lack of local building stones. One is that, throughout most of the Vale of York, the underlying solid rock is covered by appreciable thicknesses of unconsolidated Quaternary sediments, consisting mainly of tills (boulder clays), gravels, sands and clays. The second reason is that, even where the bedrock crops out, it consists of soft Triassic rocks that are virtually useless for construction. Pebbles and cobbles from the tills and gravels were used for rough flooring, road surfacing and embankments, but they are inherently unsuitable for major structural work. Some of the clays, notably those of glaciolacustrine origin, provided materials for brick and tiles, but these form a relatively minor part of the city's Roman buildings.

With the decision to build in stone, therefore, it became necessary to investigate resources beyond the Vale, in effect in a region encompassing the eastern flanks of the Pennines and the hills of eastern Yorkshire, where the rocks range in age from Carboniferous to Cretaceous (Figure 1). Four rock types were selected for large-scale usage: sandstones from the Millstone Grit, sandstones of Elland Flags-type from the Coal Measures, dolomitic limestones from the Permian Lower Magnesian Limestone, and oolitic limestones from the Middle and/or Upper Jurassic. Small quantities of other rock types were also employed, and decorative stones including marbles were imported from other parts of the Roman Empire.

Previous Work

Some comment upon building materials appears in the Royal Commission monograph on Roman York (RCHME 1962) and several excavation reports contain references to the stone recovered (Ramm 1959; 1968; Radley 1972). Systematic study of materials, however, only began with the formation of the York Archaeological Trust in 1972, when one of the present authors examined the stone of the Roman sewer in Church Street (Buckland 1976). Later work included both *in situ* Roman material and evidence for its re-use during the medieval period (Buckland 1976; 1977; 1978; 1980; 1987). The dating of the so-called Anglian Tower has also been considered in relation to its structural materials (Buckland 1984). The pattern of building stone use throughout Roman Yorkshire, with particular reference to York, was last reviewed in 1988 (Buckland 1988). More recent research has largely been carried out by Gaunt but is as yet unpublished, so this paper attempts to provide an update and summary. The rocks are dealt with in stratigraphic order.

Carboniferous Rocks

Outcrops of Carboniferous rocks are confined to the Pennines. These rocks are primarily divided into the Lower and Upper Carboniferous successions, but for descriptive purposes are here reviewed in terms of four traditional subdivisions, in ascending stratigraphic order the Carboniferous Limestone, Yoredale sequence (referred to in some recent publications as the Wensleydale Group), Millstone Grit and Coal Measures.

The Carboniferous Limestone is referable wholly to the Lower Carboniferous and consists mainly of hard, massive to thick bedded limestones, with some mudstones. Except for small occurrences near Harrogate and Pateley Bridge, the outcrops of these rocks lie well to the west of the Vale of York. The limestones are not easily dressed and their constructional use was, in the past, virtually confined to local, and generally irregular, walling. In parts of Wensleydale, Swaledale and adjacent areas, the rocks spanning the Lower to Upper Carboniferous boundary comprise the Yoredale sequence, typified by thin mudstones, limestones, fine-grained sandstones and cherts. The only outcrops close to the Vale are near Leyburn and Richmond. Some of the limestones and sandstones have been used locally for building purposes. A few scattered *tesserae* made from limestones of Carboniferous type have been found in York (Versey 1932), but, like the Carboniferous Limestone cobbles occasionally incorporated in wall footings in the fortress,

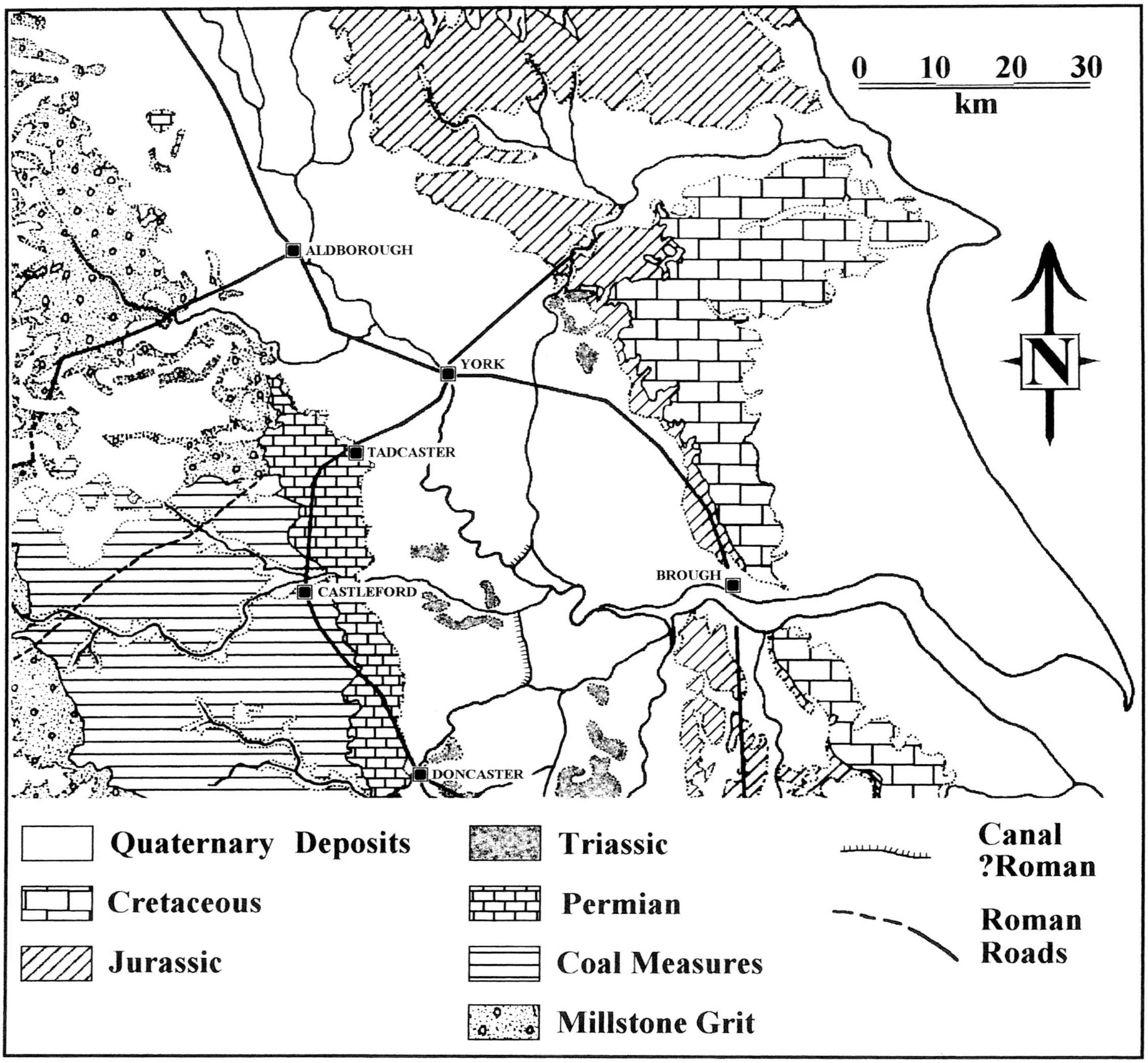

Figure 1. Geology of the York region .Drawing: P I Buckland.

these presumably derive from erratics found in local Quaternary deposits.

The Millstone Grit, which lies wholly within the Upper Carboniferous, consists of mudstones and siltstones with numerous, and in places thick, sandstones. Its outcrops occupy most of the eastern flank of the Pennines from Wensleydale southwards almost to Leeds, but farther south the outcrops are confined to the high Pennines west of Halifax, Huddersfield and Sheffield. The sandstones vary from fine to coarse grained, but the most characteristic and generally thickest are medium to at least partly coarse grained, with subangular to subrounded grains, poorly sorted and poorly compacted, yet firmly welded together into a hard rock; some grains of feldspar, flakes of muscovite and small pebbles of quartz are also variously present. In addition, the sandstones are mainly massive or thick bedded, with large-scale cross bedding in places. These lithological characters render the sandstone ideal for cereal grinding, and the manufacture of rotary querns from Millstone Grit sandstones (Gaunt nd, 10) extends back into the pre-Roman Iron Age. The characters also give the rock high load-bearing strength, making it excellent for major construction work, especially for foundations, columns, arches and other compressional situations where individual large blocks were required.

Many tens of thousands of tonnes of Millstone Grit

sandstones were taken to York by the Romans, almost certainly from the outcrops between Leeds and Masham and transported along the rivers Ure, Nidd and Wharfe. Unfortunately, it is not possible to be more specific about precise provenance, partly because virtually no Millstone Grit sandstone contains any features that distinguish it from other such sandstones, and partly because there are no certain Roman quarry sites on the outcrops. One site, adjacent to the Roman road westwards from York in Hetchells Wood, north of Thorner (Margary 1973, 403), may be of this date (Henderson 1979), but is probably only a source of the trimmed cobbles which once surfaced the road. In York, the use of some building stone of reddened Millstone Grit offers a clue to possible sources. Like most sandstones and limestones, Millstone Grit becomes reddened by fire, but in some outcrops north of Leeds the rock has been reddened naturally when it formed the floor of a hot desert in early Permian times (Turner 1970). Whereas some building stones from Roman York have, on associated evidence, clearly suffered fire damage, it is possible that others were quarried from these naturally reddened sources.

The Coal Measures form the higher part of the Upper Carboniferous and consist predominantly of mudstones and siltstones with numerous, generally thin, sandstones and coal seams. These rocks extend at outcrop from the Leeds-Bradford area south to Derby. A few of the sandstones, of limited areal extent and mainly in the basal part of the Coal Measures, are medium to coarse-grained rocks, lithologically closely comparable to those which characterise the Millstone Grit. Some have been used locally for building stones, and quern factories are known on the outcrops of some of these sandstones west of Sheffield (Butcher 1951; Wright 1988). With these exceptions, the sandstones in the Coal Measures are fine grained or, less commonly, fine to medium grained, with subangular to subrounded grains which are moderately to well sorted and compacted. The more thickly bedded of these sandstones include good building stones which are widely employed today, but they have not been positively identified from Roman remains in York, probably because of the closer availability of Millstone Grit sandstones.

Some of the finer grained and thinner bedded sandstones contain closely spaced parallel laminae rich in muscovite. These laminae allow the rock to be split readily into layers of almost constant thickness, providing both paving and roofing slabs. Such fissile, `flaggy' sandstones are present in many parts of the Coal Measures but the thickest sequence occurs in the Lower Coal Measures, being variously called the Elland Flags in western Yorkshire, the Greenmoor Rock and Grenoside Sandstone around Sheffield and the Wingfield Flags in Derbyshire. Large amounts of these fissile sandstones were utilised in *Eboracum*, mainly in slabs 15-30mm thick for roofing purposes, although some use of thicker slabs for flooring has also been noted. The rock is also used, probably largely secondarily, for *tesserae*, hones, pot lids and gaming counters. Individual flaggy sandstones from the Coal Measures are not normally identifiable in isolation and finds on archaeological sites in Yorkshire and the East Midlands are usually referred to as being of Elland Flags-type. In terms of transportation distance, however, there is little doubt that the material found at York is likely to have come from the Elland Flags *sensu stricto*, *via* the River Aire and perhaps the Calder, but comparable finds from sites in southern Yorkshire, Derbyshire and eastwards into Lincolnshire could be from a variety of comparable sandstones, using the pre-Vermuyden courses of the Don and the Idle for transport.

Permian Rocks

Outcrops of Permian rocks are present along the western margin of the Vale of York where, in effect, they form the Pennine foothills. To the north of Knaresborough these rocks are extensively covered by Quaternary deposits and consequently their outcrops are small and discontinuous, but southwards they form an almost continuous outcrop up to 10km wide, extending into Derbyshire and Nottinghamshire. The Permian rocks comprise a sixfold sequence, in ascending stratigraphic order, the impersistent and incohesive Basal Sands, the similarly impersistent Lower Marl, the Lower Magnesian Limestone, the Middle Marl, the Upper Magnesian Limestone and the Upper Marl. Only the two limestones have any constructional value although veins and nodular masses of gypsum, mainly of satinspar variety, from the Middle Marl and Upper Marl were used for plaster production in the medieval period and may have been the source of the plaster used in late Roman burials in York and elsewhere in the region (Green 1977).

The Lower Magnesian Limestone, in some recent publications referred to as the Cadeby Formation, consists predominantly of dolomite, although calcite may also be present. The rock includes a wide range of textures, the principal varieties being:

a) Oolitic, consisting of ovoid to spherical bodies called ooliths that have a concentric internal structure and are up to *c* 2mm across, any larger examples being termed pisoliths.

b) Relict oolitic, as above, but with the middle of the ooliths dissolved out to become minute voids, and the rest of the ooliths and the inter-oolith matrix partly recrystallised to the extent that the original oolitic texture is still apparent but not clearly defined.

c) Microcellular, fine grained, and seen to be finely recrystallised under the petrological microscope, but containing numerous minute voids, evidence of an original oolitic texture.

d) Medium grained or granular, consisting of dolomite crystals on which the characteristic slightly curved crystal faces may be visible. This variety looks like a medium grained sandstone, hence its alternative name `dolomite sand'. Each crystal results from recrystallisation of a single oolith.

e) Fine grained, variably i) without visible bedding structures, ii) thin-bedded, iii) with parallel planar or wavy laminae of finer or, less commonly, coarser grains, and iv) with cross-bedded laminae.

These textures may change from one variety to another across quite short distances, so that even small areas may contain two or more textures at outcrop. The limestone makes an excellent freestone, able to be worked in any direction, and it may be sculptured into intricate shapes.

Many tens of thousands of tonnes of the Lower Magnesian Limestone were used in Roman York, almost certainly taken from outcrops in the Knaresborough-Wetherby-Tadcaster area and transported along the rivers Nidd and/or Wharfe, although Buckland (1988) has discussed the possibility of a more southerly source using a canal west of Cawood, now called the Bishop Dyke, and the Ouse. Curiously, almost all of the limestone of Roman constructional date from York has oolitic, relict oolitic and microcellular textures, with a small proportion having a medium grained, 'dolomite sand' texture. This contrasts markedly with the predominately fine-grained texture of the vast amount of the limestone taken to York in the medieval period for the walls, churches and other structures. No quarries undoubtedly used for supplying Roman York[1] have been identified, possibly because any topographic evidence has been swept away by later activities. The two quarries in the Lower Magnesian Limestone of proven Roman date, at Hazel Wood south-west of Tadcaster (Ramm 1976), and Well near Masham (Gilyard Beer 1951), are too small to have been more than local sources.

In examining the inside of the fortress wall north-east of the Multangular Tower, Buckland (1988) noted massive blocks of baryte incorporated in the structure (Figure 2) and had suggested that their origin lay in the ballasting of ships returning from supplying sites in County Durham, where such minerals are recorded from veins in the Coal Measures (Buckland and Sadler 1990). More recently the same mineral has been noted clearly within Lower Magnesian Limestone from Roman deposits in Swinegate. Baryte is extremely rare in the Lower Magnesian Limestone and its presence can be used to suggest possible sources for the rock. Small mineralised nodules in the limestone at Wormald Green between Ripon and Harrogate contain baryte, but they also include other minerals (Cooper and Burgess 1993, 75) that

Figure 2. The inner face of the fortress wall, north-east of the Multangular Tower. The irregular joint indicates that the north-east wall of the Multangular Tower towards the so-called Anglian Tower has been rebuilt. The wall is of Lower Magnesian Limestone with, in the rebuilt section, scattered blocks of barytes. The rough finish would have been hidden by the rampart. (Photo: R Swift)

have not been found in York. Other occurrences were recorded by Edwards *et al* (1950, 48, 72) near Bramham and East Keswick, and a recent road cutting made during the realignment of the A1 trunk road west of Bramham exposed numerous thick veins and large nodular masses of baryte extending across several metres of limestone (Figure 3). The present village of Bramham, of medieval origin, sits on a series of flattish or gently sloping terraces, separated by areas of steep slope, suggesting several long-degraded working faces which may be the remains of a Roman quarry. Alternatively or additionally, because the baryte vein complex is aligned east-north-east and may well continue along this trend parallel to the main faults in the area, sources closer to Newton Kyme, upstream of Tadcaster on the Wharfe, may have been utilised.

As well as massive walling, the Lower Magnesian Limestone has been used, and re-used, for sculptures, memorial and other carved stones, spindle whorls and many other types of small item. Several artefacts from medieval deposits at Bedern in York, including gaming counters, mortars, lamps and seal matrices, made from Lower Magnesian Limestone with relict oolitic and microcellular textures, implying that they were shaped in medieval times from former Roman building stone.

The Upper Magnesian Limestone, referred to in some recent publications as the Brotherton Formation, is largely and in many areas wholly dolomitic, and mostly fine grained to finely crystalline, although oolitic and derived oolitic textures are present in some localities. The rock is predominantly thin bedded and splits readily along somewhat irregular bedding planes into thin but uneven layers. This `flaggy' character makes it unsuitable for massive construction use but renders it locally useful as a roofing stone. It was widely employed in the medieval period as such, for example in Doncaster (Buckland *et al* 1986, 217-19) and Kellington church, near Selby. Only one fragment is known from a Roman context in York, despite the nearest outcrop being only 15km away at Tadcaster, the Roman *Calcaria*, and Elland Flags-type sandstone was clearly preferred for roofing. The place name *Calcaria* has been interpreted as referring to either limestone quarries or lime works (Rivet and Smith 1979, 288-89) and it is possible that the rock was burnt for lime; Addyman (*pers comm*) noted calcined lime fragments, associated with Roman pottery and tile, in the Norman castle mound at Tadcaster.

Triassic Rocks

Although Triassic rocks form rockhead throughout most of the Vale of York, they are largely concealed beneath extensive Quaternary deposits and, in consequence, most of their outcrops are small and discontinuous. These rocks comprise, in ascending order, the Sherwood (formerly Bunter) Sandstone, the Mercia Mudstone (formerly Keuper Marl), the Penarth Group (formerly Rhaetic) and the basal few metres of the Lias Group, the remainder of which is referred to the Lower Jurassic. As intimated in the introduction to this paper, the Triassic rocks of the region have little or no constructional value. The Sherwood Sandstone has been utilised locally as a building stone, but it is of very poor quality. The Mercia Mudstone, which crops

Figure 3. Exposure of baryte in the A1 road cutting west of Bramham. Photo: P Buckland.

out in places in the eastern part of the Vale, readily degrades to clay upon exposure to weathering. In some localities it contains veins and nodular masses of gypsum, mainly of the satinspar variety, and like similar mineral deposits within the Permian it has been used for plaster production and may have been the source of at least some of the plaster in late Roman burials in York. The Penarth Group and basal Lias are both thin and limited to the extreme east of the Vale where, being predominantly mudstones, they rapidly weather down to clay on exposure.

Outcrops of Sherwood Sandstone are confined to the western part of the Vale of York and the rock is mainly a red or reddish brown, fine to medium-grained sandstone, which may locally be coarser. It is also invariably friable, even incohesive in places, to depths well beyond feasible quarrying and this leads to disaggregation to loose sand upon prolonged weathering. Only locally is it sufficiently cohesive to be used in construction work. Around Boroughbridge, for example, the sandstone is sufficiently hard in places for it to have been used for the third-century defences and other structures at the town of Aldborough, although Millstone Grit sandstone was used also, at least in part for the later bastions (Charlesworth 1971). Occasionally, Sherwood Sandstone was used as infill where structural integrity was not a problem, and for repair and alteration work. In the Roman sewer in Church Street at York, two blocks were used to support an Elland Flags type access cover, perhaps a late repair, yet since the nearest Sherwood Sandstone outcrops lie some 8km away near Bilborough and Long Marston, it is surprising that some more immediate source of builders' rubble was not used.

Jurassic Rocks

Outcrops of Jurassic rocks occur only to the east of York where they form the North York Moors, the Cleveland, Hambleton and Howardian Hills, and continue southwards as a narrow outcrop along the western foothills of the Yorkshire Wolds to the Humber Estuary at Brough. A much broader outcrop extends southwards through Lincolnshire. The rocks are primarily subdivided into Lower, Middle and Upper Jurassic successions and the detailed classification and nomenclature are complex, having been changed, in some cases more than once, in the last few decades. This discussion, however, will concentrate mainly upon those rocks of known or suspected use for constructional purposes in Roman York.

The Lower Jurassic, which comprises all but the basal part of the Lias Group, consists predominantly of mudstones with some thin beds and concretionary layers of limestone and, in the middle of the succession from the Howardian Hills northwards, the Staithes Sandstone and overlying Cleveland Ironstone. In places on and near its outcrop the fine to medium-grained Staithes Sandstone has been used as a building stone, for example in the twelfth-century church at Husthwaite, but it has not been identified in York. A few *tesserae* from Roman York are made from dark grey, fine-grained limestone and calcareous mudstone that are suggestive of a Lower Jurassic source.

The Middle Jurassic contains numerous sandstones and limestones with interbedded mudstones and siltstones and a few ironstones. A wide range of lithologies is present in the sandstones and limestones and, in the past, several of these rocks were utilised locally for construction on or near their outcrops. Significant quantities of one particular rock type, an oolitic limestone, were taken to York during the Roman period for use as a building stone, for which there are four possible sources in the Middle Jurassic and two others in the Upper Jurassic. In addition, small amounts of a few other Middle Jurassic rocks were employed for minor constructional needs in Roman York.

The Dogger, which forms the basal part of the Middle Jurassic, is exceptionally variable lithologically and in some areas, notably in and around upper Eskdale, it includes greenish grey calcareous and/or sideritic sandstone, the Danby facies of the Dogger. A greenish grey, fissile, flaggy unfossiliferous sandstone containing poikilitic calcite, which produces a lustre-mottled effect in strong light, has been found in two Roman structures in York, the lining of a drain adjacent to the *intervallum* road behind interval tower SW5 (Buckland 1977) and in the flooring of the Church Street sewer (Buckland 1976). These were tentatively attributed to the Dogger by Versey (in idem) from sources close to the Derwent Gorge, near Crambeck. In the Howardian Hills, however, the Dogger is generally a more massive, ferruginous and locally oolitic and/or fossiliferous sandstone, and it is not inconceivable that the York material comes from the Brandsby Roadstone, also in the Middle Jurassic (J E Hemingway and R W O'B Knox *pers comms*).

The Gormire (formerly Hydraulic) Limestone is a thin, very fine-grained, almost porcellanous-looking rock that occurs a little higher in the Middle Jurassic succession in the southern part of the Hambleton Hills and in the Howardian Hills. Although absent across the Market Weighton area, along with the entire Middle and Upper Jurassic, its stratigraphic equivalent, the Ellerker (also formerly Hydraulic) Limestone is present in the South Cave-Brough area, and across the Humber as far south as Winterton. A few scattered *tesserae* with this identical porcellanous lithology have been found at York and are assumed to have come from these sources, although lithologically comparable thin limestones occur further south in Lincolnshire in the Kirton Cementstones, also of Middle Jurassic age.

The Whitwell Oolite, present in the middle part of the Middle Jurassic in the Howardian Hills, consists of variably poor to well-graded, closely compacted ooliths, with a few scattered pisoliths and finely fragmented fossils. Its coeval stratigraphic equivalents, the Millepore Bed in the Cayton Bay area, the Cave Oolite in the South Cave-Brough area, and the Hibaldstow Limestones in north Lincolnshire (Gaunt *et al* 1992, 51-52) are all, in effect, lithologically identical. So also are two oolitic limestones in the Upper Jurassic in the Howardian Hills, the Hambleton Oolite and Malton Oolite. All these rocks are generally medium to thick bedded and make excellent freestones. A considerable quantity of this type of oolitic limestone was used as building stone in Roman York, although it does not appear to have been employed generally in the fortress where, with the exception of parts of the Church Street sewer and the so-called Anglian tower (Buckland 1984), Lower Magnesian Limestone is ubiquitous

Figure 4. The outer face of the fortress wall, north-east of the Multangular Tower, built in saxa quadrata of Lower Magnesian Limestone. The inserted Late Roman `Anglian Tower' is of Jurassic limestone. (Photo: R Swift)

in walling (Figures 4 and 5). It has been suggested (Buckland 1988) that this apparent separation between military and civil use is a reflection of the location of the Lower Magnesian Limestone quarries within the legionary *territorium*. A few *tesserae* also occur in oolitic limestone, as well as sculptural items, gaming counters and lamps.

Despite several suggested origins (cf. Versey in Wenham 1968, 44; Radley 1972; Buckland 1984), the provenance of the oolitic limestone employed in Roman York is uncertain, although the Millepore Bed can perhaps be excluded because of the distance from its coastal outcrop. Both the Whitwell Oolite and the Upper Jurassic oolitic limestones have outcrops close to the Derwent Gorge, and the latter were certainly used for Roman work in Malton. Both could have been transported down the Derwent, either to Stamford Bridge for overland cartage along the Roman road to York, or down the river to its confluence with the Ouse near Barmby on the Marsh and then upstream to the *colonia*. In this context, it seems probable that the present artificial course of the lowest part of the Derwent (Gaunt 1994, 128-30, fig 46) was cut during the Roman period to improve such a route. This would have shortened the route from the Derwent to York by some 14km and avoided the difficult currents near Kilpin Pike and the Ouse-Aire confluence. The terminology of Edgar's Charter of AD 959 (Long 1993) suggests that the diversion was old by this date, although at least part of the original natural course past Howden was still usable in the late thirteenth century. Such river modification would fit in with the suggested Roman origins of the Bickers Dyke (Buckland and Sadler 1985), Turnbrigg Dyke (Gaunt 1975; Buckland and Sadler 1985; Jones 1995) and the Bishop Dyke, all of which would have facilitated the movement of bulky commodities by water. If the Cave Oolite and/or Hibaldstow Limestones were taken to York, it would have been *via* the Humber and the Ouse, possibly at least partly as ballast (cf. Buckland and Sadler 1990). There are quite a number of very degraded quarries on both outcrops, especially on that of the Hibaldstow Limestones where it is crossed by Ermine Street south-east of Winterton, from where the Ancholme would have provided a route to the Humber; none of these quarries is of known Roman date, however.

Figure 5. The east corner tower of the Roman fortress, preserved to the base of the parapet, all in Lower Magnesian Limestone. (Photo: R Swift)

In excavations at the former Leedham's Garage site, south of the River Ouse and adjacent to the site of the Roman bridge over the river in York, a strip building fronting on to the road (Ottaway 1994) was found, built of oolitic limestone. These remains included a few large broken slabs of hard, partly oolitic limestone containing numerous pelletoid oncoliths, in effect, like very large irregular ooliths up to 5mm wide and probably of algal origin, and also bioclastic layers, that is, consisting of finely fragmented and crushed fossil debris. Thin beds of lithologically similar rocks are present in all the Middle and Upper Jurassic oolitic limestones referred to above, and these slabs were probably brought from the same source as the walling material. They are too thick and heavy for roofing purposes, but would be suitable for rough paving.

The Brandsby Roadstone, in the upper part of the Middle Jurassic in the Howardian and Hambleton Hills, is a fine-grained, strongly calcareous, thin-bedded sandstone, in which some of the contained calcite is poikilitic and produces a lustre-mottled effect. It splits readily into thin layers, some of which are uneven due to irregularities in the bedding and also to abundant burrow casts and fossil bivalve shells on the bedding planes. Other layers, however, are almost planar and of consistent thickness, where the bedding is parallel and laminated and devoid of fossils. The Brandsby Roadstone provides the best roofing stone in the entire Jurassic of Yorkshire and was used, for example, at both Roman Malton and medieval Wharram Percy. No such usage has been identified in Roman York, although the possibility exists that it was employed for the floor of the Church Street sewer and the culvert behind the rampart at Interval Tower SW5. Some facies of the Brandsby Roadstone are more thickly bedded and less fissile, and in this context a number of large blocks of sparsely fossiliferous sandstone incorporated in the church at Kirk Hammerton, near York, may be from this source. This has not been identified in Roman deposits in the city, but as the remainder of the Anglo-Saxon part of the church consists of re-used material (Morris 1976) it is probable that it comes from an unlocated Roman building in *Eboracum*.

The Upper Jurassic consists predominantly of mudstones except in the Howardian Hills, Hambleton Hills and the southern part of the North York Moors where it includes the Corallian Group, a thick sequence of mainly calcareous sandstones and a variety of limestones. Some of these Corallian rocks were used locally as building stones and at least one of them is found in Roman York. The Hambleton Oolite and, at a higher level in the Corallian sequence, the Malton (formerly Osmington) Oolite are both oolitic limestones that are lithologically closely comparable to the Whitwell Oolite and its Middle Jurassic stratigraphic equivalents. Outcrops of one or both of these Upper Jurassic oolitic limestones in the Howardian Hills almost certainly supplied much of the building stone for Roman Malton (cf. Corder 1930) and some at least of the oolitic limestone used for construction in Roman York may have come from these outcrops *via* the Derwent.

It may be mentioned that a few calcareous ooliths have been noted in mortar adhering to building stone of presumed Roman date from York. This suggests that at least some of the Roman mortar was manufactured from calcined Jurassic limestones.

The Birdsall Calcareous Grit, which occurs between the Hambleton Oolite and Malton Oolite, is largely restricted to the Birdsall area south-east of the Derwent Gorge. It is a very fine to fine-grained, locally fossiliferous, mainly medium-bedded sandstone that is calcareous at depth but is generally at least partly decalcified where weathered at outcrop and where exposed as a building stone. It provides an excellent freestone and can be intricately carved. Substantial amounts of constructional and architectural debris in this rock are present at Wharram Percy (Gaunt, in Milne and Richards 1992, 46-47), possibly re-used from a Roman villa in the area, but it has yet to be found in Roman York.

The Hildenley Limestone is a very fine-grained, locally fossiliferous, variably thin to thick-bedded limestone, not unlike Chalk in appearance but much harder, and restricted to a small well-quarried outcrop about 3km west-south-west of Malton. Its stratigraphic relationships are uncertain due to faulting, but it appears to be a local variant of the Malton Oolite, and possibly near its top, of the succeeding Coral Rag also. The rock has been used locally from Roman times for building material and artefacts, and two blocks have been recognised amongst the material from the legionary headquarters building beneath York Minster, as described by Senior (1990).

Cretaceous rocks

Outcrops of Cretaceous rocks are confined to the Yorkshire and Lincolnshire Wolds with, in the north, an extension along the cliffs near Speeton in the southern part of Filey Bay. These rocks consist, in ascending order, of the Speeton Clay which is confined to the Speeton-West Heslerton area, the Carstone, an impersistent, thin, friable and locally incohesive sandstone, and the Chalk. Only the Chalk has been put to constructional use, and that only marginally. Chalk is essentially an extremely fine-grained, medium to thick-bedded limestone, mainly white but with a red layer at the base. In the Wolds the Chalk is much harder than in southern England, but it is not sufficiently resistant to prolonged weathering to be of much value for construction, although the stone was used for sill walls on several of the east Yorkshire Roman villa sites. Because the rock is easily carved, it was used for artefacts where neither inherent strength nor resistance to weathering were required; it was extensively used also for white *tesserae*, and examples of these have been found in York. The middle part of the Chalk sequence contains nodular and tabular layers of flint, an extremely hard form of cryptocrystalline silica. Flint erratics are present in the Quaternary deposits of eastern Yorkshire and comparable nodules from deposits in southern England were widely used for building in that region, for example in the walls of Silchester (Sellwood 1984), but this usage does not appear to have occurred in York.

Quaternary clays and erratics

Clays from the widespread glaciolacustrine deposits in the York area provided abundant material for pottery, tile and brick manufacture, and workshops associated initially with the Ninth legion and later with the Sixth legion existed in the area of Aldwark, south-east of the east corner of the fortress. Much of the brick and tile recovered from Roman contexts, and still visible as bonding courses in the Multangular Tower (Figure 6), must have come from these and other sources in the Quaternary deposits of the area (Betts 1985). At Aldwark, one of the pits used as a source of clay appears to have become a pond adjacent to a large house with tessellated floor in the late Roman period (Brinklow *et al* 1986). The tills comprising parts of the moraine contain numerous erratics which were widely utilised for foundation courses for structures in the fortress (cf. Buckland 1977), as well as for the make up of roads. Local sources are also probable for the sand and gravel extensively used in the manufacture of mortar and cement in the settlement.

Quaternary tufa

Calcareous tufa is a Quaternary deposit that forms from lime-rich water in springs and streams on and adjacent to limestone and Chalk. Its texture varies from highly porous, with hollow tubercles and intertubercle voids, to more compact layered forms, but all varieties have only low inherent strength and low density and are easily crushed. Molluscs and plant fragments are not uncommon in the deposit. Despite its unlithified nature and lack of strength, tufa was used in Roman construction, notably in vaults and perhaps as insulation in bath houses. Samples from the Roman sewer and bath house in Church Street in York (Buckland 1976) are of the porous tubular variety, as also are more scattered fragments from elsewhere in the city. The largest and most impressive tufa deposits in the region occur on Carboniferous Limestone outcrops in the Pennines, notably in Gordale, but most of these are of the compacted varieties. Other tufa deposits, although generally small, are widespread on Lower Magnesian Limestone outcrops, and are mainly of the porous tubular variety. They occur from the Snape area north-east of Masham (Powell *et al* 1992, 80), near to where the Roman villa at Well is built on such a deposit (Gilyard-Beer 1951), southwards by the Ripon, Copgrove and Knaresborough areas (Cooper and Burgess 1993, 73) to localities such as Burton Salmon, south-west of York (Norris *et al* 1971). The porous tubercular variety of tufa occurs also on and adjacent to outcrops of Jurassic calcareous rocks. Examples range from the Boltby-Felixkirk-Coxwold area (Powell *et al* 1992, 81) to northern Lincolnshire (Gaunt *et al* 1992, 125). In this context, a few of the fragments of tufa from York contain calcite ooliths comparable to those that constitute the various Jurassic oolitic limestones reviewed above, so it seems likely that some at least of the York tufa came from Jurassic outcrops, most probably from locations in the Howardian

Figure 6. The interior of the south-west wall of the fortress looking towards the Multangular Tower, all in Lower Magnesian Limestone, with a brick bonding course visible in the tower. Recovered from cemeteries outside the walls and placed within the wall, are six stone coffins of Millstone Grit sandstone. (Photo: R Swift)

Hills; the low-density nature of tufa would make it possible to transport tufa from such a nearby source by cart, without recourse to river transport.

Building and decorative stones from more distant sources

A few rock fragments found at York are from sources outside the surrounding region, some from more southerly parts of England and others from elsewhere in the Roman Empire. These fragments, although diverse lithologically, all have distinctive textures and have been polished on cut or carved surfaces to a highly glossed finish, which enhances their texture for decorative purposes. A few of the fragments are from Roman levels and others, found in later deposits, are presumed to be from Roman contexts, although the possibility that some arrived in York as pilgrims' relics from Mediterranean sources cannot be precluded; Lynn (1984) has suggested such origins for Greek and Egyptian porphyritic rocks on monastic sites in Ireland. Furthermore, not all polished fragments are necessarily from wall veneers, columns or other structural features. Furniture and such items as altars, statues, bowls and vases were also made from polished stone.

In Britain true marbles, that is, of metamorphic origin, are virtually confined to the Scottish highlands and islands and to parts of Ireland, all lying outside the Roman Empire. In England, however, there are a number of sedimentary limestones with textures that, when polished, have appearances comparable to true marble; several of these 'false marbles' have been worked for decorative veneers and other items and the word marble has been incorporated in their stratigraphic name. Such is the Upper Jurassic Purbeck Marble from Dorset, a dark green to reddish limestone containing abundant freshwater gastropod shells, used from the Roman period onwards (Beavis 1970), and the Upper Carboniferous Frosterley Marble from County Durham, a dark limestone containing abundant corals, used at least since the medieval period.

Only one type of 'false marble' has been proved to have been used in Roman York, the Forest Marble, named from Wychwood Forest in Oxfordshire and part of the Middle Jurassic succession in the Cotswold region. It is a variably thin to thick-bedded, commonly cross-bedded, fossiliferous oolitic limestone, widely used from the Roman period onwards. Several large fragments, clearly from wall veneers, were found at the Bishophill site (Buckland 1978) and another fragment from a column comes from a site in Blake Street; a thin fragment from the Coppergate site, recorded as `tile', has a comparable lithology.

A few fragments of true marbles of non-British origin are recorded from York. One is off-white and coarsely crystalline with ill-defined greyish veins and irregular masses, and has a similar appearance to Marmora Marble or `Turkish White', from the island of that name in the Bosphorus; it was found in a late Roman or Anglian context at the Bishophill site. Another fragment is yellow and finely crystalline, with irregular reddish-purple-brown veins, and is comparable to Giallo Antico, a variety of Numidian marble; unfortunately it was found in a post-medieval pit. A fragment of greyish white, finely crystalline marble that is closely comparable lithologically with thin-sections of Carrara Marble, a metamorphosed Jurassic limestone in the Apennine region of Italy, was found at the medieval site in the Bedern. The two fragments of unidentified white marble figurines (RCHME 1962, 120) in the Yorkshire Museum collections may be from this source also.

Certain igneous rocks, particularly those with coarsely crystalline and/or porphyritic textures, the latter being characterised by some crystals that are markedly larger than the others, have a highly decorative finish when trimmed and polished, although their hardness makes them much more difficult to work than marbles and limestones. Only two polished fragments have been found in York, from the Coppergate site. They are virtually identical, being dark greenish grey to black, very finely crystalline rocks containing large, up to 15mm long, pale greyish green crystals, variously single, interpenetration twinned and compound, that are almost certainly sericitised plagioclase feldspars; the rock is probably a porphyritic andesite or basalt. Without thin-section analysis, however, more precise identification and provenance are uncertain.

Conclusion

From the earliest use of stone in the fortress through to its latest structures, Roman York employed a wide range of building materials, with sources extending from the Pennine fringe in the west to the North York Moors and Wolds in the east, along with occasional pieces of more exotic material. The volume moved must have amounted to many tens of thousands of tonnes, implying well maintained waterways, the cheapest means of transport in antiquity, as well as roads. Relatively little of this stonework is visible in the city at the present day. Some has been incorporated in the deposits which, at over three metres thick in places, make up the made ground upon which modern York stands, but much has been dispersed into the many medieval churches of the Vale of York and beyond. Churches like St Michael, Spurriergate, and St Mary Bishophill Junior in the city (Buckland 1987), Skipwith, Hemingborough, Howden, Holme-on-Spalding Moor and Barton-on-Humber, show evidence of the re-use of Roman material, predominantly Millstone Grit sandstone. It is probable that little quarrying of fresh stone took place between the end of the Roman period and the early Norman period when the construction of the massive cathedral by Thomas of Bayeux used up all that was readily available of the ruins of *Eboracum*.

Acknowledgements

Research in York has been facilitated by the York Archaeological Trust under its Director, Peter Addyman, and the several supervisors of excavations in the city over the past twenty years, in particular Richard Hall, Patrick Ottaway, Tony Sumpter, and Ben Whitwell. Tony Benfield first pointed out the exposure of baryte veins at Bramham and Terry O'Connor kindly commented upon the text.

Endnote

1. R Isserlin has kindly drawn our attention to an inscription dated to AD107 from York (*RIB* 665) carved from (lower) Magnesian limestone. This indicates that Lower Magnesian Limestone was being quarried at this date

This paper was completed in 1995.

Bibliography

Beavis, J, 1970 Some aspects of the use of Purbeck Marble in Roman Britain. *Proc Dorset Nat Hist Archaeol Soc* **92**, 181-204

Betts, I M, 1985 *A Scientific Investigation of the Brick and Tile Industry of York to the Mid-Eighteenth Century.* Unpubl PhD thesis, Univ Bradford

Brinklow, D, Hall, R A, Magilton, J R, and Donaghey, S, 1986 *Coney Street, Aldwark and Clementhorpe, Minor Sites and Roman Roads*. Archaeol York **6/1**. London: CBA

Buckland, P C, 1976 Geological and archaeological notes on the rocks used in the construction of the sewer. In J B Whitwell (1976) *The Church Street Sewer and an Adjacent Building.* Archaeol York **3/1**. London: CBA, 32-37

---- , 1977 Notes on the building stones. In A B Sumpter and S Coll (1977) *The Excavation of Interval Tower S.W.5, York.* Archaeol York **3/2**. London: CBA, 80-81

---- , 1978 Notes on the building stones. In M O H Carver, S Donaghey and A B Sumpter (1978) *Riverside Structures and a Well at Skeldergate and Buildings at Bishophill.* Archaeol York **4/1**. London: CBA, 40-44

---- , 1980 The building stones. In J R Magilton (1980) *The Church of St Helen-on-the Walls, Aldwark.* Archaeol York **10/1**. London: CBA, 27-29

---- , 1984 The `Anglian Tower' and the use of Jurassic Limestone in York. In P V Addyman and V Black (eds) (1984) *Archaeological Papers from York presented to M W Barley.* York: York Archaeol Trust, 51-57

---- , 1987 The building stones: petrological analysis. In L P Wenham, R A Hall, C M Briden and D A Stoker (1987) *St Mary Bishophill Junior and St Mary Castlegate.* Archaeol York **8/2**. London: CBA, 110-15

---- , 1988 The stones of York: building materials in Roman Yorkshire. In J Price and P R Wilson (eds) (1988) *Recent Research in Roman Yorkshire*. Oxford: Brit Archaeol Rep, Brit Ser **193**, 237-87

---- , Magilton, J R, and Hayfield, C, 1986 *Archaeology of Doncaster 2. The Medieval and Later Town.* Oxford: Brit Archaeol Rep, Brit Ser **202**, (2 vols)

---- , and Sadler, J P, 1985 The nature of Late Flandrian alluviation in the Humberhead Levels. *East Midland Geographer* **8**, 239-51

---- , and ---- , 1990 Ballast and building stone: a discussion. In Parsons (1990), 114-25

Butcher, L H, 1951 Archaeological remains on the Wharncliffe-Grenoside uplands, South Yorks (preliminary notice). *Trans Hunter Archaeol Soc* **7**, 38-39

Charlesworth, D, 1971 The defences of *Isurium Brigantum*. In R M Butler (ed) (1971) *Soldier and Civilian in Roman Yorkshire*. Leicester: Leicester Univ Press, 155-64

Cooper, A H, and Burgess, I C, 1993 *Geology of the Country around Harrogate*. Memoir of the British Geological Survey. Sheet **62** (England and Wales). London: HMSO

Corder, P, 1930 *The Defences of the Roman Fort at Malton*. Leeds: Roman Malton District Rep **2**

Edwards, W, Mitchell, G H, and Whitehead, T H, 1950 *Geology of the District North and East of Leeds*. Memoir of the Geological Survey. Sheet **70** (England and Wales). London: HMSO

Fulford, M, 1984 *Silchester Defences 1974-80*. London: Britannia Mono **5**

Gaunt, G D, 1975 The artificial nature of the River Don north of Thorne, Yorkshire. *Yorks Archaeol J* **47**, 15-21

----, 1994 *Geology of the Country around Goole, Doncaster and the Isle of Axholme*. Memoir of the British Geological Survey. Sheets **79** and **88** (England and Wales). London: HMSO

----, nd (1994) *A Guide to Lithological Description and Provenance of Quern Stones in Yorkshire and the North-Eastern Midlands*. Dept Archaeol Sci, Univ Bradford

---- , Fletcher, T P, and Wood, C J, 1992 *Geology of the Country around Kingston upon Hull and Brigg*. Memoir of the British Geological Survey. Sheets **80** and **89** (England and Wales). London: HMSO

Gilyard-Beer, R, 1951 *The Romano-British Baths at Well*. Leeds: Yorks Roman Antiqs Comm Rep **1**

Green, C J S, 1977 The significance of plaster burials for the recognition of Christian cemeteries. In R Reece (ed) (1977) *Burial in the Roman World*. London: CBA Res Rep **22**, 46-53

Henderson, A, 1979 Pompocali, a topographic curiosity. *Naturalist* **104**, 75-87

Jones, P, 1995 Two early Roman canals? The origins of the Turnbridgedike and Bycarrsdike. *J Railway Canal Hist Soc* **31**, 522-31

Long, M H, 1993 Howden and Old Drax. In H E J Le Patourel, M H Long and M F Pickles (eds) (1993) *Yorkshire Boundaries*. Leeds: Yorkshire Archaeological Society, 125-34

Lynn, C J, 1984 Some fragments of exotic porphyry found in Ireland. *J Irish Archaeol* **2**, 19-32

Margary, I V, 1973 *Roman Roads in Britain* (3rd edn). London: Baker

Milne, G, and Richards, J D, 1992 *Two Anglo-Saxon Buildings and Associated Finds*. Wharram: a Study of Settlement on the Yorkshire Wolds **7**. York: York Univ Archaeol Publ **9**

Morris, R, 1976 Kirk Hammerton Church: the tower and the fabric. *Archaeol J* **133**, 95-103

Norris, A, Bartley, D D, and Gaunt, G D, 1971 An account of the deposit of shell marl at Burton Salmon, West Yorkshire. *Naturalist* **96 (no 917)**, 57-63

Ottaway, P, 1994 *Roman York*. London: Batsford/English Heritage

Parsons, D, (ed) 1990 *Stone Quarrying and Building in England AD 43-1525*. Chichester: Phillimore in association with the Royal Archaeological Institute

Powell, J H, Cooper, A H, and Benfield, A C, 1992 *Geology of the Country around Thirsk*. Memoir of the British Geological Survey, Sheet **52** (England and Wales). London: HMSO

RCHME 1962 = Royal Commission on Historical Monuments England *An Inventory of the Historical Monuments in the City of York 1. Eburacum: Roman York*. London: HMSO

Radley, J, 1972 Excavations on the defences of the City of York: an early medieval stone tower and the successive earth ramparts. *Yorks Archaeol J* **44**, 38-64

Ramm, H, 1959 Notes on the use of gritstone in York. In Richardson (1959), 106-07

----, 1968 Use of Magnesian Limestone in York. *Symposium on the York Virgin*. York: RCHME (privately circulated)

----, 1976 The Roman roads west of Tadcaster. *York Hist* **1**, 3-12

Richardson, K M, 1959 Excavations in Hungate, York. *Archaeol J* **116**, 51-114

Rivet, A L F, and Smith, C, 1979 *The Place Names of Roman Britain*. London: Batsford

Sellwood, B W, 1984 The rock types represented in the town wall of Silchester. In Fulford (1984), 224-31

Senior, J R, 1990 Hildenley Limestone: a fine quality dimensional and artifact stone from Yorkshire. In Parsons (1990), 146-68

Turner, J S, 1970 Reddened Carboniferous strata on the northern side of Leeds. *J Earth Sci*, **8**, 149-58

Versey, H C, 1932 The composition of some Roman tesserae. *Naturalist* **57**, 652

Wenham, L P, 1968 *The Romano-British Cemetery at Trentholme Drive, York*. London: HMSO, Ministry Public Build Works Res Rep **5**

Wright, M E, 1988 Beehive quern manufacture in the south-east Pennines. *Scot Archaeol Rev* **5**, 65-77

INDEX